A Level

Biology
for OCR
Year 2

Series Editor
Ann Fullick

Authors
Ann Fullick
Jo Locke
Paul Bircher

OXFORD
UNIVERSITY PRESS

OXFORD
UNIVERSITY PRESS

Great Clarendon Street, Oxford, OX2 6DP, United Kingdom

Oxford University Press is a department of the University of Oxford. It furthers the University's objective of excellence in research, scholarship, and education by publishing worldwide. Oxford is a registered trade mark of Oxford University Press in the UK and in certain other countries

British Library Cataloguing in Publication Data
Data available

978-0-19-835764-3

10 9 8 7 6 5 4 3

Paper used in the production of this book is a natural, recyclable product made from wood grown in sustainable forests. The manufacturing process conforms to the environmental regulations of the country of origin.

Printed in Italy by L.E.G.O. SpA

This resource is endorsed by OCR for use with specification H420 A Level GCE Biology A. In order to gain endorsement this resource has undergone an independent quality check. OCR has not paid for the production of this resource, nor does OCR receive any royalties from its sale. For more information about the endorsement process please visit the OCR website www.ocr.org.uk

Acknowledgements

The authors would like to thank John Beazley for his reviewing, as well as Amy Johnson, Amie Hewish, Les Hopper, Clodagh Burke, Sharon Thorn for their tireless work and encouragement. In addition they would like to thank the teams at science and plants for schools (SAPS) and at the Wellcome Trust Sanger Institute for their valuable input to the project, and finally the help received from Dr Jeremy Pritchard and Jennifer Collins.

Ann Fullick would like to thank her partner Tony for his support and amazing photographs, and all of the boys William, Thomas, James, Edward, and Chris for their expert advice and for making her take time off.

Paul Bircher would like to thank his wife, Julie, and the rest of his family for their patience and support throughout the writing of this book. A special mention goes to his irrepressible grandchildren, Leo and Toby, who provided a welcome distraction. Their insanity has kept him sane.

Jo Locke would like to thank her husband Dave for all his support, encouragement, and endless cups of tea, as well as her girls Emily and Hermione who had to wait patiently for Mummy 'to just finish this paragraph'.

AS/A Level course structure

This book has been written to support students studying for OCR A Biology A. It covers the A Level Year 2 only modules from the specification. These are shown in the contents list, which also shows you the page numbers for the main topics within each chapter. There is also an index at the back to help you find what you are looking for.

AS exam

A level exam

Year 1 content

1 Development of practical skills in biology
2 Foundations in biology
3 Exchange and transport
4 Biodiversity, evolution, and disease

Year 2 content

5 Communication, homeostasis, and energy
6 Genetics, evolution, and ecosystems

A Level exams will cover content from Year 1 and Year 2 and will be at a higher demand. You will also carry out practical activities throughout your course.

Contents

This book contains many different features. Each feature is designed to support and develop the skills you will need for your examinations, as well as foster and stimulate your interest in biology.

Terms that you will need to be able to define and understand are highlighted by **bold text**.

Application features

These features contain important and interesting applications of biology in order to emphasise how scientists and engineers have used their scientific knowledge and understanding to develop new applications and technologies. There are also practical application features, with the icon ⚗, to support further development of your practical skills. There are also application features with the icon ⚙ which help support development of your understanding of scientific issues and their impact in society.

1   All application features have a question to link to material covered with the concept from the specification.

Extension features

These features contain material that is beyond the specification. They are designed to stretch and provide you with a broader knowledge and understanding and lead the way into the types of thinking and areas you might study in further education. As such, neither the detail nor the depth of questioning will be required for the examinations. But this book is about more than getting through the examinations.

1   Extension features also contain questions that link the off-specification material back to your course.

Summary Questions

1   These are short questions at the end of each topic.

2   They test your understanding of the topic and allow you to apply the knowledge and skills you have acquired.

3   The questions are ramped in order of difficulty. The icon ⚙ indicates where a question relates to scientific issues in society.

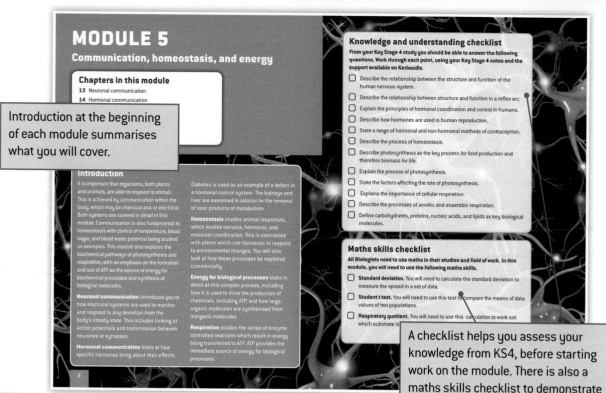

MODULE 5
Communication, homeostasis, and energy

Chapters in this module
13 Neuronal communication
14 Hormonal communication

Introduction at the beginning of each module summarises what you will cover.

Introduction

It is important that organisms, both plants and animals, are able to respond to stimuli. This is achieved by communication within the body, which may be chemical and/or electrical. Both systems are covered in detail in this module. Communication is also fundamental to homeostasis with control of temperature, blood sugar, and blood water potential being studied as examples. This module also explores the biochemical pathways of photosynthesis and respiration, with an emphasis on the formation and use of ATP as the source of energy for biochemical processes and synthesis of biological molecules.

Neuronal communication introduces you to how electrical systems are used to monitor and respond to any deviation from the body's steady state. This includes looking at action potentials and transmission between neurones at synapses.

Hormonal communication looks at how specific hormones bring about their effects.

Diabetes is used as an example of a defect in a hormonal control system. The kidneys and liver are examined in relation to the removal of toxic products of metabolism.

Homeostasis studies animal responses, which involve nervous, hormonal, and muscular coordination. This is contrasted with plants which use hormones to respond to environmental changes. You will also look at how these processes be exploited commercially.

Energy for biological processes looks in detail at this complex process, including how it is used to drive the production of chemicals, including ATP, and how large organic molecules are synthesised from inorganic molecules.

Respiration studies the series of enzyme controlled reactions which result in energy being transferred to ATP. ATP provides the immediate source of energy for biological processes.

Knowledge and understanding checklist

From your Key Stage 4 study you should be able to answer the following questions. Work through each point, using your Key Stage 4 notes and the support available on Kerboodle.

☐ Describe the relationship between the structure and function of the human nervous system.

☐ Describe the relationship between structure and function in a reflex arc.

☐ Explain the principles of hormonal coordination and control in humans.

☐ Describe how hormones are used in human reproduction.

☐ State a range of hormonal and non-hormonal methods of contraception.

☐ Describe the process of homeostasis.

☐ Describe photosynthesis as the key process for food production and therefore biomass for life.

☐ Explain the process of photosynthesis.

☐ State the factors affecting the rate of photosynthesis.

☐ Explains the importance of cellular respiration.

☐ Describe the processes of aerobic and anaerobic respiration.

☐ Define carbohydrates, proteins, nucleic acids, and lipids as key biological molecules.

Maths skills checklist

All Biologists need to use maths in their studies and field of work. In this module, you will need to use the following maths skills.

☐ **Standard deviation.** You will need to calculate the standard deviation to measure the spread in a set of data.

☐ **Student t test.** You will need to use this test to compare the means of data values of two populations.

☐ **Respiratory quotient.** You will need to use this calculation to work out which substrate is

A checklist helps you assess your knowledge from KS4, before starting work on the module. There is also a maths skills checklist to demonstrate the skills you will learn in that module.

Visual summaries show how some of the key concepts of that module interlink with other modules, across the entire A Level course.

Application task brings together some of the key concepts of the module in a new context.

Application

A major drawback of many brain imaging techniques is that individuals need to keep at least their heads completely still. In quantitative electroencephalograms (QEEGs) sensors are attached to the outside of the skull to measure the activity of the brain as people carry out different actions. It allows scientists to build up brain maps indicating which areas are used in different activities and skills. QEEGs are not as spatially accurate as fMRI can be, however, combined with other types of brain imaging they are increasing our knowledge of how the brain works.

Recent work using QEEGs to look at the changes in the brain in children affected by autism has produced some interesting findings, and a new form of therapy. They have used QEEG to show the patterns of different brain waves in autistic brains.

While QEEGs show that the activity in some regions is particularly high, the overall rate of brain activity in people affected by autism has been shown to be lower than that in unaffected people. In fact, levels of brain activity are highest in anxious people, and lowest in people with traumatic brain injuries, but autistic patients are not much above them.

A whole range of therapies is used to help people with autism cope with everyday life. A new tool is the use of neurofeedback training, which uses information from

QEEGs [...] and co[...]

There is growing evidence that for some people affected by autism this can enable them to function far more effectively and interact successfully with the people and the world around them.

1. a Far less is understood about the brain than, for example, the heart or kidney. Suggest why our understanding of the brain lags behind some other organs.
   b How have we found out what happens in the brain?
2. a What is a feedback system? Explain how they work and give examples from the various control and communication systems in the body.
   b Investigate what is meant by a neurofeedback system and discuss how this might be used to help people retrain their brains to work in different ways.
3. When QEEGs and other recordings of brain activity are taken it is often noted whether the eyes are open or closed.
   a Summarise how information from open eyes reaches the brain.
   b Suggest why it is important to record whether the eyes are open or closed.

Extension

Either investigate the main methods used for investigating the brain and make a table or poster to summarise the technology and the information it gives about the structure and function of the brain

OR investigate our current understanding of autism, including the areas and activity of the brain most

affected, the impact on functioning and examples of therapies used to help affected individuals, including at least one both drug-based intervention and neuro-feedback based on QEEGs.

Extension task bring together some key concepts of the module and develop them further, leading you towards greater understanding and further study.

Practice questions at the end of each chapter and the end of each module, including questions that cover practical and math skills.

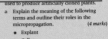
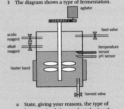

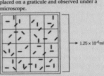
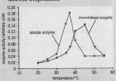

A dedicated Synoptic Concepts section at the end of the book, with help and advice on answering synoptic questions that cover multiple different topics. This section also contains further practice questions.

This book is supported by next generation Kerboodle, offering unrivalled digital support for independent study, differentiation, assessment, and the new practical endorsement.

If your school subscribes to Kerboodle, you will also find a wealth of additional resources to help you with your studies and with revision.

- Study guides
- Maths skills boosters and calculation worksheets
- On your marks activities to help you achieve your best
- Practicals and follow up activities to support the practical endorsement
- Interactive objective tests that give question-by-question feedback
- Animations and revision podcasts
- Self-assessment checklists

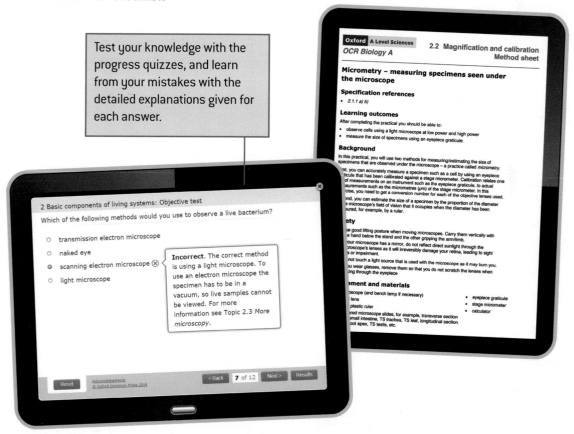

Test your knowledge with the progress quizzes, and learn from your mistakes with the detailed explanations given for each answer.

2 Basic components of living systems: Objective test

Which of the following methods would you use to observe a live bacterium?

- ○ transmission electron microscope
- ○ naked eye
- ⊙ scanning electron microscope ⊗
- ○ light microscope

Incorrect. The correct method is using a light microscope. To use an electron microscope the specimen has to be in a vacuum, so live samples cannot be viewed. For more information see Topic 2.3 *More microscopy.*

Reset   Acknowledgements   < Back **7** of 12 Next > Results
© Oxford University Press 2014

Oxford A Level Sciences
OCR Biology A   2.2 Magnification and calibration
                           Method sheet

Micrometry – measuring specimens seen under the microscope

Specification references
- 2.1.1 a) b)

Learning outcomes
After completing the practical you should be able to:
- observe cells using a light microscope at low power and high power
- measure the size of specimens using an eyepiece graticule.

Background
In this practical, you will use two methods for measuring/estimating the size of specimens that are observed under the microscope – a practice called *micrometry*.

First, you can accurately measure a specimen such as a cell by using an eyepiece graticule that has been calibrated against a stage micrometer. Calibration relates one set of measurements on an instrument such as the eyepiece graticule, to actual measurements such as the micrometres (μm) of the stage micrometer. In this exercise, you need to get a conversion number for each of the objective lenses used.

Second, you can estimate the size of a specimen by the proportion of the diameter of the microscope's field of vision that it occupies when the diameter has been measured, for example, by a ruler.

Safety
Use good lifting posture when moving microscopes. Carry them vertically with one hand below the stand and the other gripping the arm/limb.

If your microscope has a mirror, do not reflect direct sunlight through the microscope's lenses as it will irreversibly damage your retina, leading to sight loss or impairment.

Do not touch a light source that is used with the microscope as it may burn you. If you wear glasses, remove them so that you do not scratch the lenses when looking through the eyepiece

Equipment and materials
- microscope (and bench lamp if necessary)
- lens
- plastic ruler
- prepared microscope slides, for example, transverse section of small intestine, TS trachea, TS leaf, longitudinal section of root apex, TS testis, etc.
- eyepiece graticule
- stage micrometer
- calculator

For teachers, Kerboodle also has plenty of further assessment resources, answers to the questions in the book, and a digital markbook along with full teacher support for practicals and the worksheets, which include suggestions on how to support and stretch students. All of the resources are pulled together into teacher guides that suggest a route through each chapter.

MODULE 5
Communication, homeostasis, and energy

Introduction

It is important that organisms, both plants and animals, are able to respond to stimuli. This is achieved by communication within the body, which may be chemical and/or electrical. Both systems are covered in detail in this module. Communication is also fundamental to homeostasis with control of temperature, blood sugar, and blood water potential being studied as examples. This module also explores the biochemical pathways of photosynthesis and respiration, with an emphasis on the formation and use of ATP as the source of energy for biochemical processes and synthesis of biological molecules.

Neuronal communication introduces you to how electrical systems are used to monitor and respond to any deviation from the body's steady state. This includes looking at action potentials and transmission between neurones at synapses.

Hormonal communication looks at how specific hormones bring about their effects.

Diabetes is used as an example of a defect in a hormonal control system. The kidneys and liver are examined in relation to the removal of toxic products of metabolism.

Homeostasis studies animal responses, which involve nervous, hormonal, and muscular coordination. This is contrasted with plants which use hormones to respond to environmental changes. You will also look at how these processes be exploited commercially.

Energy for biological processes looks in detail at this complex process, including how it is used to drive the production of chemicals, including ATP, and how large organic molecules are synthesised from inorganic molecules.

Respiration studies the series of enzyme controlled reactions which result in energy being transferred to ATP. ATP provides the immediate source of energy for biological processes.

Knowledge and understanding checklist

From your Key Stage 4 study you should be able to answer the following questions. Work through each point, using your Key Stage 4 notes and the support available on Kerboodle.

☐ Describe the relationship between the structure and function of the human nervous system.

☐ Describe the relationship between structure and function in a reflex arc.

☐ Explain the principles of hormonal coordination and control in humans.

☐ Describe how hormones are used in human reproduction.

☐ State a range of hormonal and non-hormonal methods of contraception.

☐ Describe the process of homeostasis.

☐ Describe photosynthesis as the key process for food production and therefore biomass for life.

☐ Explain the process of photosynthesis.

☐ State the factors affecting the rate of photosynthesis.

☐ Explains the importance of cellular respiration.

☐ Describe the processes of aerobic and anaerobic respiration.

☐ Define carbohydrates, proteins, nucleic acids, and lipids as key biological molecules.

Maths skills checklist

All Biologists need to use maths in their studies and field of work. In this module, you will need to use the following maths skills.

☐ **Standard deviation.** You will need to calculate the standard deviation to measure the spread in a set of data.

☐ **Student t test.** You will need to use this test to compare the means of data values of two populations.

☐ **Respiratory quotient.** You will need to use this calculation to work out which substrate is being metabolised.

MyMaths.co.uk
Bringing Maths Alive

13 NEURONAL COMMUNICATION
13.1 Coordination
Specification reference: 5.1.1

Learning outcomes

Demonstrate knowledge, understanding, and application of:

→ the need for communication systems in multicellular organisms

→ the communication between cells by cell signalling.

▲ **Figure 1** *Complex coordination systems are needed to enable your body to cope with the sort of changes you encounter when moving from a warm, lit house to a dark, frosty outside world*

Synoptic link

You will find out about the nervous system throughout Chapter 13, Neuronal communication, the hormonal system in Chapter 14, Hormonal communication, and coordination in plants in Chapter 16, Plant responses.

When changes occur in an organism's internal or external environment, the organism must respond to these changes in order to survive. Examples of these changes are shown in Table 1.

▼ Table 1

Internal environment	External environment
blood glucose concentration	humidity
internal temperature	external temperature
water potential	light intensity
cell pH	new or sudden sound

Animals and plants respond to these changes in a variety of ways. Animals react through electrical responses (via neurones), and through chemical responses (via hormones). Plant responses are based on a number of chemical communication systems including plant hormones. These communication systems must be coordinated to produce the required response in an organism.

Why coordination is needed

As species have evolved, cells within organisms have become specialised to perform specific functions. As a result organisms need to coordinate the function of different cells and systems to operate effectively. Few body systems can work in isolation (apart from a few exceptions, for example, a heart can continue to beat if placed in the right bathing solution). For example, red blood cells transport oxygen effectively, but have no nucleus. This means that these cells are not able to replicate – a constant supply of red blood cells to the body is maintained by haematopoietic stem cells. In order to contract, muscle cells must constantly respire, and thus require a consistent oxygen supply. As these cells cannot transport oxygen, they are dependent on red blood cells for this function. In plants flowering needs to coordinate with the seasons, and pollinators must coordinate with the plants. In temperate climates light-sensitive chemicals enable plants to coordinate the development of flower buds with the lengthening days that signal the approach of spring and summer.

◀Figure 2 *Everything from the fresh spring leaves to the reproductive behaviour of these birds and of the insects with which they feed their young is controlled by chemicals and/or nervous coordination*

Homeostasis

In many relatively large multicellular animals, different organs have different functions in the body. Therefore, the functions of organs must be coordinated in order to maintain a relatively constant internal environment. This is known as **homeostasis**. For example, the digestive organs such as the exocrine pancreas, duodenum, and ileum along with the endocrine pancreas and the liver work together to maintain a constant blood glucose concentration.

Cell signalling

Nervous and hormonal systems coordinate the activities of whole organisms. This coordination relies on communication at a cellular level through cell signalling. This occurs through one cell releasing a chemical which has an effect on another cell, known as a target cell. Through this process, cells can:

- transfer signals locally, for example, between neurones at synapses. Here the signal used is a neurotransmitter.
- transfer signals across large distances, using hormones. For example, the cells of the pituitary gland secrete antidiuretic hormone (ADH), which acts on cells in the kidneys to maintain water balance in the body.

Coordination in plants

Plants do not have a nervous system like animals. However, to survive they still must respond to internal and external changes to their environment. For example, plant stems grow towards a light source to maximise their rate of photosynthesis. This is achieved through the use of plant hormones.

Synoptic link

You will find out about homeostasis in Chapter 15, Homeostasis.

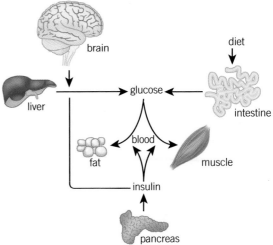

▲ **Figure 3** *To maintain blood glucose concentration, the function of many organs must be coordinated*

Synoptic link

You will find out more about cell signalling in Topic 14.5, Coordinated responses.

▲ **Figure 4** *When aphids attack certain types of wheat, the wheat plant cells produce chemicals both to alert other cells within the plant to the attack, and to signal to other aphids, putting them off landing on the plant and attacking it further*

Synoptic link

You will find out more about plant hormones in Chapter 16, Plant responses.

Summary questions

1 State one internal factor which causes a response in: *(2 marks)*
   a a plant
   b an animal

2 Describe how cells are able to communicate with one another. *(2 marks)*

3 Using examples, explain how and why coordination is required in a multicellular organism.   *(6 marks)*

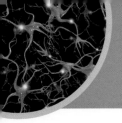

13.2 Neurones

Specification reference: 5.1.3

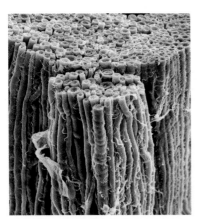

▲ **Figure 1** *Nerve fibres (axons). They are grouped together into bundles called nerves, which transmit electrical impulses to and from the central nervous system approx ×800 magnification*

The nervous system is responsible for detecting changes in the internal and external environment. These changes are known as a **stimulus**. This information then needs to be processed and an appropriate **response** triggered.

Both the nervous system and hormonal system play a role in reacting to stimuli, but they do so in very different ways. In this chapter you will focus on neuronal communication. This is generally a much faster and more targeted response than that produced by hormonal communication.

Neurones

You have already learnt that the nervous system is made up billions of specialised nerve cells called **neurones**. The role of neurones is to transmit electrical impulses rapidly around the body so that the organism can respond to changes in its internal and external environment. There are several different types of neurone found within a mammal. They work together to carry information detected by a sensory receptor to the effector, which in turn carries out the appropriate response.

Structure of a neurone

Mammalian neurones have several key features:

- Cell body – this contains the nucleus surrounded by cytoplasm. Within the cytoplasm there are also large amounts of endoplasmic reticulum and mitochondria which are involved in the production of **neurotransmitters**. These are chemicals which are used to pass signals from one neurone to the next. You will find out more about the important role of neurotransmitters in the nervous system in Topic 13.5, Synapses.

- Dendrons – these are short extensions which come from the cell body. These extensions divide into smaller and smaller branches known as dendrites. They are responsible for transmitting electrical impulses towards the cell body.

- Axons – these are singular, elongated nerve fibres that transmit impulses away from the cell body. These fibres can be very long, for example, those that transmit impulses from the tips of toes and fingers to the spinal cord. The fibre is cylindrical in shape consisting of a very narrow region of cytoplasm (in most cases approximately 1 μm) surrounded by a plasma membrane.

Types of neurone

Neurones can be divided into three groups according to their function. As a result they have slightly different structures:

- Sensory neurones – these neurones transmit impulses from a sensory receptor cell to a relay neurone, motor neurone, or the brain. They have one dendron, which carries the impulse to the cell body, and one axon, which carries the impulse away from the cell body.

- Relay neurones – these neurones transmit impulses between neurones. For example, between sensory neurones and motor neurones. They have many short axons and dendrons.

- Motor neurones – these neurones transmit impulses from a relay neurone or sensory neurone to an effector, such as a muscle or a gland. They have one long axon and many short dendrites.

In most nervous responses the electrical impulse follows the pathway:

Receptor → sensory neurone → relay neurone → motor neurone → effector cell

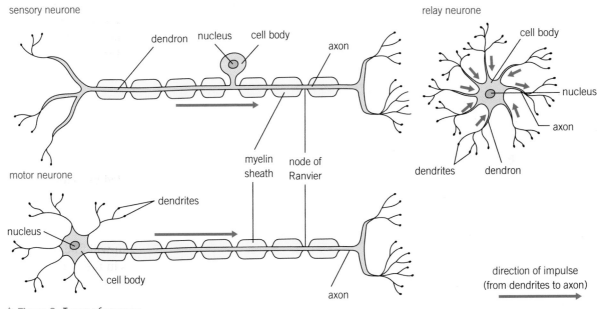

▲ Figure 2 *Types of neurone*

Myelinated neurones

The axons of some neurones are covered in a **myelin sheath**, made of many layers of plasma membrane. Special cells, called Schwann cells, produce these layers of membrane by growing around the axon many times. Each time they grow around the axon, a double layer of phospholipid bilayer is laid down. When the Schwann cell stops growing there may be more than 20 layers of membrane. The myelin sheath acts as an insulating layer and allows these myelinated neurones to conduct the electrical impulse at a much faster speed than unmyelinated neurones. Myelinated neurones can transmit impulses at up to 100 metres per second. In comparison, non-myelinated neurones can only conduct impulses at approximately 1 metre per second.

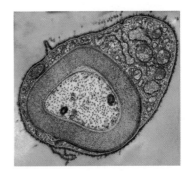

▲ Figure 3 *Transverse section through an axon showing the myelin sheath created by the Schwann cell's membrane wrapping around the axon many times, approx ×25 000 magnification*

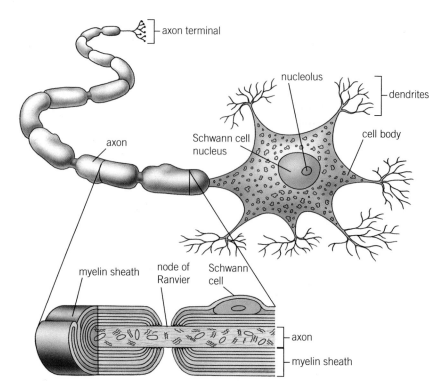

▲ **Figure 4** *Structure of a myelinated motor neurone*

Summary questions

1 State the difference between the function of a motor and a sensory neurone. (*1 mark*)

2 Draw and annotate a diagram of a motor neurone. (*4 marks*)

3 Describe the difference in structure between a myelinated and a non-myelinated neurone and how this affects the speed a nerve impulse is transmitted. (*4 marks*)

Between each adjacent Schwann cell there is a small gap (2–3 μm) known as a node of Ranvier. This creates gaps in the myelin sheath. In humans these occur every 1–3 mm. The myelin sheath is an electrical insulator. In myelinated neurones, the electrical impulse 'jumps' from one node to the next as it travels along the neurone. This allows the impulse to be transmitted much faster. In non-myelinated neurones the impulse does not jump – it transmits continuously along the nerve fibre, so is much slower. You will find out more detail about the transmission of impulses along axons in Topic 13.4, Nervous transmission.

Multiple sclerosis

◀**Figure 5** *Computer artwork comparing a healthy myelinated nerve fibre (bottom) with one from a person with multiple sclerosis (top)*

Multiple sclerosis (MS) is a neurological condition which affects around 100 000 people in the UK. Most people are diagnosed between the ages of 20 and 40.

MS affects nerves in the brain and spinal cord, causing a wide range of symptoms, including problems with muscle movement, balance, and vision.

MS is known to be an autoimmune disease. This is where the immune system mistakenly attacks healthy body tissue. This results in a thinning or complete loss of the myelin sheath and, as the disease advances, results in the breakdown of the axons of neurones. It is not known what triggers this disorder but is thought to be a combination of genetic and environmental factors, such as a viral infection.

1 The population of the UK is around 65 000 000. Calculate the proportion of the population who suffer from MS.

2 State what is meant by an autoimmune disease.

3 Describe the role of myelin in the body.

4 One of the symptoms of MS is the loss of vision, normally in only one eye. Suggest why a damaged myelin sheath could prevent a person from being able to see.

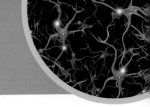

13.3 Sensory receptors

Specification reference: 5.1.3

The body is able to detect changes in its environment using groups of specialised cells known as sensory receptors. These are often located in the sense organs, such as the ear and eye.

Sensory receptors convert the stimulus they detect into a nerve impulse. The information is then passed through the nervous system and on into the central nervous system (CNS) – normally to the brain. The brain coordinates the required response and sends an impulse to an **effector** (normally a muscle or gland) to result in the desired response.

Features of sensory receptors

All sensory receptors have two main features:

- They are specific to a single type of stimulus.
- They act as a transducer – they convert a stimulus into a nerve impulse.

There are four main types of sensory receptor present in an animal, shown in Table 1.

▼ Table 1

Type of sensory receptor	Stimulus	Example of receptor	Example of sense organ
mechanoreceptor	pressure and movement	Pacinian corpuscle (detects pressure)	skin
chemoreceptor	chemicals	olfactory receptor (detects smells)	nose
thermoreceptor	heat	end-bulbs of Krause	tongue
photoreceptors	light	cone cell (detects different light wavelengths)	eye

Role as a transducer

Sensory receptors detect a range of different stimuli including light, heat, sound, or pressure. The receptor converts the stimulus into a nervous impulse, called a generator potential. For example, a rod cell (found in your eye) responds to light and produces a generator potential.

Pacinian corpuscle

Pacinian corpuscles are specific sensory receptors that detect mechanical pressure. They are located deep within your skin and are most abundant in the fingers and the soles of the feet. They are also found within joints, enabling you to know which joints are changing direction.

Learning outcomes

Demonstrate knowledge, understanding, and application of:

→ the roles of mammalian sensory receptors in converting different types of stimuli into nerve impulses.

Synoptic link

Plants also respond to stimuli, but their receptor cells produce chemicals rather than nerve impulses. You will find out more about plant responses in Chapter 16, Plant responses.

▲ Figure 1 A Pacinian corpuscle. These sensory receptors are found mostly in the skin of the feet, hands, genitals, and nipples. They respond to vibration and deep pressure

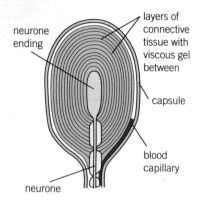

▲ Figure 2 *The structure of a Pacinian corpuscle*

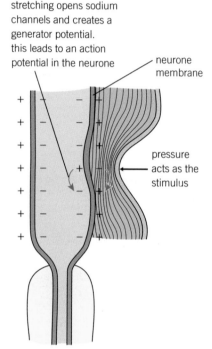

▲ Figure 3 *A generator potential is created when mechanical pressure causes a Pacinian corpuscle to change shape*

Figure 2 shows the structure of a Pacinian corpuscle. The end of the sensory neurone is found within the centre of the corpuscle, surrounded by layers of connective tissue. Each layer of tissue is separated by a layer of gel.

Within the membrane of the neurone there are sodium ion channels. These are responsible for transporting sodium ions across the membrane. The neurone ending in a Pacinian corpuscle has a special type of sodium channel called a stretch-mediated sodium channel. When these channels change shape, for example, when they stretch, their permeability to sodium also changes.

You learnt about the structure of different neurones in Topic 13.2, Neurones and you will learn more about the transmission of nerve impulses in Topic 13.4, Nervous transmission.

The following steps explain how a Pacinian corpuscle converts mechanical pressure into a nervous impulse:

1   In its normal state (known as its resting state), the stretch-mediated sodium ion channels in the sensory neurone's membrane are too narrow to allow sodium ions to pass through them. The neurone of the Pacinian corpuscle has a **resting potential**.

2   When pressure is applied to the Pacinian corpuscle, the corpuscle changes shape. This causes the membrane surrounding its neurone to stretch.

3   When the membrane stretches, the sodium ion channels present widen. Sodium ions can now diffuse into the neurone.

4   The influx of positive sodium ions changes the potential of the membrane – it becomes *depolarised*. This results in a generator potential.

5   In turn, the generator potential creates an action potential (a nerve impulse) that passes along the sensory neurone.

The **action potential** will then be transmitted along neurones to the CNS.

Summary questions

1   Describe the role of a sensory receptor in the body.   *(2 marks)*

2   State the transformation that takes place in a cone cell.   *(1 mark)*

3   Explain how your body detects that your finger has touched a pin.   *(6 marks)*

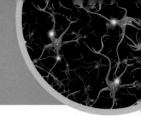

13.4 Nervous transmission

Specification reference: 5.1.3

After the sensory receptor has detected a change in the environment, an impulse is sent along the neurone by temporarily changing the voltage (potential difference) across the axon's membrane. As a result, the axon membrane switches between two states – a resting potential and an action potential. You will find out how the impulse travels between neurones in Topic 13.5, Synapses.

Resting potential

When a neurone is not transmitting an impulse, the potential difference across its membrane (difference in charge between the inside and outside of the axon) is known as a resting potential. In this state, the outside of the membrane is more positively charged than the inside of the axon. The membrane is said to be *polarised* as there is a potential difference across it. It is normally about −70 mV.

The resting potential occurs as a result of the movement of sodium and potassium ions across the axon membrane. The phospholipid bilayer prevents these ions from diffusing across the membrane and, therefore, they have to be transported via channel proteins. Some of these channels are gated – they must be opened to allow specific ions to pass through them. Other channels remain open all of the time allowing sodium and potassium ions to simply diffuse through them.

The following events result in the creation of a resting potential:

- Sodium ions (Na⁺) are actively transported *out* of the axon whereas potassium ions (K⁺) are actively transported *into* the axon by a specific intrinsic protein known as the sodium–potassium pump. However, their movement is not equal. For every three sodium ions that are pumped out, two potassium ions are pumped in.

- As a result there are more sodium ions outside the membrane than inside the axon cytoplasm, whereas there are more potassium ions inside the cytoplasm than outside the axon. Therefore, sodium ions diffuse back into the axon down its electrochemical gradient (this is the name given to a concentration gradient of ions), whereas potassium ions diffuse out of the axon.

- However, most of the 'gated' sodium ion channels are closed, preventing the movement of sodium ions, whereas many potassium ion channels are open, thus allowing potassium ions to diffuse out of the axon. Therefore, there are more positively charged ions outside the axon than inside the cell. This creates the resting potential across the membrane of −70 mV, with the inside negative relative to the outside.

Learning outcomes

Demonstrate knowledge, understanding, and application of:

→ the generation and transmission of nerve impulses in mammals.

Synoptic link

Look back at Chapter 5, Plasma membranes to ensure you fully understand the structure of a plasma membrane and the role of intrinsic proteins.

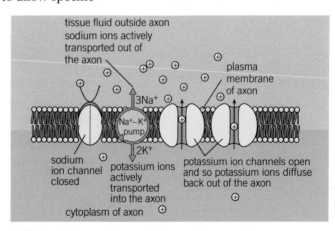

▲ Figure 1 *Axon membrane during a resting potential*

Action potential

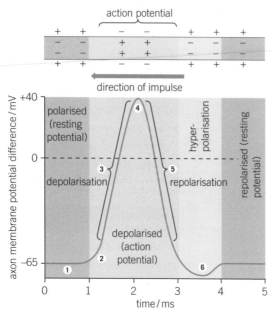

▲ **Figure 2** *Change in potential difference across an axon membrane during an action potential*

When a stimulus is detected by a sensory receptor, the energy of the stimulus temporarily reverses the charges on the axon membrane. As a result the potential difference across the membrane rapidly changes and becomes positively charged at approximately +40 mV. This is known as **depolarisation** – a change in potential difference from negative to positive. As the impulse passes **repolarisation** then occurs – a change in potential difference from positive back to negative. The neurone returns to its resting potential.

An action potential occurs when protein channels in the axon membrane change shape as a result of the change of voltage across its membrane. The change in protein shape results in the channel opening or closing. These channels are known as *voltage-gated ion channels*.

Figure 2 shows the changes in potential difference which occur across the axon membrane during an action potential.

The numbers on the graph correspond to the sequence of events that take place during an action potential:

1   The neurone has a resting potential – it is not transmitting an impulse. Some potassium ion channels are open (mainly those that are not voltage-gated) but sodium voltage-gated ion channels are closed.

2   The energy of the stimulus triggers some sodium voltage-gated ion channels to open, making the membrane more permeable to sodium ions. Sodium ions therefore diffuse into the axon down their electrochemical gradient. This makes the inside of the neurone less negative.

3   This change in charge causes more sodium ion channels to open, allowing more sodium ions to diffuse into the axon. This is an example of *positive feedback*.

4   When the potential difference reaches approximately +40 mV the voltage-gated sodium ion channels close and voltage-gated potassium ion channels open. Sodium ions can no longer enter the axon, but the membrane is now more permeable to potassium ions.

5   Potassium ions diffuse out of the axon down their electrochemical gradient. This reduces the charge, resulting in the inside of the axon becoming more negative than the outside.

6   Initially, lots of potassium ions diffuse out of the axon, resulting in the inside of the axon becoming more negative (relative to the outside) than in its normal resting state. This is known as *hyperpolarisation*. The voltage-gated potassium channels now close. The sodium-potassium pump causes sodium ions to move out of the cell, and potassium ions to move in. The axon returns to its resting potential – it is now repolarised.

Propagation of action potentials

A nerve impulse is an action potential that starts at one end of the neurone and is propagated along the axon to the other end of the neurone.

The initial stimulus causes a change in the sensory receptor which triggers an action potential in the sensory receptor, so the first region of the axon membrane is depolarised. This acts as a stimulus for the depolarisation of the next region of the membrane. The process continues along the length of the axon forming a wave of depolarisation. Once sodium ions are inside the axon, they are attracted by the negative charge ahead and the concentration gradient to diffuse further along inside the axon, triggering the depolarisation of the next section.

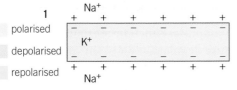

1 At resting potential the concentration of sodium ions outside the axon membrane is high relative to the inside, whereas that of the potassium ions is high inside the membrane relative to the outside. The overall concentration of positive ions is, however, greater on the outside, making this positive compared with the inside. The axon membrane is polarised.

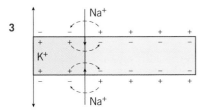

2 A stimulus causes a sudden influx of sodium ions and hence a reversal of charge on the axon membrane. This is the action potential and the membrane is depolarised.

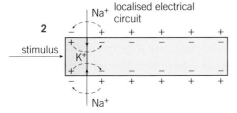

3 The localised electrical circuits established by the influx of sodium ions cause the opening of sodium voltage-gated channels a little further along the axon. The resulting influx of sodium ions in this region causes depolarisation. Behind this new region of depolarisation, the sodium voltage-gated channels close and the potassium ones open. Potassium ions begin to leave the axon along their electrochemical gradient.

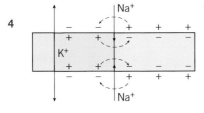

4 The action potential (depolarisation) is propagated in the same way further along the axon. The outward movement of the potassium ions has continued to the extent that the axon membrane behind the action potential has returned to its original charged state (positive outside, negative inside), that is, it has been repolarised.

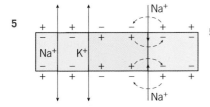

5 Following repolarisation the axon membrane returns to its resting potential in readiness for a new stimulus if it comes.

▲ Figure 3 *Propagation of an action potential along a non-myelinated neurone*

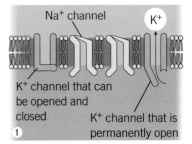

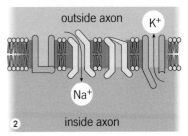

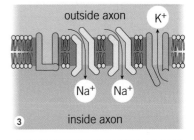

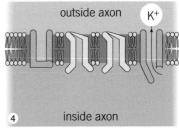

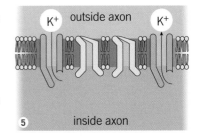

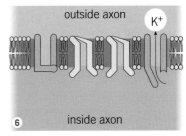

▲ Figure 4 *Changes in channel proteins in axon membrane during an action potential*

The region of the membrane which has been depolarised as the action potential passed along now undergoes repolarisation to return to its resting potential.

After an action potential there is a short period of time when the axon cannot be excited again, this is known as the *refractory period*. During this time, the voltage-gated sodium ion channels remain closed, preventing the movement of sodium ions into the axon.

A refractory period is important because it prevents the propagation of an action potential backwards along the axon as well as forwards. The refractory period makes sure action potentials are unidirectional. It also ensures that action potentials do not overlap and occur as discrete impulses.

Saltatory conduction

Myelinated axons transfer electrical impulses much faster than non-myelinated axons. This is because depolarisation of the axon membrane can only occur at the nodes of Ranvier where no myelin is present. Here the sodium ions can pass through the protein channels in the membrane. Longer localised circuits therefore arise between adjacent nodes. The action potential then 'jumps' from one node to another in a process known as saltatory conduction. This is much faster than a wave of depolarisation along the whole length of the axon membrane. Every time channels open and ions move it takes time, so reducing the number of places where this happens speeds up the action potential transmission. Long-term, saltatory conduction is also more energy efficient. Repolarisation uses ATP in the sodium pump, so by reducing the amount of repolarisation needed, saltatory conduction makes the conduction of impulses more efficient.

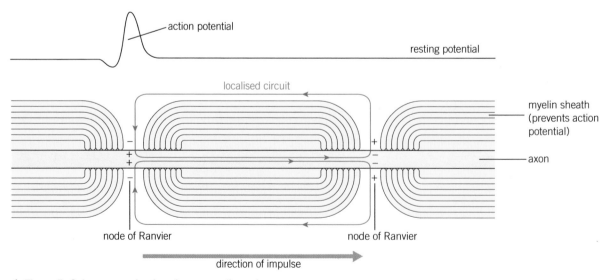

▲ Figure 5 *Saltatory conduction along a myelinated axon*

Apart from myelination, two other factors affect the speed at which an action potential travels:

- Axon diameter – the bigger the axon diameter, the faster the impulse is transmitted. This is because there is less resistance to the flow of ions in the cytoplasm, compared with those in a smaller axon.

- Temperature – the higher the temperature, the faster the nerve impulse. This is because ions diffuse faster at higher temperatures. However, this generally only occurs up to about 40 °C as higher temperatures cause the proteins (such as the sodium–potassium pump) to become denatured.

All-or-nothing principle

Nerve impulses are said to be all-or-nothing responses. A certain level of stimulus, the *threshold value*, always triggers a response. If this threshold is reached an action potential will always be created. No matter how large the stimulus is, the same sized action potential will always be triggered. If the threshold is not reached, no action potential will be triggered.

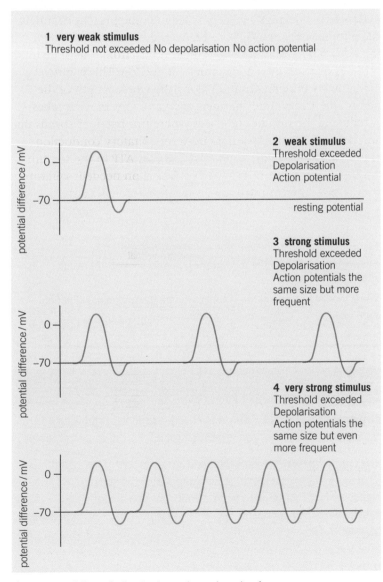

▲ Figure 6 *Effect of stimulus intensity on impulse frequency*

The size of the stimulus, however, does affect the number of action potentials that are generated in a given time. The larger the stimulus the more frequently the action potentials are generated.

The effect of the size of the stimulus on the frequency of nerve impulses can be seen in Figure 6.

Measuring action potentials

The presence and frequency of action potentials can be recorded using an oscilloscope. The diagram below shows some sample data collected in this manner, showing two action potentials:

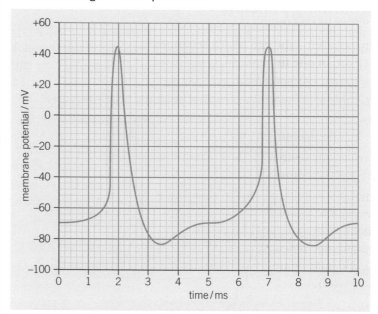

▲ Figure 7

1 State what occurs in the neurone between 1 ms and 2 ms.
2 State and explain how the membrane potential changes between 5.5 ms and 7 ms.
3 Using the data shown in the graph, calculate the frequency of action potentials.

Summary questions

1 State how the body detects the difference between a small and a large stimulus. *(1 mark)*

2 State the difference between depolarisation, repolarisation, and hyperpolarisation. *(2 marks)*

3 Describe what would happen if a refractory period did not exist. *(2 marks)*

4 Describe how the movement of ions establishes the resting potential in an axon. *(4 marks)*

5 Explain how temperature receptors in the hand generate an action potential in the sensory neurone to tell the body that you are touching a hot object. *(6 marks)*

13.5 Synapses

Specification reference: 5.1.3

In Topic 13.4, Nervous transmission, you learnt about how impulses travel along each neurone in the form of an action potential. However, to reach the CNS or an effector, the impulse often needs to be passed between several neurones. The junction between two neurones (or a neurone and effector) is called a **synapse**. Impulses are transmitted across the synapse using chemicals called **neurotransmitters**.

Learning outcomes

Demonstrate knowledge, understanding, and application of:

→ the structure and roles of synapses in neurotransmission.

Synapse structure

All synapses have a number of key features:

- Synaptic cleft – the gap which separates the axon of one neurone from the dendrite of the next neurone. It is approximately 20–30 nm across.

- Presynaptic neurone – neurone along which the impulse has arrived.

- Postsynaptic neurone – neurone that receives the neurotransmitter

- Synaptic knob – the swollen end of the presynaptic neurone. It contains many mitochondria and large amounts of endoplasmic reticulum to enable it to manufacture neurotransmitters (in most cases).

- Synaptic vesicles – vesicles containing neurotransmitters. The vesicles fuse with the presynaptic membrane and release their contents into the synaptic cleft.

- Neurotransmitter receptors – receptor molecules which the neurotransmitter binds to in the postsynaptic membrane.

Types of neurotransmitter

Neurotransmitters can be grouped into two categories:

1 Excitatory – these neurotransmitters result in the depolarisation of the postsynaptic neurone. If the threshold is reached in the postsynaptic membrane an action potential is triggered. Acetylcholine is an example of an excitatory neurotransmitter.

2 Inhibitory – these neurotransmitters result in the hyperpolarisation of the postsynaptic membrane. This prevents an action potential being triggered. Gamma-aminobutyric acid (GABA) is an example of an inhibitory neurotransmitter that is found in some synapses in the brain.

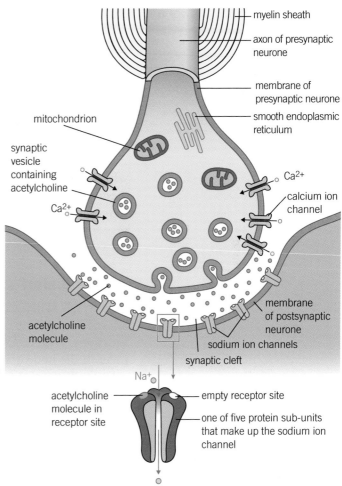

▲ Figure 1 *Structure of a synapse. This synapse is known as a cholinergic synapse as the neurotransmitter which passes across the synapse is acetylcholine*

Transmission of impulses across synapses

Synaptic transmission occurs as a result of the following:

- The action potential reaches the end of the presynaptic neurone
- Depolarisation of the presynaptic membrane causes calcium ion channels to open
- Calcium ions diffuse into the presynaptic knob
- This causes synaptic vesicles containing neurotransmitters to fuse with the presynaptic membrane. Neurotransmitter is released into the synaptic cleft by exocytosis
- Neurotransmitter diffuses across the synaptic cleft and binds with its specific receptor molecule on the postsynaptic membrane
- This causes sodium ion channels to open
- Sodium ions diffuse into the postsynaptic neurone
- This triggers an action potential and the impulse is propagated along the postsynaptic neurone.

Once a neurotransmitter has triggered an action potential in the postsynaptic neurone, it is important that it is removed so the stimulus is not maintained, and so another stimulus can arrive at and affect the synapse. Any neurotransmitter left in the synaptic cleft is removed. Acetycholine is broken down by enzymes, which also releases them from the receptors on the postsynaptic membrane. The products are taken back into the presynaptic knob. Removing the neurotransmitter from the synaptic cleft prevents the response from happening again and allows the neurotransmitter to be recycled.

Transmission across cholinergic synapses

Cholinergic synapses use the neurotransmitter acetylcholine. They are common in the CNS of vertebrates and at neuromuscular junctions – where a motor neurone and a muscle cell (an effector) meet. If the neurotransmitter reaches the receptors on a muscle cell, it will cause the muscle to contract. Acetylcholine is released from the vesicles in the presynaptic knob (Figure 2). It then diffuses across the synaptic cleft where it binds with specific receptors in the postsynaptic membrane. This triggers an action potential in the postsynaptic neurone or muscle cell. Once an action potential has been triggered, acetylcholine is hydrolysed by a specific enzyme – acetylcholinesterase. This enzyme is also situated on the postsynaptic membrane. Acetylcholine is hydrolysed to give choline and ethanoic acid. One molecule of acetylcholinesterase can break down around 25 000 molecules of acetylcholine per minute. The breakdown products are taken back into the presynaptic knob to be reformed into acetylcholine, and the postsynaptic membrane is ready to receive another impulse.

1 The arrival of an action potential at the end of the presynaptic neurone causes calcium ion channels to open and calcium ions (Ca^{2+}) enter the synaptic knob.

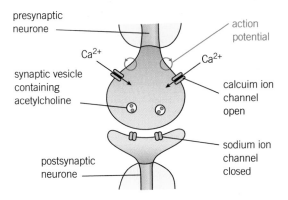

presynaptic neurone

action potential

Ca^{2+}

Ca^{2+}

synaptic vesicle containing acetylcholine

calcuim ion channel open

sodium ion channel closed

postsynaptic neurone

4 The influx of sodium ions generates a new action potential in the postsynaptic neurone.

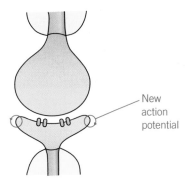

New action potential

2 The influx of calcium ions into the presynaptic neurone causes synaptic vesicles to fuse with the presynaptic membrane, so releasing acetylcholine into the synaptic cleft.

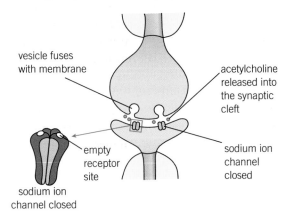

vesicle fuses with membrane

acetylcholine released into the synaptic cleft

empty receptor site

sodium ion channel closed

sodium ion channel closed

5 Acetylcholinesterase hydrolyses acetylcholine into choline and ethanoic acid (acetyl), which diffuse back across the synaptic cleft into the presynaptic neurone (= recycling). In addition to recycling the choline and ethanoic acid, the breakdown of acetylcholine also prevents it from continuously generating a new action potential in the postsynaptic neurone.

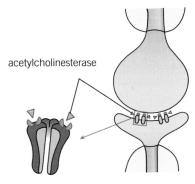

acetylcholinesterase

3 Acetylcholine molecules fuse with receptor sites on the sodium ion channel in the membrane of the postsynaptic neurone. This causes the sodium ion channels to open, allowing sodium ions (Na^+) to diffuse in rapidly along a concentration gradient.

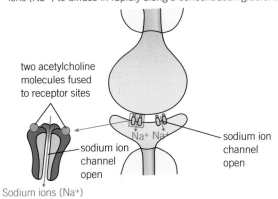

two acetylcholine molecules fused to receptor sites

sodium ion channel open

Na^+ Na^+

sodium ion channel open

Sodium ions (Na^+)

6 ATP released by mitochondria is used to recombine choline and ethanoic acid into acetycholine. This is stored in synaptic vesicles for future use. Sodium ion channels close in the absence of acetylcholine in the receptor sites.

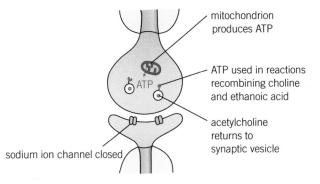

mitochondrion produces ATP

ATP

ATP used in reactions recombining choline and ethanoic acid

acetylcholine returns to synaptic vesicle

sodium ion channel closed

▲ **Figure 2** *Mechanism of transmission across a cholinergic synapse – only essential structures are shown*

Role of synapses

In simple models a neurone with a single synapse is shown. In the body, however, one neurone may make thousands of synapses using the dendron, dendrites, and axons. Synapses play an important role in the nervous system:

- They ensure impulses are unidirectional. As the neurotransmitter receptors are only present on the postsynaptic membrane, impulses can only travel from the presynaptic neurone to the postsynaptic neurone.
- They can allow an impulse from one neurone to be transmitted to a number of neurones at multiple synapses. This results in a single stimulus creating a number of simultaneous responses.
- Alternatively, a number of neurones may feed in to the same synapse with a single postsynaptic neurone. This results in stimuli from different receptors interacting to produce a single result.

Summation and control

Each stimulus from a presynaptic neurone causes the release of the same amount of neurotransmitter into the synapse. In some synapses, however, the amount of neurotransmitter from a single impulse is not enough to trigger an action potential in the postsynaptic neurone, as the threshold level is not reached. However, if the amount of neurotransmitter builds up sufficiently to reach the threshold then this will trigger an action potential. This is known as **summation**. There are two ways this can occur:

- *Spatial summation* – this occurs when a number of presynaptic neurones connect to one postsynaptic neurone. Each releases neurotransmitter which builds up to a high enough level in the synapse to trigger an action potential in the single postsynaptic neurone (Figure 3).
- *Temporal summation* – this occurs when a single presynaptic neurone releases neurotransmitter as a result of an action potential several times over a short period. This builds up in the synapse until the quantity is sufficient to trigger an action potential (Figure 4).

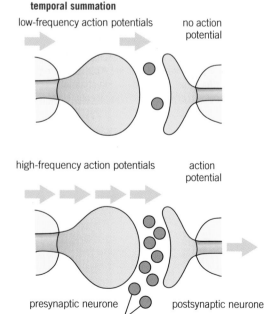

spatial summation

presynaptic neurone A

no action potential

presynaptic neurone B

no action potential

action potential

▲ Figure 3 *On their own, neurone A and neurone B do not produce enough neurotransmitter to trigger an action potential. However, together there is enough neurotransmitter to reach the threshold and trigger an action potential in the postsynaptic neurone*

temporal summation

low-frequency action potentials

no action potential

high-frequency action potentials

action potential

presynaptic neurone

postsynaptic neurone

neurotransmitter

◀Figure 4 *If neurotransmitter is released several times in quick succession from a presynaptic neurone, the level builds up in the synapse and triggers an action potential in the postsynaptic neurone*

Effects of drugs on synapses

Many recreational and medical drugs cause their effects by acting on synapses. This will result in the nervous system being stimulated or inhibited.

Drugs that stimulate the nervous system create more action potentials in postsynaptic neurones, resulting in an enhanced response. For example, if the targeted synapse is with a neurone that transmits an impulse from a sound receptor, the body will perceive a louder sound. These drugs may work by:

- Mimicking the shape of the neurotransmitter – nicotine is the same shape as acetylcholine. It can therefore bind to acetylcholine receptors on the postsynaptic membrane and trigger action potentials in the postsynaptic neurone.

- Stimulating the release of more neurotransmitter. For example, amphetamines.

- Inhibiting the enzyme responsible for breaking down the neurotransmitter in the synapse. For example, nerve gases stop acetylcholine being broken down. This can result in a loss of muscle control.

Drugs that inhibit the nervous system create fewer action potentials in postsynaptic neurones, resulting in a reduced response. For example, if the targeted synapse is with a neurone that transmits an impulse from a sound receptor, the body will perceive a quieter sound. These drugs may work by:

- Blocking receptors – this means the neurotransmitter can no longer bind and activate the receptor. For example, curare blocks acetylcholine receptors at neuromuscular junctions. The muscle cells cannot therefore be stimulated, and the person suffers from paralysis.

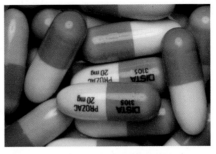

▲ Figure 5 *Prozac is used to treat depression. It acts on neurotransmitters in synapses in the brain*

- Binding to specific receptors on the post-synaptic membrane of some neurones and changing the shape of the receptor such that binding of the neurotransmitter increases. This therefore increases activity. An example of this is alcohol binding to $GABA_A$ receptors.

1. State the difference between the result of an inhibitory and a stimulatory drug that acts on the nervous system.
2. Explain how amphetamines will affect the nervous system.
3. Explain how alcohol affects the nervous system.
4. Serotonin is a neurotransmitter involved in the regulation of sleep and other emotional states. Low levels of serotonin are found in patients suffering from depression. Prozac is an example of a drug used to treat depression. It works be blocking the reuptake of serotonin into the presynaptic neurone. Using your knowledge of synapses, explain how Prozac causes its effects.

Summary questions

1. State what is meant by a synapse. *(1 mark)*

2. Explain how synapses ensure impulses are only transmitted in one direction. *(2 marks)*

3. Describe one similarity and one difference between temporal and spatial summation. *(2 marks)*

4. Explain in detail how a motor neurone causes a postsynaptic neurone to depolarise. *(6 marks)*

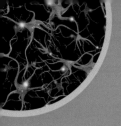

13.6 Organisation of the nervous system

Specification reference: 5.1.5

In the last few topics, you have looked in detail at how nervous impulses are transmitted around the nervous system – how are the billions of neurones in your nervous system organised?

Structural organisation

The mammalian nervous system is organised structurally into two systems:

- **Central nervous system** (CNS) – this consists of your brain and spinal cord.
- **Peripheral nervous system** (PNS) – this consists of all the neurones that connect the CNS to the rest of the body. These are the sensory neurones which carry nerve impulses from the receptors to the CNS, and the motor neurones which carry nerve impulses away from the CNS to the effectors.

Functional organisation

The nervous system is also functionally organised into two systems:

- **Somatic nervous system** – this system is under conscious control – it is used when you voluntarily decide to do something. For example, when you decide to move a muscle to move your arm. The somatic nervous system carries impulses to the body's muscles.
- **Autonomic nervous system** – this system works constantly. It is under subconscious control and is used when the body does something automatically without you deciding to do it – it is involuntary. For example, to cause the heart to beat, or to digest food. The autonomic nervous system carries nerve impulses to glands, smooth muscle (for example, in the walls of the intestine), and cardiac muscle.

The autonomic nervous system is then further divided by function into the sympathetic and parasympathetic nervous system. Generally, if the outcome increases activity it involves the sympathetic nervous system – for example, an increase in heart rate. If the outcome decreases activity it involves the parasympathetic nervous system – for example, a decrease in heart or breathing rate after a period of exercise.

Study tip

Although primarily unconscious, many aspects of the autonomic nervous system can come under conscious control. For example, people can choose to hold their breath or swallow rapidly. When people do not actively choose to control these functions, the autonomic nervous system takes over and controls them. This frees up the conscious areas of the brain – it would be hard to think of much else if you had to concentrate on breathing and keeping your heart beating.

▶ Table 1 *Examples of the effects of sympathetic and parasympathetic stimulation. Note that most of these result in opposite effects on the body*

Structure	Sympathetic stimulation	Parasympathetic stimulation
salivary glands	saliva production reduced	saliva production increased
lung	bronchial muscle relaxed	bronchial muscle contracted
kidney	decreased urine secretion	increased urine secretion
stomach	peristalsis reduced	gastric juice secreted
small intestine	peristalsis reduced	digestion increased

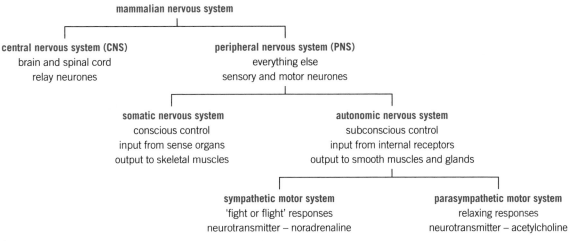

▲ Figure 1 *Summary of the organisation of the mammalian nervous system*

▼ Table 2 *Comparison of autonomic and somatic motor systems, where ACh = acetylcholine and NA = noradrenaline*

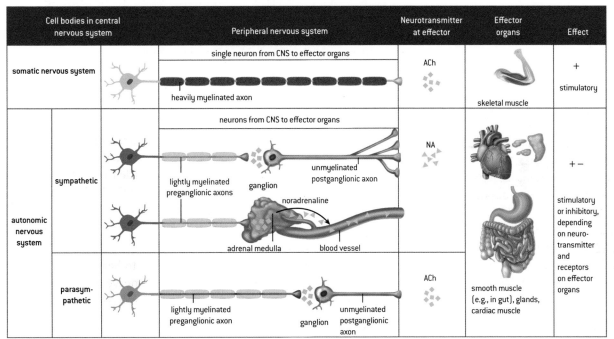

Summary questions

1 State the difference between the peripheral and the central nervous system. *(1 mark)*

2 Sort the following activities into those which are controlled by the somatic nervous system and those which are controlled by the autonomic nervous system
   a pupil dilation   b blood pressure
   c throwing a ball   d walking. *(2 marks)*

3 State and explain one reason why many autonomic functions can also be controlled by the somatic nervous system. *(3 marks)*

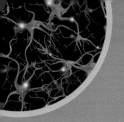

13.7 Structure and function of the brain

Specification reference: 5.1.5

An adult human brain contains approximately 86 billion neurones. The brain is responsible for processing all the information collected by receptor cells about changes in the internal and external environment. It also receives and processes information from the hormonal system through molecules in the blood. It must then produce a coordinated response.

The advantage of having a central control centre for the whole body is that communication between the billions of neurones involved is much faster than if control centres for different functions were distributed around the body. With the exception of reflex actions, all other nervous reactions are processed by the brain. You will find out about reflex actions in Topic 13.8, Reflexes.

Gross structure

The brain is protected by the skull. It is also surrounded by protective membranes (called meninges). The human brain is extremely complex, but the structures you need to know about are shown in Figure 1.

There are five main areas. They are distinguishable by their shape, colour, or microscopic structure:

- Cerebrum – controls voluntary actions, such as learning, memory, personality, and conscious thought.
- Cerebellum – controls unconscious functions such as posture, balance, and non-voluntary movement.
- Medulla oblongata – used in autonomic control, for example, it controls heart rate and breathing rate.
- Hypothalamus – regulatory centre for temperature and water balance.
- Pituitary gland – stores and releases hormones that regulate many body functions

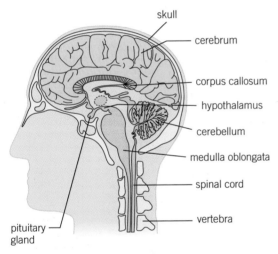

▲ Figure 1 *Main structures in the brain*

Different images of the brain

Many different techniques are used to study the brain in order to understand its function. Figures 2, 3, and 4 all show a cross-section through the brain. Can you identify the main structures?

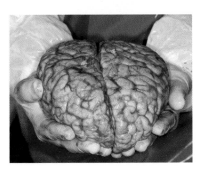

▲ **Figure 2** *Photo of the brain. Images like this have been taken during autopsies. The position of a lesion caused as a result of an accident, tumour, or stroke can be linked to observed changes in a patient's behaviour or capabilities before their death*

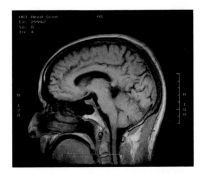

▲ **Figure 3** *Magnetic resonance imaging (MRI) of the brain. MRI is used to investigate the structure of the brain. A specialised version of MRI, called functional magnetic resonance imaging (fMRI), has been developed which allows the brain to be studied during activity. Active areas of the brain can be identified due to increased blood flow*

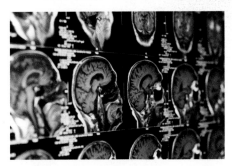

▲ **Figure 4** *Computerised tomography (CT) scan of the brain. A CT scan uses a series of X-rays to create detailed three-dimensional images of the inside of the body*

Cerebrum

The cerebrum receives sensory information, interprets it with respect to that stored from previous experiences, and then sends impulses along motor neurones to effectors to produce an appropriate response. It is responsible for coordinating all of the body's voluntary responses as well as some involuntary ones.

The cerebrum is highly convoluted, which increases its surface area considerably and therefore its capacity for complex activity. It is split into left and right halves known as the cerebral hemispheres. Each hemisphere controls one half of the body, and has discrete areas which perform specific functions – these areas are mirrored in each hemisphere. The outer layer of the cerebral hemispheres is known as the cerebral cortex. It is 2–4 mm thick. The most sophisticated processes such as reasoning and decision-making occur in the frontal and prefrontal lobe of the cerebral cortex.

Each sensory area within the cerebral hemispheres receives information from receptor cells located in sense organs. The size of the sensory area allocated is in proportion to the relative number of receptor cells present in the body part. The information is then passed on to other areas of the brain, known as association areas, to be analysed and acted upon. Impulses come into the motor areas where motor neurones send out impulses, for example, to move skeletal muscles. The size of the motor area allocated is in proportion to the relative number of motor endings in it. The main region which controls movement is the primary motor cortex located at the back of the frontal lobe.

In the base of the brain, impulses from each side of the body cross – therefore the left hemisphere receives impulses from the right-hand side of the body, and the right hemisphere receives impulses from the left-hand side of the body. For example, inputs from the eye pass to the visual area in the occipital lobe. Impulses from the right side of the field of vision in each eye are sent to the visual cortex in the left hemisphere, whereas impulses from the left side of the field of vision are sent to the right hemisphere. Through the integration of these inputs the brain is able to judge distance and perspective.

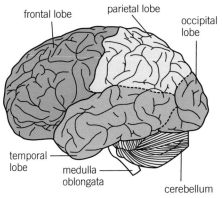

▲ **Figure 5** *The folded structure of the cerebral cortex viewed from the left side*

▲ **Figure 6** *A sensory homunculus. Neurobiologists have constructed models of the body in which the body part is made in proportion to the number of sensory inputs received from it. The hands and lips are very sensitive, and therefore they are drawn as particularly large in relation to other body parts*

Synoptic link

You will find out more about controlling heart rate in Topic 14.6, Controlling heart rate.

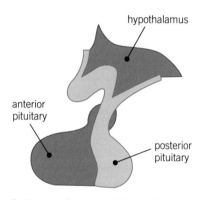

▲ **Figure 7** *Structure of the pituitary gland*

Cerebellum

This area of the brain is concerned with the control of muscular movement, body posture, and balance – it does not initiate movement, but coordinates it. Therefore, if this area of the brain is damaged, a person suffers from jerky and uncoordinated movement. The cerebellum receives information from the organs of balance in the ears and information about the tone of muscles and tendons. It then relays this information to the areas of the cerebral cortex that are involved in motor control.

Medulla oblongata

The medulla oblongata contains many important regulatory centres of the autonomic nervous system. These control reflex activities such as ventilation (breathing rate) and heart rate. It also controls activities such as swallowing, peristalsis, and coughing.

Hypothalamus

This is the main controlling region for the autonomic nervous system. It has two centres – one for the parasympathetic and one for the sympathetic nervous system. It has a number of functions, which include:

- controlling complex patterns of behaviour, such as feeding, sleeping, and aggression
- monitoring the composition of blood plasma, such as the concentration of water and blood glucose – therefore it has a very rich blood supply
- producing hormones – it is an endocrine gland, that is, it produces hormones.

Pituitary gland

This is found at the base of the hypothalamus. It is approximately the size of a pea but it controls most of the glands in the body. It is divided into two sections:

- Anterior pituitary (front section) – produces six hormones including follicle-stimulating hormone (FSH), which is involved in reproduction and growth hormones.
- Posterior pituitary (back section) – stores and releases hormones produced by the hypothalamus, such as ADH involved in urine production.

Summary questions

1   State the difference between the function of the anterior pituitary and the posterior pituitary. (*1 mark*)

2   Sounds are interpreted by the auditory area in the temporal lobe. State the pathway followed by a nervous impulse produced by a sound wave. (*3 marks*)

3   A patient displays three symptoms: *asynergia* – a lack of coordination in their motor movement, *adiadochokinesia* – an inability to perform rapid movements, and *ataxic gait* – staggering movements. Suggest and explain which part of the brain may have been damaged to cause these symptoms. (*2 marks*)

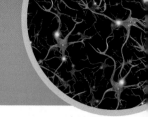

13.8 Reflexes

Specification reference: 5.1.5

When the body is in danger, it can respond to situations without conscious thought. This causes a faster response, preventing or minimising damage to the body. This is known as a **reflex action**. A reflex is an involuntary response to a sensory stimulus.

Reflex arc

The pathway of neurones involved in a reflex action is known as a reflex arc. Most reflexes follow the same steps between the stimulus and the response:

- Receptor – detects stimulus and creates an action potential in the sensory neurone.
- Sensory neurone – carries impulse to spinal cord.
- Relay neurone – connects the sensory neurone to the motor neurone within the spinal cord or brain.
- Motor neurone – carries impulse to the effector to carry out the appropriate response.

Figure 1 illustrates what happens when you touch a hot candle – this is known as a withdrawal reflex. Before your brain registers that your hand is hot, the muscles in your arm have already pulled your hand away from the danger, minimising damage to your hand.

Learning outcomes

Demonstrate knowledge, understanding, and application of:

→ the reflex actions.

5 motor neurone passes impulses to the muscle

7 response hand is moved quickly away from flame

6 effector contracts

transverse section through spinal cord (magnified five times in relation to man)

1 stimulus heat from candle flame

2 thermoreceptor in skin detects heat

3 sensory neurone passes nerve impulses to spinal cord

4 relay neurone passes impulses across the spinal cord

▲ Figure 1 *Reflex arc involved in the withdrawal of the hand from a heat stimulus*

Spinal cord

The spinal cord is a column of nervous tissues running up the back. It is surrounded by the spine for protection. At intervals along the spinal cord pairs of neurones emerge, as shown in Figure 2.

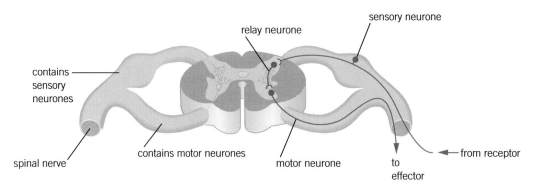

▲ Figure 2 *Section through a spinal cord showing the neurones in a reflex arc*

Knee-jerk reflex

The knee-jerk reflex is a reflex commonly tested by doctors. It is a spinal reflex – this means that the neural circuit only goes up to the spinal cord, not the brain.

When the leg is tapped just below the kneecap (patella), it stretches the patellar tendon and acts as a stimulus. This stimulus initiates a reflex arc that causes the extensor muscle on top of the thigh to contract. At the same time, a relay neurone inhibits the motor neurone of the flexor muscle, causing it to relax. This contraction, coordinated with the relaxation of the antagonistic flexor hamstring muscle, causes the leg to kick.

After the tap of a hammer, the leg is normally extended once and comes to rest. The absence of this reflex may indicate nervous problems and multiple oscillation of the leg may be a sign of a cerebellar disease.

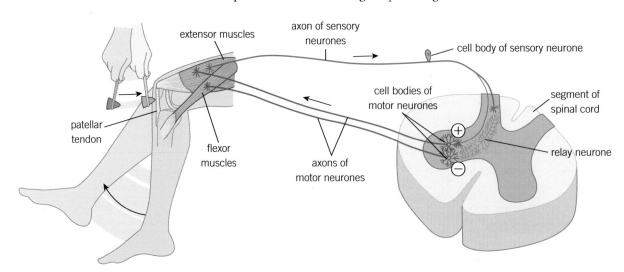

▲ Figure 3 *The knee-jerk reflex*

This reflex is used by the body to help maintain posture and balance, allowing you to remain balanced with little effort or conscious thought.

Blinking reflex

The blinking reflex is an involuntary blinking of the eyelids (Figure 4). It occurs when the cornea is stimulated, for example, by being touched. Its purpose is to keep the cornea safe from damage due to foreign bodies such as dust or flying insects entering the eye – this type of response is known as the corneal reflex. A blink reflex also occurs when sounds greater than 40–60 dB are heard, or as a result of very bright light. Blinking as a reaction to over-bright light (to protect the lens and retina) is known as the optical reflex. The blinking reflex is a cranial reflex – it occurs in the brain, not the spinal cord.

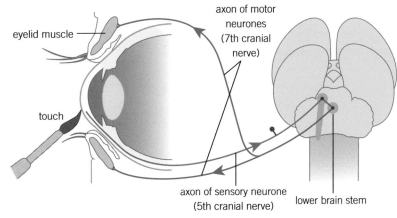

▲ Figure 4 *The blinking reflex*

When the cornea of the eye is irritated by a foreign body, the stimulus triggers an impulse along a sensory neurone (the fifth cranial nerve). The impulse then passes through a relay neurone in the lower brain stem. Impulses are then sent along branches of the motor neurone (the seventh cranial nerve) to initiate a motor response to close the eyelids. The reflex initiates a consensual response – this means that both eyes are closed in response to the stimulus.

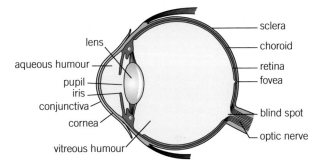

▲ Figure 5 *Structure of the eye*

The blinking reflex is very rapid – it occurs in around one tenth of a second.

Doctors test for the blinking reflex when examining unconscious patients. If this reflex is present, it indicates that the lower brain stem is functioning. This procedure is therefore used as part of an assessment to determine whether or not a patient is brain-dead – if the corneal reflex is present the person cannot be diagnosed as brain-dead.

  Measuring reaction time

When a person catches a falling object, at least part of this response is a reflex reaction. Measuring the time taken to catch a falling object can therefore be used to measure a person's reaction time.

To measure the reaction time, a suitable scale is placed onto the ruler which converts the distance dropped by the ruler into a reaction time. One investigation which can be carried out using this approach is to measure the effect of caffeine concentration on a person's reaction time.

1 State two variables which should be controlled when carrying out this investigation.
2 Explain why, in this investigation, only 'part of this response is a reflex action'.
3 Plan an investigation into how the concentration of caffeine affects a person's reaction time.

When carrying out this investigation, a researcher may choose to give a placebo caffeine drink to some of the people being tested. This is a drink labelled as a caffeine drink, but contains no caffeine.

> 4 Explain why a researcher may choose to give some people being tested a placebo.

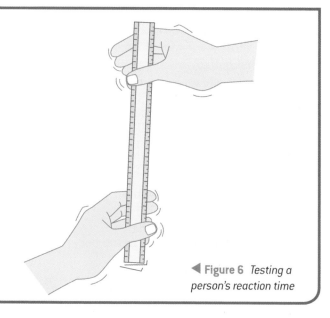

◄ Figure 6 *Testing a person's reaction time*

Survival importance

Reflexes are essential for survival as they avoid the body being harmed, or reduce the severity of any damage. For example, the iris contracts the pupil in bright light to prevent damage to the retina. In dim light, the reverse occurs to enable you to see as much as possible. Reflexes increase your chances of survival by:

- Being involuntary responses – the decision-making regions of the brain are not involved, therefore the brain is able to deal with more complex responses. It prevents the brain from being overloaded with situations in which the response is always the same.

- Not having to be learnt – they are present at birth and therefore provide immediate protection.

- Extremely fast – the reflex arc is very short. It normally only involves one or two synapses, which are the slowest part of nervous transmission.

- Many reflexes are what we would consider everyday actions, such as those which keep us upright (and thus not falling over), and those which control digestion.

Summary questions

1 State the reflex arc which occurs when a doctor tests the knee-jerk reflex. *(1 mark)*

2 Which of the following actions are reflexes
   a gagging   b speaking
   c jumping   d pupil dilation. *(1 mark)*

3 State and explain how a reflex action can improve an organism's chances of survival. *(2 marks)*

4 ⚙ State and explain the considerations a researcher should take into account when planning an investigation into the effect of drugs, such as caffeine, on a group of human volunteers. *(4 marks)*

13.9 Voluntary and involuntary muscles

Specification reference: 5.1.5

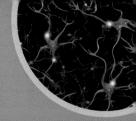

There are around 650 muscles in the body, making up roughly half of the body's weight. The contraction of many muscle cells causes the body to move (Topic 13.10, Sliding filament theory). However, there are many muscle cells in the body whose contractions you are largely unaware of.

Types of muscle

There are three types of muscle in the body:

- Skeletal muscle – skeletal muscles make up the bulk of body muscle tissue. These are the cells responsible for movement, for example, the biceps and triceps.

- Cardiac muscle – cardiac muscle cells are found only in the heart. These cells are myogenic, meaning they contract without the need for a nervous stimulus, causing the heart to beat in a regular rhythm.

- Involuntary muscle (also known as smooth muscle) – involuntary muscle cells are found in many parts of the body – for example, in the walls of hollow organs such as the stomach and bladder. They are also found in the walls of the blood vessels and the digestive tract, where through peristalsis they move food along the gut.

Learning outcomes

Demonstrate knowledge, understanding, and application of:

→ the structure of mammalian muscle and the mechanism of muscular contraction

→ the examination of stained sections or photomicrographs of skeletal muscle.

▼ Table 1 *Important differences in the structure and function of the different types of muscle*

Type of muscle	Skeletal	Cardiac	Involuntary
Fibre appearance	striated	specialised striated	non-striated
Control	conscious (voluntary)	involuntary	involuntary
Arrangement	regularly arranged so muscle contracts in one direction	cells branch and interconnect resulting in simultaneous contraction	no regular arrangement – different cells can contract in different directions
Contraction speed	rapid	intermediate	slow
Length of contraction	short	intermediate	can remain contracted for a relatively long time
Structure	Muscles showing cross striations are known as striated or striped muscles. Fibres are tubular and multinucleated.	Cardiac muscle does show striations but they are much fainter than those in skeletal muscle. Fibres are branched and uninucleated.	Muscles showing no cross striations are called non-striated or unstriped muscles. Fibres are spindle shaped and uninucleated.

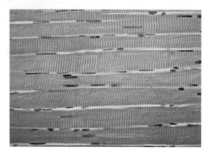

▲ Figure 1 *Skeletal muscle, ×2 magnification*

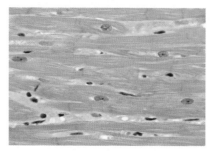

▲ Figure 2 *Cardiac muscle, approx ×300 magnification*

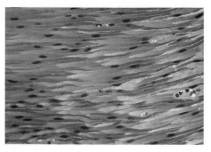

▲ Figure 3 *Involuntary muscle, approx ×157 magnification*

Structure of skeletal muscle

Muscle fibres

Skeletal muscles are made up of bundles of muscle fibres. These are enclosed within a plasma membrane known as the *sarcolemma*.

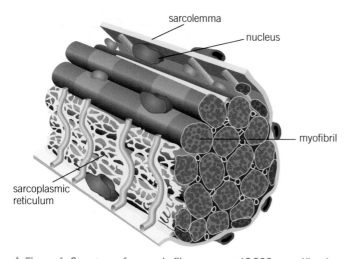

▲ Figure 4 *Structure of a muscle fibre, approx ×10 000 magnification*

The muscle fibres contain a number of nuclei and are much longer than normal cells, as they are formed as a result of many individual embryonic muscle cells fusing together. This makes the muscle stronger, as the junction between adjacent cells would act as a point of weakness. The shared cytoplasm within a muscle fibre is known as *sarcoplasm*.

Parts of the sarcolemma fold inwards (known as transverse or T tubules) to help spread electrical impulses throughout the sarcoplasm. This ensures that the whole of the fibre receives the impulse to contract at the same time.

Muscle fibres have lots of mitochondria to provide the ATP that is needed for muscle contraction. They also have a modified version of the endoplasmic reticulum, known as the sarcoplasmic reticulum. This extends throughout the muscle fibre and contains calcium ions required for muscle contraction.

Myofibrils

Each muscle fibre contains many **myofibrils**. These are long cylindrical organelles made of protein and specialised for contraction. On their own they provide almost no force but collectively they are very powerful. Myofibrils are lined up in parallel to provide maximum force when they all contract together. Myofibrils are made up of two types of protein filament:

- Actin – the thinner filament. It consists of two strands twisted around each other.

- Myosin – the thicker filament. It consists of long rod-shaped fibres with bulbous heads that project to one side.

> **Study tip**
>
> When thinking about the structure of muscles think of a rope. A rope is made up of lots of strings (muscle fibres), which themselves are made up of lots of threads (myofibrils), acting together to give it its strength.

Myofibrils have alternating light and dark bands – these result in their striped appearance:

- Light bands – these areas appear light as they are the region where the actin and myosin filaments do not overlap. (They are also known as isotropic bands or I-bands.)

- Dark bands – these areas appear dark because of the presence of thick myosin filaments. The edges are particularly dark as the myosin is overlapped with actin. (They are also known as anisotropic bands or A-bands.)

- Z-line – this is a line found at the centre of each light band. The distance between adjacent Z-lines is called a **sarcomere**. The sarcomere is the functional unit of the myofibril. When a muscle contracts the sarcomere shortens.

- H-zone – this is a lighter coloured region found in the centre of each dark band. Only myosin filaments are present at this point. When the muscle contracts the H-zone decreases.

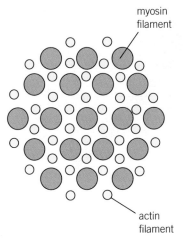

▲ Figure 5 *Transverse section through a myofibril, approx × 50 000 magnification*

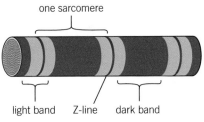

▲ Figure 6 *The structure of a myofibril*

Drawing a labelled diagram of a sarcomere

Key points you should remember when drawing and labelling a sarcomere are:

- Show two Z-lines to demonstrate your understanding of the length of a sarcomere.

- Ensure there are heads present on the myosin filaments.

- Connect actin filaments to the Z-line.

- Clearly label the light and dark bands.

- Show the position of the H-zone.

It is useful to note the position of the A band, I band, H zone, and Z-line but you do not need to learn them.

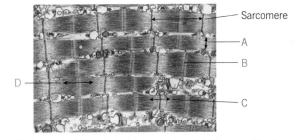

▲ Figure 7 *TEM of skeletal muscle, ×5 000 magnification*

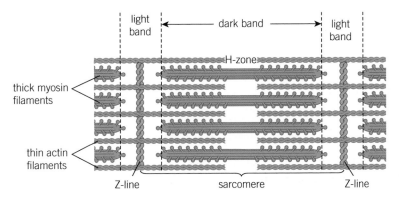

▲ Figure 8 *The structure of a sarcomere*

Why are there light bands within a sarcomere?

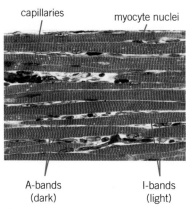

capillaries   myocyte nuclei

A-bands   I-bands
(dark)   (light)

▲ **Figure 9** *Stained skeletal muscle as seen through a microscope, ×570 magnification*

Histology of skeletal muscle

Figure 9 is a stained section of skeletal muscle viewed through a microscope. You should be able to identify the following features:

- Individual muscle fibres – long and thin multinucleated fibres that are crossed with a regular pattern of fine red and white lines.
- The highly structured arrangement of sarcomeres which appear as dark (A-bands) and light (I-bands) bands.
- Streaks of connective and adipose tissue.
- Capillaries running in between the fibres.

➕ Slow-twitch and fast-twitch muscles

There are two types of muscle fibres found in your body. Different muscles in the body have different proportions of each fibre.

Properties of slow-twitch fibres:

- fibres contract slowly
- provide less powerful contractions but over a longer period
- used for endurance activities as they do not tire easily
- gain their energy from aerobic respiration
- rich in myoglobin, a bright red protein which stores oxygen – this makes the fibres appear red
- rich supply of blood vessels and mitochondria.

Slow-twitch fibres are found in large proportions in muscles which help to maintain posture such as those in the back and calf muscles which have to contract continuously to keep the body upright.

Properties of fast-twitch fibres:

- fibres contract very quickly
- produce powerful contractions but only for short periods
- used for short bursts of speed and power as they tire easily
- gain their energy from anaerobic respiration
- pale coloured as they have low levels of myoglobin and blood vessels
- contain more, and thicker, myosin filaments
- store creatine phosphate – a molecule that can rapidly generate ATP from ADP in anaerobic conditions.

Fast-twitch fibres are found in high proportions in muscles which need short bursts of intense activity, such as biceps and eyes.

1 🧪 State and explain how you could tell the difference between areas of fast-twitch and slow-twitch muscle when observing skeletal muscle under the microscope.

2 State and explain any differences in composition which may exist in the skeletal muscles of marathon runners and sprinters.

Summary questions

1 Describe simply the structure of skeletal muscle. (*3 marks*)

2 🧪 Figure 7 shows an image of skeletal muscle viewed under an electron microscope. Name the structures labelled A–D. (*4 marks*)

3 Describe the similarities and differences in the structure and function of cardiac and involuntary muscle. (*4 marks*)

4 The drawings in Figure 10 show myofibrils in transverse section.
   a Describe the difference between taking a transverse and a longitudinal section of muscle. (*1 mark*)
   b State and explain which section through a sarcomere is represented by each image in Figure 10. (*4 marks*)

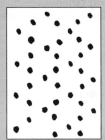 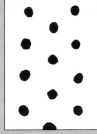 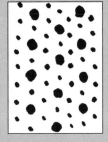

▲ Figure 10 *Transverse sections of striated muscle*

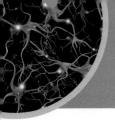

In the previous topic, you looked in detail at the structure of skeletal muscle fibres. In order to contract and cause movement, the actin and myosin filaments within the myofibrils have to slide past each other. Muscle contraction is usually described using the **sliding filament model**.

Sliding filament model

During contraction the myosin filaments pull the actin filaments inwards towards the centre of the sarcomere. This results in:

- the light band becoming narrower
- the Z lines moving closer together, shortening the sarcomere
- the H-zone becoming narrower.

The dark band remains the same width, as the myosin filaments themselves have not shortened, but now overlap the actin filaments by a greater amount.

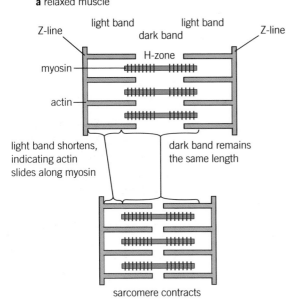

a relaxed muscle

b contracted muscle

▲ Figure 2 *Comparison of a relaxed and contracted sarcomere*

▲ Figure 1 *Electron micrograph of relaxed and contracted sarcomeres*

The simultaneous contraction of lots of sarcomeres means that the myofibrils and muscle fibres contract. This results in enough force to pull on a bone and cause movement. When sarcomeres return to their original length the muscle relaxes.

Structure of myosin

Myosin filaments have globular heads that are hinged which allows them to move back and forwards. On the head is a binding site for each of actin and ATP. The tails of several hundred myosin molecules are aligned together to form the myosin filament.

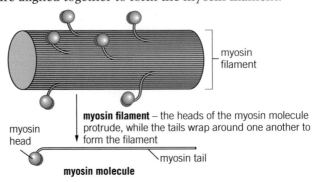

myosin filament – the heads of the myosin molecule protrude, while the tails wrap around one another to form the filament

▶ Figure 3 *Myosin structure*

Structure of actin

Actin filaments have binding sites for myosin heads. These are called actin–myosin binding sites. However, these binding sites are often blocked by the presence of another protein called tropomyosin which is held in place by the protein troponin.

When a muscle is in a resting state (relaxed) the actin–myosin sites are blocked by tropomyosin. The myosin heads can therefore not bind to the actin, and the filaments cannot slide past each other.

When a muscle is stimulated to contract, the myosin heads form bonds with actin filaments known as actin–myosin cross-bridges. The myosin heads then flex (change angle) in unison, pulling the actin filament along the myosin filament. The myosin then detaches from the actin and its head returns to its original angle, using ATP. The myosin then reattaches further along the actin filament and the process occurs again. This is repeated up to 100 times per second.

How muscle contraction occurs

Neuromuscular junction

Muscle contraction is triggered when an action potential arrives at a neuromuscular junction – this is the point where a motor neurone and a skeletal muscle fibre meet (Topics 13.4, Nervous transmission and 13.5, Synapses). There are many neuromuscular junctions along the length of a muscle to ensure that all the muscle fibres contract simultaneously. If only one existed, the muscle fibres would not contract together therefore the contraction of the muscle would not be as powerful. It would also be much slower, as a wave of contraction would have to travel across the muscle to stimulate the individual fibres to contract.

All the muscle fibres supplied by a single motor neurone are known as a motor unit – the fibres act as a single unit. If a strong force is needed, a large number of motor units are stimulated, whereas only a small number are stimulated if a small force is required.

When an action potential reaches the neuromuscular junction, it stimulates calcium ion channels to open. Calcium ions then diffuse from the synapse into the synaptic knob, where they cause synaptic vesicles to fuse with the presynaptic membrane. Acetylcholine is released into the synaptic cleft by exocytosis and diffuses across the synapse. It binds to receptors on the postsynaptic membrane (the sarcolemma), opening sodium ion channels, and resulting in depolarisation.

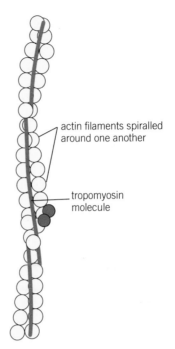

▲ Figure 4 *Actin structure*

actin filaments spiralled around one another

tropomyosin molecule

Study tip

Try to visualise the sliding filament model as a rowing boat containing several rowers. When the oars (myosin heads) are dipped into the river (bind to actin filament), the oars change angle (myosin heads flex), then are removed (myosin heads detach). This is then repeated further along the river. The rowers work in unison, and so the boat and water move relative to one another.

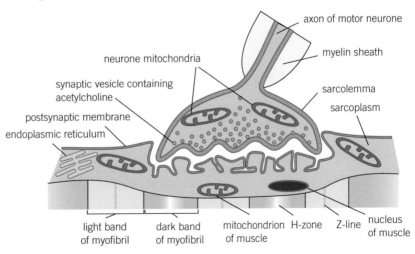

▲ Figure 5 *Neuromuscular junction*

axon of motor neurone

myelin sheath

neurone mitochondria

synaptic vesicle containing acetylcholine

postsynaptic membrane

endoplasmic reticulum

sarcolemma

sarcoplasm

light band of myofibril

dark band of myofibril

mitochondrion of muscle

H-zone

Z-line

nucleus of muscle

Acetylcholine is then broken down by acetylcholinesterase into choline and ethanoic acid. This prevents the muscle being overstimulated. Choline and ethanoic acid diffuse back into the neurone, where they are recombined into acetylcholine, using the energy provided by mitochondria.

Sarcoplasm

The depolarisation of the sarcolemma travels deep into the muscle fibre by spreading through the T-tubules. These are in contact with the sarcoplasmic reticulum. The sarcoplasmic reticulum contains stored calcium ions which it actively absorbs from the sarcoplasm.

When the action potential reaches the sarcoplasmic reticulum it stimulates calcium ion channels to open. The calcium ions diffuse down their concentration gradient flooding the sarcoplasm with calcium ions.

The calcium ions bind to troponin causing it to change shape. This pulls on the tropomyosin moving it away from the actin–myosin binding sites on the actin filament. Now that the binding sites have been exposed the myosin head binds to the actin filament forming an actin–myosin cross-bridge.

Once attached to the actin filament the myosin head flexes, pulling the actin filament along. The molecule of ADP bound to the myosin head is released. An ATP molecule can now bind to the myosin head. This causes the head to detach from the actin filament.

The calcium ions present in the sarcoplasm also activate the ATPase activity of the myosin. This hydrolyses the ATP to ADP and phosphate, releasing energy which the myosin head uses to return to its original position.

The myosin head can now attach itself to another actin–myosin binding site further along the actin filament and the cycle is repeated. The cycle continues as long as the muscle remains stimulated. During the period of stimulation many actin–myosin bridges form and break rapidly, pulling the actin filament along. This shortens the sarcomere and causes the muscle to contract.

Figure 6 summarises what takes place in the sarcoplasm.

Energy supply during muscle contraction

Muscle contraction requires large quantities of energy. This is provided by the hydrolosis of ATP into ADP and phosphate. The energy is required for the movement of the myosin heads and to enable the sarcoplasmic reticulum to actively reabsorb calcium ions from the sarcoplasm. The three main ways ATP is generated are described here – many activities use a combination of these processes.

Aerobic respiration

Most of the ATP used by muscle cells is regenerated from ADP during oxidative phosphorylation. This chemical reaction takes place inside the mitochondria which are plentiful in the muscle. However, this can only occur in the presence of oxygen. Aerobic respiration is therefore used for long periods of low-intensity exercise.

1 Tropomyosin molecule prevents myosin head from attaching to the binding site on the actin molecule.

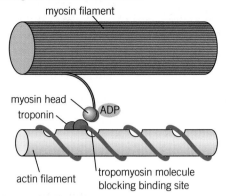

5 ATP molecule fixes to myosin head, causing it to detach from the actin filament.

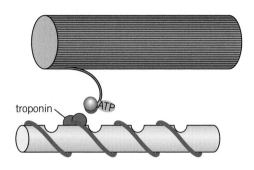

2 Calcium ions released from the endoplasmic reticulum cause the tropomyosin molecule to pull away from the binding sites on the actin molecule.

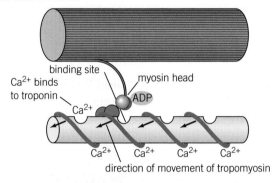

6 Hydrolysis of ATP to ADP by myosin provides the energy for the myosin head to resume its normal position.

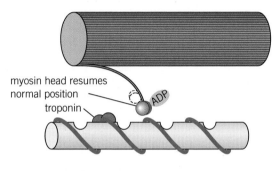

3 Myosin head now attaches to the binding site on the actin filament.

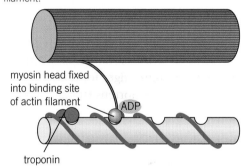

7 Head of myosin reattaches to a binding site further along the actin filament and the cycle is repeated.

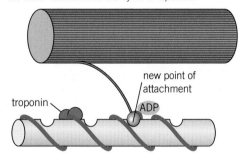

4 Head of myosin changes angle, moving the actin filament along as it does so. The ADP molecule is released.

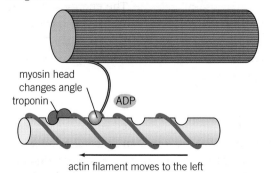

▲ Figure 6 *Interaction of myosin and actin during muscle contraction. Only one myosin head is shown for simplicity*

Anaerobic respiration

In a very active muscle, oxygen is used up more quickly than the blood supply can replace it. Therefore, ATP has to be generated anaerobically. ATP is made by glycolysis but, as no oxygen is present, the pyruvate which is also produced is converted into lactate (lactic acid). This can quickly build up in the muscles resulting in muscle fatigue. Anaerobic respiration is used for short periods of high-intensity exercise, such as sprinting.

Creatine phosphate

Another way the body can generate ATP is by using the chemical *creatine phosphate* which is stored in muscle. To form ATP, ADP has to be phosphorylated – a phosphate group has to be added. Creatine phosphate acts as a reserve supply of phosphate, which is available immediately to combine with ADP, reforming ATP. This system generates ATP rapidly, but the store of phosphate is used up quickly. As a result this is used for short bursts of vigorous exercise, such as a tennis serve. When the muscle is relaxed, the creatine phosphate store is replenished using phosphate from ATP.

> **Synoptic link**
>
> You will study in detail the chemical reactions which take place during respiration in Chapter 18, Respiration.

Monitoring muscle activity with sensors

Sensors can be used to monitor the electrical activity in a muscle. These can be used to measure the strength of a muscle contraction, or to track muscle fatigue levels.

The resultant trace, an electromyogram (EMG), is a record of the electrical activity in a muscle during an activity.

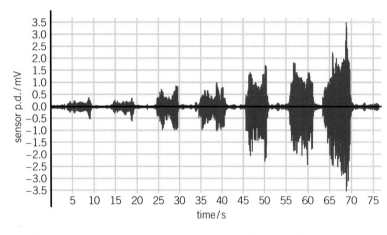

◀ **Figure 7** *EMG sensor trace from a biceps muscle. The subject is applying a gradually increasing force, which results in a larger output trace on the EMG, p.d = partial discharge*

Muscle fatigue is a long-lasting reduction of the ability to contract and exert force. It is normally localised and occurs after prolonged, relatively strong muscle activity. Occasionally, this can be beneficial, through promoting muscle growth (as seen in bodybuilders). However, it is usually harmful – serious injury is most likely to occur when the level of fatigue in a muscle is high.

The detection and classification of muscle fatigue is important in research into human–computer interactions, sport injuries and performance, ergonomics, and prosthetics.

A typical experiment in muscle fatigue research involves a subject performing a set task such as moving a limb in a specified manner. A signal is acquired using sensors attached to the skin, which is recorded and processed to reveal the characteristics of the muscle during that particular exercise.

To gain the signal trace, electromyography uses electrodes which detect the electrical currents created when muscles contract. Changes in this signal are used to identify fatigue, which may include:

- an increase in the mean amplitude of the signal
- a decrease in the frequency of the signal
- disruption to the overall pattern existing within the signal.

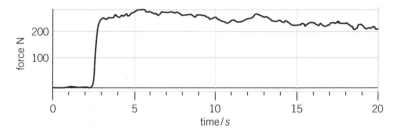

◄ **Figure 8** *At 2.5s the subject grasped a grip strength meter as hard as possible. Notice how the strength of the subject's grip peaked at 5.5s and then gradually declined until recording ended at 20s*

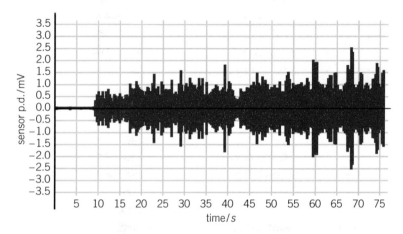

◄ **Figure 9** *EMG sensor trace from a muscle undergoing fatigue. Notice that the electrical activity in the muscle increases over time while, as shown in Figure 8, the maximum force produced by a muscle decreases*

1   Plan an investigation into muscle fatigue, using an EMG to monitor the activity of the muscle.

2   Suggest how the outcomes from the investigation could be used to identify when the muscle became fatigued.

Summary questions

1   State two differences in the appearance of a sarcomere in a relaxed and contracted muscle when observed through a microscope. (*2 marks*)

2   Professional sprinters have high levels of creatine phosphate in their muscle cells. Describe why this is advantageous. (*2 marks*)

3   After a person's death, their body can no longer produce ATP. This results in the stiffening of muscles (rigor mortis). Explain why a lack of ATP prevents muscles relaxing. (*3 marks*)

4   Bepridil is a drug that can be used to treat angina, a form of heart disease. It works by partially blocking calcium ion channels. Explain the effect bepridil will have on heart muscle contraction. (*4 marks*)

Practice questions

1 Tension in muscles is created as the myosin filaments pull the actin filaments towards each other as cross bridges form between the filaments. The resting length of muscle fibres determines, along with the frequency of stimulation, the magnitude of this tension.

Elastic proteins, such as titins, present in muscle resist the overstretching of muscle fibres.

a (i) State the name of cross bridges that form between the filaments. (*1 mark*)

(ii) State the term that describes the fact that both the frequency of stimulation and arrival of action potentials determine the size of the response. (*1 mark*)

The diagram shows the changes in tension as a muscles contracts.

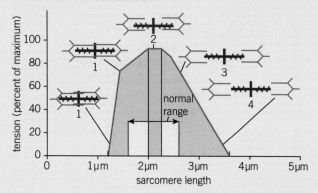

b (i) Describe how muscle tension changes with changing sarcomere length. (*4 marks*)

(ii) State the optimal resting length of a sarcomere. (*2 marks*)

(iii) Explain, using the diagram, why muscle fibres have a small range of optimal resting lengths. (*5 marks*)

2 Multiple sclerosis (MS) is a condition that affects the nervous system. The immune system of people with MS treats parts of the nervous system as foreign, and launches an immune response damaging neurons. An example of the damage caused is shown in the diagram.

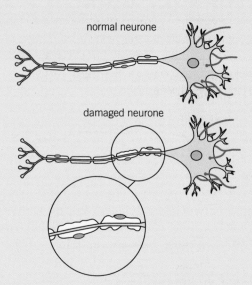

a (i) State the name of the specific part of the nervous system damaged in people with multiple sclerosis. (*1 mark*)

(ii) Describe and explain how this damage would affect the functioning of the nervous system. (*4 marks*)

Some of the symptoms of MS are described here:

temporary loss of vision

loss of balance and co-ordination

incontinence / constipation

temporary loss of sensation in limbs

sensory impairment

b Explain why people with MS may get these symptoms. (*4 marks*)

3 An imbalance in the neurotransmitters in the brain, particularly a shortage of serotonin, is now considered to be one of the causes of depression.

The diagram shows a synapse in the brain which uses serotonin as a neurotransmitter.

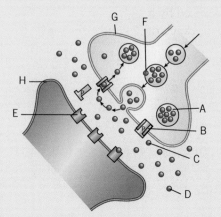

a (i) Name the structures labelled A, F, G, and H. (*4 marks*)

(ii) State the name of the process that results in the release of F into the synaptic cleft. (*1 mark*)

Structure B is a serotonin transporter (SERT) that transports serotonin back into the synaptic knob. Structure P is a drug that is used to treat depression.

b (i) Explain why serotonin needs to be removed from a synapse after it has been released. (*2 marks*)

(ii) Suggest the action of drug P on the synapse in the diagram and explain how this would improve the symptoms of depression. (*4 marks*)

Fluoxetine is an example of a drug used in the treatment of depression. Fluoxetine only inhibits the reuptake of serotonin. It is known as a selective serotonin reuptake inhibitor (SSRI).

Cocaine, a drug of misuse, also inhibits the reuptake of serotonin as well as the neurotransmitters dopamine and noradrenaline.

c Discuss why cocaine is not used to treat depression. (*4 marks*)

4 When patients arrive at hospital with a suspected stroke they are usually given an MRI or CT brain scan. Doctors need to find out as much information as possible to help with the diagnosis.

Strokes are often diagnosed by studying images of the brain produced during these brain scans. New guidelines have suggested that doctors should use MRI scans to diagnose strokes instead of CT scans.

MRI scans are better at detecting stroke damage caused by a lack of blood flow (usually due to a blockage or a blood clot) in the brain compared to CT scans. The majority of strokes are caused by a lack of blood flow in the brain. There is only a short time that treatment used to reverse the damage is effective.

In diffusion MRI, the movement of water in brain tissue is measured. The movement of water is restricted in damaged tissue.

a Suggest why the movement of water is restricted in damaged tissue. (*2 marks*)

CT scans use X-rays to produce multiple images building up a detailed, three-dimensional picture of the brain. An injection of a dye into one of the veins in the arm can be administered during the scan to help improve the clarity of the image.

b Explain how the dye improves the clarity of the images produced. (*2 marks*)

MRI scans use a strong magnetic field and radio waves to produce a detailed picture of the brain. A dye can also be used to improve scan images.

The photos show two brain scans performed on patients after admission to hospital with suspected strokes.

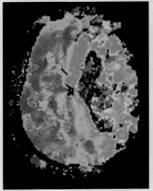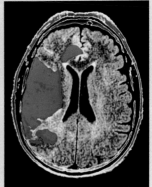

The photo on the left shows a functional magnetic resonance imaging (fMRI) scan of the brain of a 32 year old woman after a massive stroke. This scan shows the amount of blood flow received by areas of the brain. Green and blue areas are receiving normal blood flow, while yellow, red, and black are receiving abnormal blood flow. The lack of blood flow in the right hemisphere is due to a blocked right internal carotid artery. The photo on the right shows a coloured computed tomography (CT) scan of a section through a patient's brain showing internal bleeding (red) due to a stroke.

c Describe the differences between the two images. (*3 marks*)

d Outline the advantages and disadvantages of MRI and CT scans. (*4 marks*)

e Suggest why a CT scan may still be recommended if an MRI scanner is not immediately available for patients that require an emergency injection to break up blood clots. (*2 marks*)

Learning outcomes

Demonstrate knowledge, understanding, and application of:

→ endocrine communication by hormones

→ the structure and functions of the adrenal glands.

In the previous chapter you looked in detail at how the nervous system detects and responds to changes in the internal and external environment. The body has a second system, the endocrine system, which works alongside the neuronal system to react to changes. The endocrine system uses **hormones** to send information about changes in the environment around the body to bring about a designated response.

The endocrine system

Endocrine glands

The endocrine system is made up of **endocrine glands**. An endocrine gland is a group of cells which are specialised to secrete chemicals – these chemicals are known as hormones, and are secreted directly into the bloodstream. Examples of endocrine glands include the pancreas and adrenal glands.

Figure 1 shows the positions of the major endocrine glands in the body and the hormones they secrete. The pituitary gland at the base of the brain makes several hormones, which in turn control the release of other hormones. The close proximity of the pituitary gland to the hypothalamus ensures that the nervous and hormonal responses of the body are closely linked and coordinated.

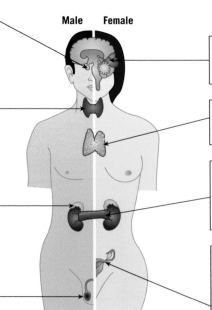

Pituitary gland – produces growth hormone, which controls growth of bones and muscles; anti-diuretic hormone, which increases reabsorption of water in kidneys; and gonadotrophins, which control development of ovaries and testes.

Thyroid gland – produces thyroxine which controls rate of metabolism and rate that glucose is used up in respiration, and promotes growth.

Adrenal gland – produces adrenaline which increases heart and breathing rate and raises blood sugar level.

Testis – produces testosterone which controls sperm production and secondary sexual characteristics.

Male Female

Pineal gland – produces melatonin which affects reproductive development and daily cycles.

Thymus – produces thymosin which promotes production and maturation of white blood cells.

Pancreas – produces insulin which converts excess glucose into glycogen in the liver; and glucagon, which converts glycogen back to glucose in the liver.

Ovary – produces oestrogen, which controls ovulation and secondary sexual characteristics; and progesterone, oestrogen, which controls ovulation and secondary sexual characteristics; and progesterone, which prepares the uterus lining for receiving an embryo.

▲ Figure 1 *Position of the major endocrine glands in the body, hormones are highlighted in green, white boxes are features common to both sexes*

(Remember that, by contrast, exocrine glands, such as those in the digestive system, secrete chemicals through ducts into organs, or to the surface of the body.)

Hormones

Hormones are often referred to as chemical messengers because they carry information from one part of the body to another. They can be steroids, proteins, glycoproteins, polypeptides, amines, or tyrosine derivatives. Although they are chemically different, they share many characteristics.

Hormones are secreted directly into the blood when a gland is stimulated. This can occur as a result of a change in concentration of a particular substance, such as blood glucose concentration. It can also occur as the result of another hormone or a nerve impulse.

Once secreted, the hormones are transported in the blood plasma all over the body. The hormones diff⸻ ⸻ of the blood and bind to specific receptors for that hormone, f⸻ ⸻ membranes, or in the cytoplasm of cells in the target organ⸻ ⸻ as **target cells**. Once bound to their receptors the ⸻ ⸻ target cells to produce a response. You wi⸻ ⸻ d glucose concentration is maintained ⸻ ⸻ ose concentration.

The ty⸻ ⸻ effect on a target cell. ⸻

- *Ster⸻
  throu⸻
  memb⸻
  receptors⸻
  complex. T⸻
  cytoplasm or⸻
  hormone. The ⸻
  formed acts as a t⸻
  turn facilitates or in⸻
  a specific gene. Oestro⸻
  hormone which works i⸻

- *Non-steroid hormones* are hy⸻ ⸻ ot
  pass directly through the cell ⸻ ⸻ .
  Instead they bind to specific rec⸻ ⸻ rs on
  the cell surface membrane of the target cell.
  This triggers a cascade reaction mediated
  by chemicals called second messengers.
  Adrenaline is an example of a hormone which
  works in this way.

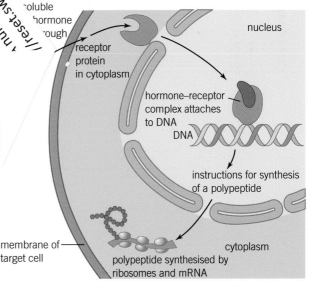

▲ Figure 2 *Mechanism of action of steroid hormones*

Labels: nucleus; receptor protein in cytoplasm; hormone–receptor complex attaches to DNA; DNA; instructions for synthesis of a polypeptide; membrane of target cell; polypeptide synthesised by ribosomes and mRNA; cytoplasm; soluble hormone through

Hormonal versus neuronal communication

As hormones are not released directly onto their target cells, this results in a slower and less specific form of communication than neuronal communication. However, as hormones are not broken down as quickly as neurotransmitters, it can result in a much longer lasting and widespread effect. For example, the hormones insulin and

Synoptic link

You will find out how the second messenger model works in Topic 15.6, The kidney and osmoregulation.

glucagon are responsible for controlling blood glucose concentration. A number of organs are involved in this response.

Table 1 summarises the main differences between the actions of the hormonal and nervous systems.

▼ Table 1 *Comparison of the hormonal and nervous systems*

Hormonal system	Nervous system
communication is by chemicals called hormones	communication is by nerve impulses
transmission is by the blood system	transmission is by neurones
transmission is usually relatively slow	transmission is very rapid
hormones travel to all parts of the body, but only target organs respond	nerve impulses travel to specific parts of the body
response is widespread	response is localised
response is slow	response is rapid
response is often long-lasting	response is short-lived
effect may be permanent and irreversible	effect is temporary and reversible

Adrenal glands

The adrenal glands are two small glands that measure approximately 3 cm in height and 5 cm in length. They are located on top of each kidney and are made up of two distinct parts surrounded by a capsule:

- The adrenal cortex – the outer region of the glands. This produces hormones that are vital to life, such as cortisol and aldosterone.
- The adrenal medulla – the inner region of the glands. This produces non-essential hormones, such as adrenaline which helps the body react to stress.

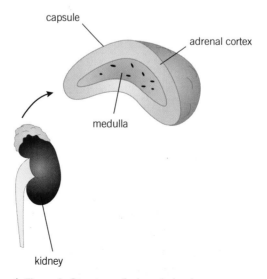

▲ Figure 3 *Structure of adrenal glands*

Adrenal cortex

The production of hormones by the adrenal cortex is itself controlled by hormones released from the pituitary gland in the brain. There are three main types of hormones produced by the adrenal cortex:

- *Glucocorticoids.* These include cortisol which helps regulate metabolism by controlling how the body converts fats, proteins, and carbohydrates to energy. It also helps regulate blood pressure and cardiovascular function in response to stress. Another glucocorticoid hormone released is corticosterone. This works with cortisol to regulate immune response and suppress inflammatory reactions. The release of these hormones is controlled by the hypothalamus.

- *Mineralocorticoids.* The main one produced is aldosterone which helps control blood pressure by maintaining the balance between salt and water concentrations in the blood and body fluids. Its release is mediated by signals triggered by the kidney.

- *Androgens.* Small amounts of male and female sex hormones are released – their impact is relatively small compared with the larger amounts of hormones, such as oestrogen and testosterone, released by the ovaries or testes after puberty, but they are still important, especially in women after the menopause.

Adrenal medulla

The hormones of the adrenal medulla are released when the sympathetic nervous system is stimulated. This occurs when the body is stressed. You can find out more about the fight or flight response in Topic 14.5, Coordinated responses.

The hormones secreted by the adrenal medulla are:

- *Adrenaline.* This increases the heart rate sending blood quickly to the muscles and brain. It also rapidly raises blood glucose concentration levels by converting glycogen to glucose in the liver.

- *Noradrenaline.* This hormone works with adrenaline in response to stress, producing effects such as increased heart rate, widening of pupils, widening of air passages in the lungs, and the narrowing of blood vessels in non-essential organs (resulting in higher blood pressure).

Summary questions

1   Using a named example, explain the function of an endocrine gland.
    (2 marks)

2   Describe the pathway triggered by a stimulus in hormonal communication.   *(2 marks)*

3   Bright light causes the iris muscles in your eyes to contract, constricting the pupil and preventing damage to the eye. State and explain whether hormonal or neuronal communication would be used in this response.   *(2 marks)*

4   A person falls into a fast-flowing river. State and explain the changes that may occur in the body and increase the person's chances of survival in this situation.   *(6 marks)*

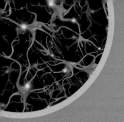

14.2 Structure and function of the pancreas

Specification reference: 5.1.4

The pancreas is found in the upper abdomen, behind the stomach (Figure 1). It plays a major role in controlling blood glucose concentration, and in digestion. It is a glandular organ – its role is to produce and secrete hormones and digestive enzymes.

Function of the pancreas

The pancreas has two main functions in the body, as an:

● exocrine gland – to produce enzymes and release them via a duct into the duodenum

● endocrine gland – to produce hormones and release them into the blood.

Role as an exocrine gland

Most of the pancreas is made up of exocrine glandular tissue. This tissue is responsible for producing digestive enzymes and an alkaline fluid known as pancreatic juice. The enzymes and juice are secreted into ducts which eventually lead to the pancreatic duct. From here they are released into the duodenum, the top part of the small intestine. The pancreas produces three important types of digestive enzymes:

● Amylases – break down starch into simple sugars. For example, pancreatic amylase.

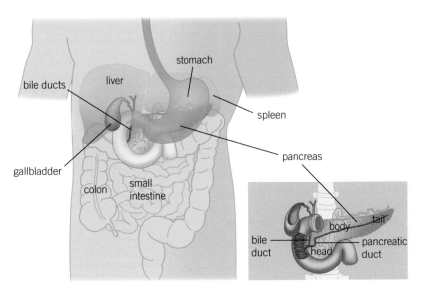

▲ Figure 1 Position of the pancreas in the body

● Proteases – break down proteins into amino acids. For example, trypsin.

● Lipases – break down lipids into fatty acids and glycerol. For example, pancreatic lipase.

Role as an endocrine gland

The pancreas is responsible for producing insulin and glucagon. These two hormones play an essential role in controlling blood glucose concentration, which you will read about in Topic 14.3, Regulation of blood glucose concentration. Within the exocrine tissue there are small regions of endocrine tissue called **islets of Langerhans**. The cells of the islets of Langerhans are responsible for producing insulin and glucagon, and secreting these hormones directly into the bloodstream.

Histology of the pancreas

When viewed under a microscope, you can clearly see the differences between endocrine and exocrine pancreatic tissue. The main differences are summarised in Table 1.

▼ Table 1

Structure	Appearance	Shape	Type of tissue	Function
islets of Langerhans	lightly stained	large, spherical clusters	endocrine pancreas	produce and secrete hormones
pancreatic acini (singular – acinus)	darker stained	small, berry-like clusters	exocrine pancreas	produce and secrete digestive enzymes

Islets of Langerhans

Within the islets of Langerhans are different types of cell. They are classified according to the hormone they secrete:

- α (alpha) cells – these produce and secrete glucagon
- β (beta) cells – these produce and secrete insulin

Alpha cells are larger and more numerous than beta cells within an islet.

Using standard staining techniques, it is often very difficult to distinguish between the cell types within an islet of Langerhans. In Figure 2, a differential stain has been used. The β cells of the islets that produce insulin are stained blue, and the α cells that produce glucagon are stained pink.

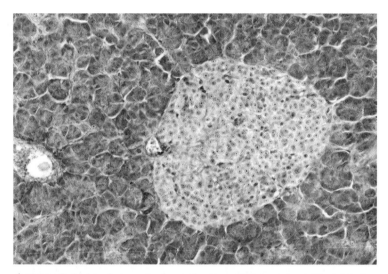

▲ **Figure 2** *Light micrograph of a section through the pancreas. The islet of Langerhans (centre right) is composed of groups of secretory cells. The main secretions from these cells are the hormones insulin and glucagon, which control blood sugar. These cells are endocrine – their secretions go straight into the bloodstream. The cells surrounding the islet are packed into secretory acini (pink) which secrete digestive enzymes. This part of the pancreas is exocrine – the enzymes pass straight out into the gut, via ducts. The structure on the left is a branch of the pancreatic duct, ×195 magnification*

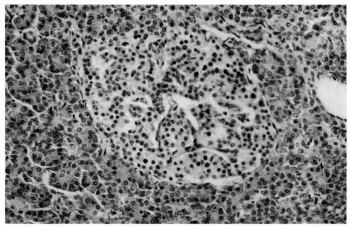

▲ **Figure 3** *Islet of Langerhan cell in human pancreas, ×300 magnification*

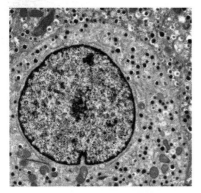

▲ **Figure 4** *A colour transmission electron micrograph of a pancreatic alpha cell. Alpha cells make up 15–20% of the cells in the islets of Langerhans. They produce glucagon, a hormone that regulates blood sugar levels by increasing the amount of glucose available. The glucagon is carried into the bloodstream by granules (dark red). The nucleus (round, blue) and mitochondria (brown) can also be seen, ×5 000 magnification*

Summary questions

1 State the difference between endocrine and exocrine glandular tissue.   (*1 mark*)

2 State two features which would enable you to identify the islets of Langerhans through a cross-section of pancreatic material, when viewed under a light microscope.   (*2 marks*)

3 Figure 6 shows pancreatic tissue which has been stained for insulin. Areas which contain insulin appear brown. Describe the structure and function of the pancreatic tissue shown on this slide.   (*6 marks*)

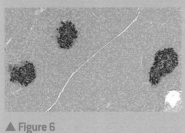

▲ Figure 6

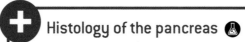

Histology of the pancreas

The following activity will allow you to view the structures within the pancreas.

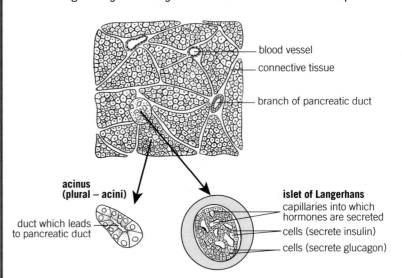

▲ **Figure 5** *Cross-section of pancreatic material, as viewed through a light microscope*

Place a stained slide of pancreatic tissue on the stage of a light microscope.

Begin by viewing under low power. Compare this with Figure 5 – can you identify the following structures?

a Exocrine tissue, which secretes digestive enzymes
b Blood vessels
c Branches of the pancreatic duct.

Select one of the groups of exocrine cells and view under high power. You will be viewing an acinus – can you identify the following features?

d Small group of cells formed in a cluster
e Clearly visible nuclei within these cells
f Central duct (which leads to the pancreatic duct).

Make a scientific drawing of an acinus. Remember to include a scale on your drawing and label all of the key features.

Return to low power and look for small groups of cells which look different to the rest of the material. These are the islets of Langerhans. These groups of cells are likely to appear lighter than the surrounding material.

View an islet of Langerhans under high power. Can you spot the following?

g Capillaries, which transport the secreted hormones
h α cells, which secrete glucagon
i β cells, which secrete insulin.

You may find that your slide has been prepared using a differential stain, which colours the α and β cells differently to enable you to identify these structures more readily.

Make a drawing of an islet of Langerhans. Remember to include a scale on your drawing and label all of the key features.

14.3 Regulation of blood glucose concentration

Specification reference: 5.1.4

During respiration the body uses glucose to produce ATP. To remain healthy it is important that the concentration of glucose in your blood is kept constant. Without control, blood glucose concentration would range from very high levels after a meal, to very low levels several hours later. At these very low levels cells would not have enough glucose for respiration. Blood glucose concentration is kept constant by the action of the two hormones – insulin and glucagon.

Learning outcomes

Demonstrate knowledge, understanding, and application of:

→ how blood glucose concentration is regulated.

Increasing blood glucose concentration

Glucose is a small, soluble molecule that is carried in the blood plasma. Blood glucose is normally maintained at a concentration of around $90\,\mathrm{mg\,cm^{-3}}$ of blood. Blood glucose concentration can increase as a result of:

- Diet – when you eat carbohydrate-rich foods such as pasta and rice (which are rich in starch) and sweet foods such as cakes and fruit (which contain high levels of sucrose), the carbohydrates they contain are broken down in the digestive system to release glucose. The glucose released is absorbed into the bloodstream, and the blood glucose concentration rises.

- **Glycogenolysis** – glycogen stored in the liver and muscle cells is broken down into glucose which is released into the bloodstream increasing blood glucose concentration.

- **Gluconeogenesis** – the production of glucose from non-carbohydrate sources. For example, the liver is able to make glucose from glycerol (from lipids) and amino acids. This glucose is released into the bloodstream and causes an increase in blood glucose concentration.

Decreasing blood glucose concentration

Blood glucose concentration can be decreased by:

- Respiration – some of the glucose in the blood is used by cells to release energy. This is required to perform normal body functions. However, during exercise, more glucose is needed as the body needs to generate more energy in order for muscle cells to contract. The higher the level of physical activity, the higher the demand for glucose and the greater the decrease of blood glucose concentration.

- **Glycogenesis** – the production of glycogen. When blood glucose concentration is too high, excess glucose taken in through the diet is converted into glycogen which is stored in the liver.

Role of insulin

Insulin is produced by the β cells of the islets of Langerhans in the pancreas. If the blood glucose concentration is too high, the

Study tip

There are a lot of new terms in this topic – if you remember the root of the words it will help you remember the chemical process they are referring to:

- **lysis** – means splitting
- neo – means new
- genesis – means birth/origin

Therefore, glycogeno**lysis** means the splitting of glycogen (to produce glucose) and gluconeogenesis means the formation of new glucose.

β cells detect this rise in blood glucose concentration and respond by secreting insulin directly into the bloodstream.

Virtually all body cells have insulin receptors on their cell surface membrane (an exception being red blood cells). When insulin binds to its glycoprotein receptor, it causes a change in the tertiary structure of the glucose transport protein channels. This causes the channels to open allowing more glucose to enter the cell. Insulin also activates enzymes within some cells to convert glucose to glycogen and fat.

Insulin therefore lowers blood glucose concentration by:

- increasing the rate of absorption of glucose by cells, in particular skeletal muscle cells
- increasing the respiratory rate of cells – this increases their need for glucose and causes a higher uptake of glucose from the blood
- increasing the rate of glycogenesis – insulin stimulates the liver to remove glucose from the blood by turning the glucose into glycogen and storing it in the liver and muscle cells
- increasing the rate of glucose to fat conversion
- inhibiting the release of glucagon from the α cells of the islets of Langerhans.

Insulin is broken down by enzymes in the cells of the liver. Therefore, to maintain its effect it has to be constantly secreted. Depending on the food eaten, insulin secretion can begin within minutes of the food entering the body and may continue for several hours after eating.

As blood glucose concentration returns to normal, this is detected by the β cells of the pancreas. When it falls below a set level, the β cells reduce their secretion of insulin. This is an example of *negative feedback*. Negative feedback ensures that, in any control system, changes are reversed and returned back to the set level.

Role of glucagon

Glucagon is produced by the α cells of the islets of Langerhans in the pancreas. If the blood glucose concentration is too low, the α cells detect this fall in blood glucose concentration and respond by secreting glucagon directly into the bloodstream.

Unlike insulin, the only cells in the body which have glucagon receptors are the liver cells and fat cells – therefore these are the only cells that can respond to glucagon.

Glucagon raises blood glucose concentration by:

- glycogenolysis – the liver breaks down its glycogen store into glucose and releases it back into the bloodstream
- reducing the amount of glucose absorbed by the liver cells
- increasing gluconeogenesis – increasing the conversion of amino acids and glycerol into glucose in the liver.

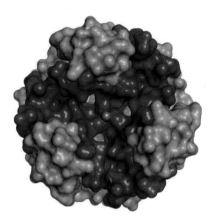

▲ Figure 1 *Computer model showing the structure of a molecule of the hormone insulin. Insulin is a globular protein made up of 51 amino acids which are arranged into two chains*

Synoptic link

You will find out more about the structure and function of the liver in Topic 15.4, Excretion, homeostasis, and the liver.

▲ Figure 2 *Computer model showing the structure of glucagon. The secondary structure of the hormone as a coiled ribbon can be seen and the atoms it is made up of. Atoms are colour-coded spheres (carbon: grey, nitrogen: blue, and oxygen: red)*

As blood glucose concentration returns to normal, this is detected by the α cells of the pancreas. When it rises above a set level, the α cells reduce their secretion of glucagon. This is another example of negative feedback. The feedback causes the corrective measures to be switched off, returning the system to its original (normal) level.

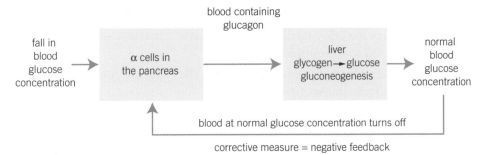

▲ **Figure 3** *Negative feedback in control of blood glucose concentration*

Interaction of insulin and glucagon

Figure 4 shows how insulin and glucagon work together to maintain a constant blood glucose concentration. Insulin and glucagon are antagonistic hormones, that is, they work against each other.

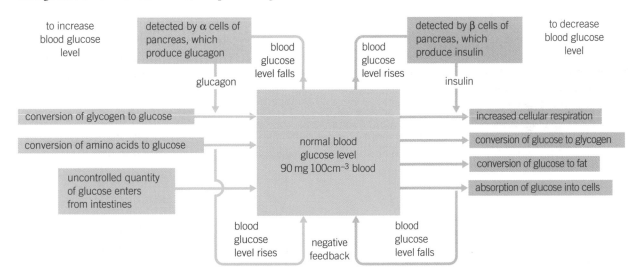

▲ **Figure 4** *Summary of blood glucose concentration regulation*

The system of maintaining blood glucose concentration is said to be self-regulating, as it is the level of glucose in the blood that determines the quantity of insulin and glucagon that is released. Blood glucose concentration is not constant, but fluctuates around a set point as the result of negative feedback. In times of stress adrenaline is released by the body. One of the effects of this hormone is to raise the blood glucose concentration to allow more respiration to occur. You will find out more about the fight and flight response in Topic 14.5, Coordinated responses.

Control of insulin secretion

When blood glucose concentration rises above the set level, this is detected by the β cells in the islets of Langerhans and insulin is released. The mechanism by which this occurs is as follows:

1 At normal blood glucose concentration levels, potassium channels in the plasma membrane of β cells are open and potassium ions diffuse out of the cell. The inside of the cell is at a potential of −70 mV with respect to the outside of the cell.

2 When blood glucose concentration rises, glucose enters the cell by a glucose transporter.

3 The glucose is metabolised inside the mitochondria, resulting in the production of ATP.

4 The ATP binds to potassium channels and causes them to close. They are known as ATP-sensitive potassium channels.

5 As potassium ions can no longer diffuse out of the cell, the potential difference reduces to around −30 mV and depolarisation occurs.

6 Depolarisation causes the voltage-gated calcium channels to open.

7 Calcium ions enter the cell and cause secretory vesicles to release the insulin they contain by exocytosis.

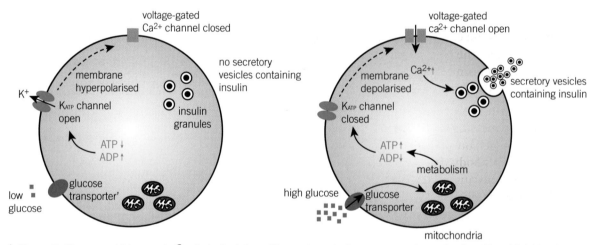

▲ Figure 5 Changes which occur in β cells in the islets of Langerhans in the pancreas when stimulated by a high blood glucose concentration

Summary questions

1 Describe what is meant by negative feedback. (1 mark)

2 Describe the role glucagon plays in the control of blood glucose concentration. (3 marks)

3 Describe the changes which take place inside a β cell to cause the release of insulin in the presence of high blood glucose concentration. (4 marks)

4 Explain how hormones return blood glucose concentration to normal after a meal. (6 marks)

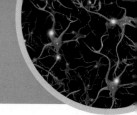

To keep blood glucose concentration constant the body relies on the interaction between glucagon and insulin. However, for over 300 million people in the world this system of regulation does not work properly. They suffer from the chronic disease, diabetes mellitus (usually referred to as diabetes). This means they are unable to metabolise carbohydrates properly, in particular glucose.

Types of diabetes

If you suffer from diabetes your pancreas either does not produce enough insulin, or your body cannot effectively respond to the insulin produced. This means that blood glucose concentration remains high. Hyperglycaemia, or raised blood sugar, is a common effect of uncontrolled diabetes. Over time this can lead to serious damage of many body systems, especially the nerves and blood vessels.

There are two main types of diabetes:

- Type 1 diabetes. Patients with type 1 diabetes are unable to produce insulin. The β cells in the islets of Langerhans do not produce insulin. The cause of type 1 diabetes is not known and so, at the moment, the disease cannot be prevented or cured. It is possible, however, to treat the symptoms. Evidence suggests that in many cases the condition arises as a result of an autoimmune response where the body's own immune system attacks the β cells. This condition normally begins in childhood, and people develop symptoms of the disease quickly.

- Type 2 diabetes. Patients with type 2 diabetes cannot effectively use insulin and control their blood sugar levels. This is either because the person's β cells do not produce enough insulin *or* the person's body cells do not respond properly to insulin. This is often because the glycoprotein insulin receptor on the cell membrane does not work properly. The cells lose their responsiveness to insulin, and therefore do not take up enough glucose, leaving it in the bloodstream. Globally, approximately 90% of people with diabetes have type 2 diabetes. This is largely as a result of excess body weight, physical inactivity, and habitual, excessive overeating of (refined) carbohydrates. Symptoms are similar to those of type 1 diabetes, but are often less severe and develop slowly. As a result, the disease is often only diagnosed after complications have already arisen. Risk of type 2 diabetes increases with age. Until recently, this type of diabetes was seen only in adults (normally over the age of 40), but it is now also occurring in children.

▼ Table 1 *Symptoms of diabetes*

Common symptoms of diabetes
• High blood glucose concentration
• Glucose present in urine
• Excessive need to urinate (polyuria)
• Excessive thirst (polydipsia)
• Constant hunger
• Weight loss
• Blurred vision
• Tiredness

▲ Figure 1 *People with diabetes have to measure their blood glucose levels regularly. This is normally done using a finger prick test*

Type 2 diabetes

There has been a significant increase in the number of cases of diabetes diagnosed in the UK, rising from 1.4 million in 1996 to around 3 million in 2014. It is estimated that there may be as many as 5 million sufferers of diabetes by 2025, with around 85% having type 2 diabetes.

A clear causal link has been established between obesity and the onset of type 2 diabetes. This information has been used to launch initiatives to promote healthy eating and exercise. These include, for example, the Change4Life campaign. This is an example of where scientific evidence has been used to inform decision-making at a national level.

Discussion

1 To what extent is a government responsible for the health of its citizens?
2 What steps would you take to minimise the risk of diabetes amongst the UK population?
3 Should universal benefits – for example, free healthcare – be available to those whose lifestyle choices cause the onset of a medical condition?

Study tip

Most of the symptoms of diabetes are logical. Think about what would happen if your blood glucose concentration remained high, but the level of glucose in your cells was low.

▲ **Figure 2** *People with type 1 diabetes need regular insulin injections. As insulin is a protein, it cannot be taken by mouth because it will be digested*

Diabetes treatment

Diabetes is not a curable disease, but it can be controlled successfully, allowing sufferers to lead a normal life. Treatment differs for both types of diabetes.

Type 1 diabetes

Type 1 diabetes is controlled by regular injections of insulin and is therefore said to be insulin-dependent.

People with the condition have to regularly test their blood glucose concentration, normally by pricking their finger. The drop of blood is then analysed by a machine, which tells the person their blood glucose concentration. Based on this concentration, the person can work out the dose of insulin they need to inject. The insulin administered increases the amount of glucose absorbed by cells and causes glycogenesis to occur, resulting in a reduction of blood glucose concentration.

If a person with diabetes injects himself or herself with too much insulin, they may experience hypoglycaemia (very low blood glucose concentrations) that can result in unconsciousness. However, too low an insulin dose results in hyperglycaemia, which can also result in unconsciousness and death if left untreated. Careful monitoring and dose regulation is therefore required.

Figure 3 shows how blood glucose concentration and insulin levels vary in a person with type 1 diabetes, and a person without diabetes.

If the person with diabetes injects himself or herself with insulin, there will be a surge of insulin in their blood which will cause their blood glucose level to drop quickly.

Type 2 diabetes

The first line of control in type 2 diabetes is to regulate the person's carbohydrate intake through their diet and matching this to their exercise levels. This often involves increasing exercise levels. Overweight people are also encouraged to lose weight.

In some cases, diet and exercise are not enough to control blood glucose concentration so drugs also have to be used. These can include drugs that stimulate insulin production, drugs that slow down the rate at which the body absorbs glucose from the intestine, and ultimately even insulin injections.

Medically produced insulin

Originally, insulin was obtained from the pancreas of cows and pigs which had been slaughtered for food. This process was difficult and expensive. The insulin extracted could also cause allergic reactions as it differed slightly from human insulin.

In 1955, the structure of human insulin was identified and it is now made by genetically modified bacteria. This has a number of advantages:

- Human insulin is produced in a pure form – this means it is less likely to cause allergic reactions.

- Insulin can be produced in much higher quantities.

- Production costs are much cheaper.

- People's concerns over using animal products in humans, which may be religious or ethical, are overcome.

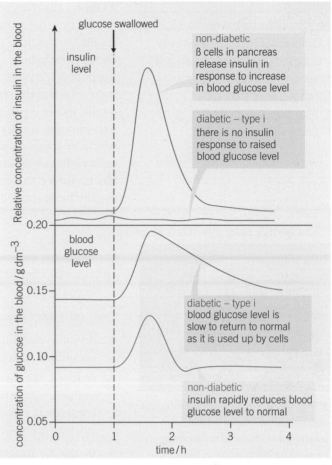

▲ **Figure 3** *Comparison of blood glucose and insulin levels in a person with type 1 diabetes and a person without diabetes after each has swallowed a glucose tablet*

Potential use of stem cells in diabetes treatment

For decades, diabetes researchers have been searching for ways to replace the faulty β cells in the pancreatic islets of diabetic sufferers. Each year, over 1000 people with type 1 diabetes receive a pancreas transplant. After a year, over 80% of these patients have no symptoms of diabetes and do not have to take insulin. However, the demand for transplantable pancreases far outweighs their availability. The risk of having a transplant can also be a greater health risk than the diabetes itself – immunosuppressant drugs are required to ensure the body accepts the transplanted pancreas, which can leave a person susceptible to infection.

Doctors have attempted to cure diabetes by injecting patients with pancreatic β islet cells, but fewer than 8% of cell transplants performed have been successful. The immunosuppressant drugs used to prevent rejection of these cells increases the metabolic demand on insulin-producing cells. Eventually this exhausts their capacity to produce insulin.

Synoptic link

You will find out how bacteria are genetically engineered to produce human insulin in Topic 21.4, Genetic engineering, and about the industrial production of insulin in 22.5, Microorganisms, medicines, and bioremediation.

Synoptic link

Look back at Topic 6.5, Stem cells to remind yourself of the different types of stem cell that exist and the sources of stem cells.

As type 1 diabetes results from the loss of a single cell type, and there is evidence that a relatively small number of islet cells can restore insulin production, the disease is a perfect candidate for stem cell therapy. Totipotent stem cells have the potential to grow into any of the body's cell types. Scientists have been researching the best type of stem cells and the signals required to promote their differentiation into β cells, either directly in the patient or in the laboratory before being transplanted. It is likely that the stem cells used in diabetes treatment would be taken from embryos. To obtain the stem cells, the early embryo has to be destroyed. This means destroying a potential human life. However, the embryos used as a source for these stem cells would usually be destroyed anyway – they are 'spare' embryos from infertility treatments or from terminated pregnancies.

Stem cells lines formed from a small number of embryos can be used to treat many patients – each treatment does not require a separate embryo. An alternative to using embryonic matter is that of using preserved umbilical stem cells.

Stem cells offer many advantages over current therapies:

- donor availability would not be an issue – stem cells could produce an unlimited source of new β cells
- reduced likelihood of rejection problems as embryonic stem cells are generally not rejected by the body (although some evidence contradicts this). Stem cells can also be made by somatic cell nuclear transfer (SCNT)
- people no longer have to inject themselves with insulin.

However, because our ability to control growth and differentiation in stem cells is still limited, a major consideration is whether any precursor or stem-like cells transplanted into the body might induce the formation of tumours as a result of unlimited cell growth.

▲ **Figure 4** *A human embryo. Embryonic stem cells could lead to the discovery of new medical treatments (such as the replacement of β cells in diabetes patients) that would alleviate the suffering of many people. However, many people argue against using the cells, as an embryo has to be destroyed, approx ×600 magnification*

Summary questions

1   Copy and complete the table to compare and contrast the differences between the causes of type 1 and type 2 diabetes.   *(3 marks)*

	Type 1	Type 2
Cause		
When does it develop		
Period of development		

2   Explain why people with type 1 have to constantly monitor their blood glucose concentration.   *(3 marks)*

3   Explain why some who suffer from diabetes mellitus who can produce insulin cannot control their blood glucose concentration.   *(3 marks)*

4   Discuss the advantages and disadvantages of the different treatments for diabetes.   *(6 marks)*

14.5 Coordinated responses

Specification reference: 5.1.5

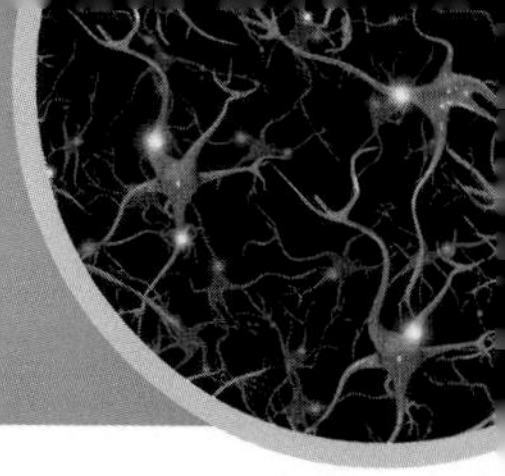

On many occasions the body responds to changes in its internal and external environment through a coordinated response. The nervous and endocrine systems work together to detect and respond appropriately to stimuli. One example of coordination between these two systems is the mammalian 'fight or flight' response.

Learning outcomes

Demonstrate knowledge, understanding, and application of:

→ the coordination of responses by the nervous and endocrine systems.

Fight or flight response

The fight or flight response is an instinct that all mammals possess. When a potentially dangerous situation is detected, the body automatically triggers a series of physical responses. These are intended to help mammals survive by preparing the body to either run or fight for life, hence the name of the response.

Once a threat is detected by the autonomic nervous system, the hypothalamus communicates with the sympathetic nervous system and the adrenal–cortical system. The sympathetic nervous system uses neuronal pathways to initiate body reactions whereas the adrenal–cortical system uses hormones in the bloodstream. The combined effects of these two systems results in the fight or flight response. The overall process is summarised in Figure 1.

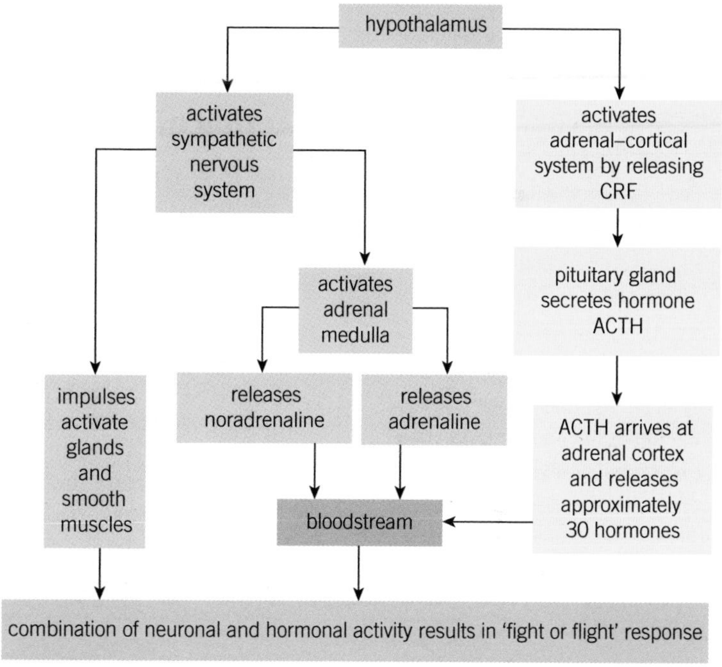

▲ **Figure 1** *Summary of coordination of 'fight or flight' response*

The sympathetic nervous system sends out impulses to glands and smooth muscles and tells the adrenal medulla to release adrenaline and noradrenaline into the bloodstream. These 'stress hormones' cause several changes in the body, including an increased heart rate.

The release of other stress hormones which have a longer-term action from the adrenal cortex is controlled by hormones produced by the pituitary gland in the brain. The hypothalamus stimulates the pituitary gland to secrete adrenocorticotropic hormone (ACTH). This travels in the bloodstream to the adrenal cortex, where it activates the release of many hormones that prepare the body to deal with a threat. Look back at Topic 14.1, Hormonal communication to remind yourself of the structure and function of the adrenal glands.

The physiological responses which occur as part of the fight or flight response are summarised in Table 1.

▼ **Table 1** *Fight or flight physiological responses*

Physical response	Purpose
heart rate increases	to pump more oxygenated blood around the body
pupils dilate	to take in as much light as possible for better vision
arterioles in skin constrict	more blood to major muscle groups, brain, heart, and muscles of ventilation
blood glucose level increases	increase respiration to provide energy for muscle contraction
smooth muscle of airways relaxes	to allow more oxygen into lungs
non-essential systems (like digestion) shut down	to focus resources on emergency functions
difficulty focusing on small tasks	brain solely focused only on where threat is coming from

Study tip

Even though the fight or flight response is automatic it is normally triggered as a false alarm, such as the panic you may feel when sitting an exam – there is no threat to your survival. The amygdala (part of the brain that initiates the fight or flight response) can't distinguish between a real threat and a perceived threat.

Action of adrenaline

One of adrenaline's main functions during the fight and flight response is to trigger the liver cells to undergo glycogenolysis so that glucose is released into the bloodstream. This allows respiration to increase so more energy is available for muscle contraction.

Adrenaline is a hormone. It is hydrophilic therefore cannot pass through cell membranes. Adrenaline binds with receptors on the surface of a liver cell membrane and triggers a chain reaction inside the cell:

● When adrenaline binds to its receptor, the enzyme *adenylyl cyclase* (which is also present in the cell membrane) is activated.

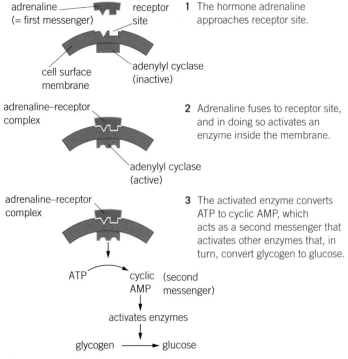

▲ **Figure 2** *Second messenger model of hormone action*

- Adenylyl cyclase triggers the conversion of ATP into *cyclic adenosine mono-phosphate* (cAMP) on the inner surface of the cell membrane in the cytoplasm.

- The increase in cAMP levels activates specific enzymes called *protein kinases* which phosphorylate, and hence activate, other enzymes. In this example, enzymes are activated which trigger the conversion of glycogen into glucose.

This model of hormone action is known as the *second messenger model*. The hormone is known as the first messenger (in this example, adrenaline) and cAMP is the second messenger. One hormone molecule can cause many cAMP molecules to be formed. At each stage, the number of molecules involved increases so the process is said to have a cascade effect (Figure 3).

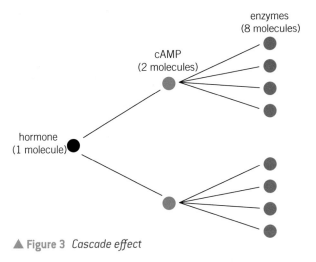

▲ Figure 3 *Cascade effect*

Summary questions

1 State and explain two physical responses which occur as a result of the 'fight and flight' response. *(2 marks)*

2 Explain why people often feel cold in times of stress. *(2 marks)*

3 Explain how the nervous and endocrine systems work together to enable the body to respond to danger. *(6 marks)*

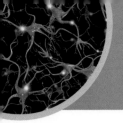

14.6 Controlling heart rate

Specification reference: 5.1.5

The human heart beats at approximately 70 beats per minute at rest. However, when you exercise, or in times of danger, it is essential that the heart rate increases to provide the extra oxygen required for increased respiration.

Controlling heart rate

Heart rate is involuntary and controlled by the autonomic nervous system.

The medulla oblongata in the brain is responsible for controlling heart rate and making any necessary changes. There are two centres within the medulla oblongata, linked to the sinoatrial node (SAN) in the heart by motor neurones:

- one centre increases heart rate by sending impulses through the sympathetic nervous system, these impulses are transmitted by the accelerator nerve
- one centre decreases heart rate by sending impulses through the parasympathetic nervous system, these impulses are transmitted by the vagus nerve.

Which centre is stimulated depends on the information received by receptors in the blood vessels. There are two types of receptors which provide information that affects heart rate:

- **baroreceptors** (pressure receptors) – these receptors detect changes in blood pressure. For example, if a person's blood pressure is low, the heart rate needs to increase to prevent fainting. Baroreceptors are present in the aorta, vena cava, and carotid arteries.
- **chemoreceptors** (chemical receptors) – these receptors detect changes in the level of particular chemicals in the blood such as carbon dioxide. Chemoreceptors are located in the aorta, the carotid artery (a major artery in the neck that supplies the brain with blood), and the medulla.

Chemoreceptors

Chemoreceptors are sensitive to changes in the pH level of the blood. If the carbon dioxide level in the blood increases, the pH of the blood decreases because carbonic acid is formed when the carbon dioxide interacts with water in the blood. If the chemoreceptors detect a decrease in blood pH, a response is triggered to increase heart rate – blood therefore flows more quickly to the lungs so the carbon dioxide can be exhaled.

This process is summarised in Figure 2

Learning outcomes

Demonstrate knowledge, understanding, and application of:

→ the effects of hormones and nervous mechanisms on heart rate.

Synoptic link

Look back at Topic 8.5, The heart to remind yourself of the heart's structure and the intrinsic rhythmicity of the heart.

Synoptic link

Look back at Topic 13.6, Organisation of the nervous system to remind yourself of the two different functional systems of the autonomic nervous system – the parasympathetic and sympathetic nervous system.

Synoptic link

Look back at Topic 7.4, Ventilation and gas exchange in other organisms to remind yourself of transport of carbon dioxide in the blood.

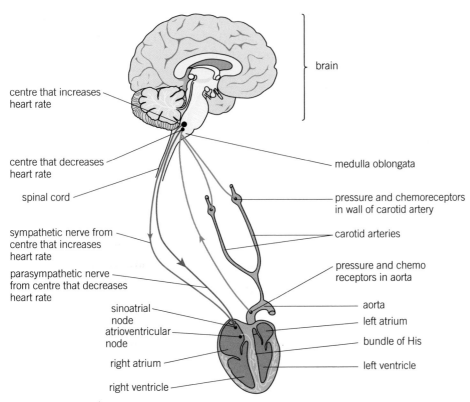

▲ Figure 1 *Control of heart rate*

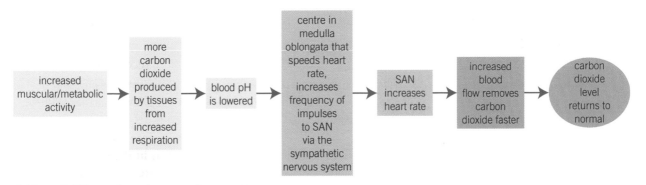

▲ Figure 2 *Effects of exercise on cardiac output*

When the carbon dioxide level in the blood decreases, the pH of the blood rises. This is detected by the chemoreceptors in the wall of the carotid arteries and the aorta. This results in a reduction in the frequency of the nerve impulses being sent to the medulla oblongata. In turn, this reduces the frequency of impulses being sent to the SAN via the sympathetic nervous system, and thus heart rate decreases back to its normal level.

Baroreceptors

Baroreceptors present in the aorta and carotid artery wall detect changes in pressure. If blood pressure is too high, impulses are sent to the medulla oblongata centre which decreases heart rate. The medulla oblongata sends impulses along parasympathetic neurones to the SAN which decreases the rate at which the heart beats. This reduces blood pressure back to normal.

If blood pressure is too low, impulses are sent to the medulla oblongata centre which increases heart rate. The medulla oblongata sends impulses along sympathetic neurones to the SAN which increases the rate at which the heart beats. This increases blood pressure back to normal.

Hormonal control

Heart rate is also influenced by the presence of hormones. For example, in times of stress adrenaline and noradrenaline are released. These hormones affect the pacemaker region of the heart itself – they speed up your heart rate by increasing the frequency of impulses produced by the SAN. Look back at Topic 14.1, Hormonal control and Topic 14.5, Coordinated responses to remind yourself how adrenaline is released and its importance in the 'fight or flight' response.

Monitoring heart rate

Heart rate can be determined by taking a pulse. Each time the heart beats a surge of blood travels around the body, which can be felt as a pulse in your blood vessels. The most common places to take a pulse are at the wrist (radial artery) or in the neck (carotid artery). The number of pulses felt per minute is equivalent to the number of times your heart beats per minute (bpm). Many people who exercise regularly and have a keen interest in monitoring their fitness levels wear a pulse monitor.

A group of students wanted to investigate how their heart rate was affected by exercise (Table 1).

1 Devise a practical procedure to collect these data.

The following were collected from a sample of 10 students. The students' pulse rates were measured at rest, during exercise, and two minutes after completing exercise.

2 State why it was important for the students to know the resting pulse rates.

3 a Calculate the mean resting heart rate of the students.

b Calculate the standard deviation of the students' heart rate during exercise.

$$\sigma = \sqrt{\frac{\Sigma(x - \overline{x})^2}{n - 1}}$$

4 a Calculate the percentage increase in the average heart rate two minutes after exercise, compared with the resting heart rate. (2 marks)

b Calculate the Spearman's rank correlation coefficient between the students' resting pulse rates, and their pulse rates during exercise. (6 marks)

c Evaluate the strength of the correlation between the two sets of data. (2 marks)

Study tip

You can find a full table of values of Spearman's rank correlation coefficient in the appendix.

Study tip

Look back at Topic 10.6, Representing variation graphically, to see a worked example of how to calculate the standard deviation of a set of data, and how to calculate a Spearman's rank correlation coefficient.

▼ Table 1

Student	Resting pulse / bpm	Pulse rate during exercise / bpm	Pulse rate after exercise / bpm
1	57	84	62
2	64	89	66
3	54	84	58
4	78	101	80
5	74	98	75
6	68	92	72
7	65	90	68
8	60	86	66
9	64	95	67
10	72	94	75

Synoptic link

Look back at Topic 10.6, Representing variation graphically to see a worked example of how to calculate the standard deviation of a set of data.

Summary questions

1 Copy and complete the table to summarise the control of heart rate:

Stimulus	Receptor	Nervous system involved	Effect on heart rate
high blood pressure			
low blood pressure			
low blood CO_2 concentration			
high blood CO_2 concentration			

(4 marks)

2 Explain why an athlete's heart rate may increase just before a race has begun. (2 marks)

3 a Explain why blood pH values vary during exercise. (2 marks)
  b Explain why an increase in the pH of blood leads to a decrease in heart rate. (3 marks)

Practice questions

1 Insulin-dependent diabetes most often appears during childhood. It is caused by the autoimmune destruction of β cells in the pancreas. In this autoimmune response, specific white blood cells respond to proteins on the surface of the β cells.

   a (i) Name the hormone that will become deficient due to the autoimmune destruction of the β cells. *(1 mark)*

      (ii) Suggest what happens to the proteins on the cell surface membrane of the β cells to stimulate the autoimmune response. *(1 mark)*

   b In individuals with insulin-dependent diabetes, there is excessive secretion of glucagon. Increased glucagon concentration in the blood results in additional metabolic changes to those cause by damaged β cells. Describe the effects of increased glucagon concentration on the liver. *(3 marks)*

   c The graph shows the changes in blood glucose concentration following a meal in a diabetic individual and in a healthy individual.

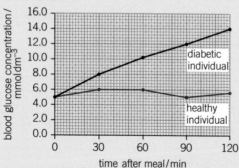

      (i) **Describe** the changes in blood glucose concentration in the **diabetic** individual for the 120 minutes following the meal. *(3 marks)*

      (ii) **Explain** the changes in blood glucose concentration in the **healthy** individual for the 120 minutes following the meal. *(3 marks)*

                                                  OCR Jan 2010

2 One risk factor that increases the risk of developing Type 2 diabetes is obesity.

The graph shows how prevalence of diabetes and mean body mass changed in one population from 1990 to 2000.

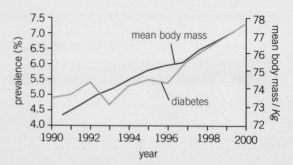

   a Explain why the data in the graph indicates a correlation and does not prove that obesity is a risk factor in diabetes. *(2 marks)*

   b Describe how the change in prevalence of diabetes changes with mean body mass in the graph. *(3 marks)*

   c Explain why the following symptoms may be seen in diabetics. *(5 marks)*

      sudden weight loss, increased need to urinate, dehydration

   d It is believed by some scientists that insulin resistance, the loss of sensitivity of receptors to insulin, is a result of an 'energy surplus' within cells.

      Some drugs used in the treatment of type 2 diabetes have been found to be inhibitors of ATP synthase of protein complexes in the electron transport chains.

      Suggest why this activity would lead to an increase in receptor sensitivity to insulin. *(3 marks)*

3 The pancreas acts as both an endocrine and exocrine gland. The diagram shows the arrangement of some of the different types of cells present in the pancreas.

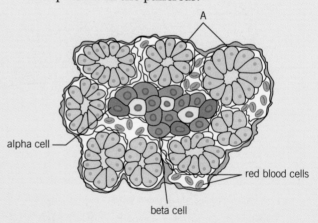

a **(i)** State the name given to a group of cells that work together to carry out a function. *(1 mark)*

   (ii) State the name of the structure that contains the alpha and beta cells in the pancreas. *(1 mark)*

   (iii) Outline the roles of the alpha and beta cells. *(4 marks)*

b Describe the role of the red blood cells with regard to the function of the alpha and beta cells. *(3 marks)*

c **(i)** State the name of the cell labelled A. *(1 mark)*

   (ii) State the function of cell A. *(1 mark)*

   (iii) Explain why cells, of which cell A is an example, are arranged in a different way to the alpha and beta cells. *(5 marks)*

4 A student carried out an investigation to study the effect of caffeine on daphnia heart rate. The following procedure was followed:

1 Daphnia, *transparent* aquatic organisms, of *the same species and size* were *placed on ice for 20 minutes* before the experiment.

2 A single daphnia was then placed in a beaker containing the solution under investigation *for 5 minutes*.

3 A small quantity of *cotton wool was placed on a microscope slide*.

4 A few *drops of pond water were placed on the cotton wool* on the microscope slide.

5 The daphnia was then placed on a microscope slide using a piece of filter paper. *A cover slip was not used.*

6 A cavity slide containing cold water was put under the microscope slide.

7 The daphnia was viewed under a *low power lens* of light microscope focused on the heart.

8 The number of heart beats every 30 seconds was then recorded.

9 This was *repeated* seven times for each concentration of caffeine.

10 The same procedure was repeated for solutions with different concentrations of caffeine.

The table shows the results that the student obtained.

Caffeine	1%	0.50%	0.25%	0.10%
1	401	392	308	232
2	400	380	276	228
3	380	360	284	204
4	404	400	268	232
5	360	332	292	248
6	428	368	260	256
7	440	340	280	240
8	368	320	308	216
Average	398	362	285	232

a Describe the significance of the terms in italics at each step in the procedure with regard to the following:

   (i) Reliability *(3 marks)*

   (ii) Experimental errors *(3 marks)*

   (iii) Ethics *(2 marks)*

b Plot a graph using the results in the table *(3 marks)*

c Describe the trends shown in the graph you have drawn for the different concentrations of caffeine. *(3 marks)*

The standard deviation, calculated as the square root of the variance, is used in the t-test calculation to compare two sets of data.

d Describe what standard deviation shows. *(1 mark)*

The variance is calculated in the following way:

Subtract the mean from each number in a column and square the difference. Add all of these squared differences together for each column and divide by $n - 1$ (one less than the number of results obtained). t is calculated using the following formula:

$$t = \frac{(\overline{x_1} - \overline{x_2})}{\sqrt{\dfrac{\sigma_1^2}{N_1} + \dfrac{\sigma_2^2}{N_2}}}$$

Where:

X_1 is the mean of the first data set
X_2 is the mean of the second data set

σ_1 is the standard deviation of the first data set
σ_2 is the standard deviation of the second data set
N_1 is the number of elements in the first data set
N_2 is the number of elements in the second data set

e Calculate t for the 1% and 0.5% solutions of caffeine and state whether there is a significant difference between these two sets of results. *(4 marks)*

15 HOMEOSTASIS
15.1 The principles of homeostasis
Specification reference: 5.1.1

The enzyme-controlled reactions of life can only take place if the conditions are right. The concentration of chemicals such as glucose and sodium ions must be kept within a narrow range, as must the pH and water balance of the body fluids, and the core temperature of the body. Organisms use both chemical and electrical systems to monitor and respond to any changes from the steady state of the body, and use the information to maintain a dynamic equilibrium.

Receptors and effectors

It is impossible to maintain a living mammal in a completely stable state because everything causes minute changes. Instead, the body maintains a dynamic equilibrium, with small fluctuations over a narrow range of conditions. This is known as **homeostasis**.

▲ Figure 1 *Mammals maintain the conditions inside their body within very narrow limits wherever they live and whatever they do*

Receptors and effectors are vital for the body to maintain this dynamic equilibrium. As you have seen, sensory receptors detect changes in the internal and external environment of an organism. In homeostasis, it is essential to monitor changes in the internal environment, for example, the pH of the blood, core body temperature, and concentrations of urea and sodium ions in the blood.

Information from the sensory receptors is transmitted to the brain and impulses are sent along the motor neurones to the effectors to bring about changes to restore the equilibrium in the body. Effectors are the muscles or glands that react to the motor stimulus to bring about a change in response to a stimulus. Both are vital in a homeostatic system – detecting change is no use without the means to react to that change, but effectors cause chaos unless responding to a need.

Synoptic link

You learnt about sensory receptors, sensory and motor neurones, and coordination in the brain in Chapter 13, Neuronal communication.

Feedback systems

Homeostasis depends on sensory receptors detecting small changes in the body, and effectors working to restore the status quo. These precise control mechanisms in the body are based on feedback systems that enable the maintenance of a relatively steady state around a narrow range of conditions.

Negative feedback systems

Most of the feedback systems in the body involve negative feedback. A small change in one direction is detected by sensory receptors. As a result, effectors work to reverse the change and restore conditions to their base level. Negative feedback systems work to reverse the initial stimulus. You have seen negative feedback in action in the control of blood sugar levels by insulin and glucagon. Negative feedback systems

Synoptic link

You learnt about negative feedback in Chapter 14, Hormonal communication.

are also important in many other aspects of homeostasis including temperature control and the water balance of the body. The general principles of negative feedback systems are shown in Figure 2.

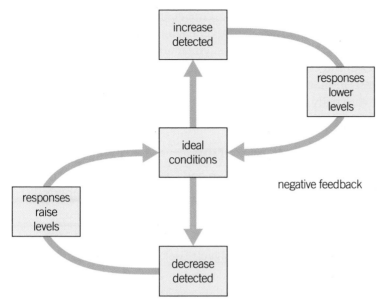

▲ **Figure 2** *The principles of negative feedback*

Positive feedback systems

There are relatively few positive feedback systems in the body. In a positive feedback system, a change in the internal environment of the body is detected by sensory receptors, and effectors are stimulated to reinforce that change and increase the response. One example occurs in the blood clotting cascade. When a blood vessel is damaged, platelets stick to the damaged region and they release factors that initiate clotting and attract more platelets. These platelets also add to the positive feedback cycle and it continues until a clot is formed. Another example of a positive feedback mechanism is seen during childbirth. The head of the baby presses against the cervix, stimulating the production of the hormone oxytocin. Oxytocin stimulates the uterus to contract, pushing the head of the baby even harder against the cervix and triggering the release of more oxytocin. This continues until the baby is born.

Synoptic link

You learnt about the clotting cascade in Topic 12.5, Non-specific animal defences against pathogens.

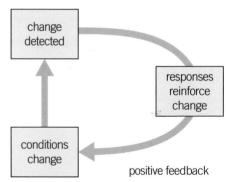

▲ **Figure 3** *The principles of positive feedback*

Summary questions

1   a   Suggest three different types of receptors explaining what changes they detect.   *(3 marks)*

     b   Suggest two different types of effector and give an example of what they do.   *(4 marks)*

2   a   What is homeostasis?   *(2 marks)*

     b   Why are both receptors and effectors important in homeostasis?   *(3 marks)*

3   Suggest why effective homeostasis depends on negative rather than positive feedback systems.   *(6 marks)*

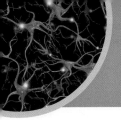

15.2 Thermoregulation in ectotherms

Specification reference: 5.1.1

Synoptic link

You learnt about enzymes in Chapter 4, Enzymes.

The enzymes controlling the rates of the chemical reactions needed for life are very temperature sensitive. Each enzyme has an optimum temperature at which it works most efficiently. If the temperature gets too high the enzymes are denatured. An important aspect of homeostasis in many animals is the maintenance of a relatively constant core body temperature to maintain optimum enzyme activity. This process is known as **thermoregulation**.

Endotherms and ectotherms

Organisms are constantly heating up and cooling down as a result of their surroundings. These changes depend on a number of physical processes. These include:

- Exothermic chemical reactions.
- Latent heat of evaporation – objects cool down as water evaporates from a surface.
- Radiation – the transmission of electromagnetic waves to and from the air, water, or ground.
  - Convection – the heating and cooling by currents of air or water, warm air or water rises and cooler air or water sinks setting up convection currents around an organism.
  - Conduction – heating as a result of the collision of molecules. Air is not a good conductor of heat but the ground and water are.

In many cases, the balance between heating and cooling determines the core temperature of the organism. Animals can be classified as **ectotherms** or **endotherms** depending on how they maintain and control their body temperature.

gains in heat		losses of heat
waste heat from cell respiration		evaporation of water
conduction from surroundings	organism	conduction to surroundings
convection from surroundings		convection to surroundings
radiation from surroundings		radiation to surroundings

▲ **Figure 1** *Ways in which animals warm up and cool down*

Ectotherms

Most animals are ectotherms and use their surroundings to warm their bodies (ectotherm literally means 'outside heat'). Their core body temperature is heavily dependent on their environment. Ectotherms include all the invertebrate animals, along with fish, amphibians, and reptiles.

Many ectotherms living in water do not need to thermoregulate. The high heat capacity of water means that the temperature of their environment does not change much. Ectotherms that live on land have a much bigger problem with temperature regulation. The temperature of the air can vary dramatically both between seasons and even over a 24-hour period from the middle of the day to the end of the night. As a result ectotherms have evolved a range of strategies that enable them to cool down or warm up.

Synoptic link

You learnt about the properties of water in Topic 3.2, Water.

Endotherms

Mammals and birds are endotherms. They rely on their metabolic processes to warm up and they usually maintain a very stable core body temperature regardless of the temperature of the environment (endotherm literally means 'inside heat'). They have adaptations which enable them to maintain their body temperature and to take advantage of warmth from the environment. As a result, endotherms survive in a wide range of environments. Keeping warm in cold conditions and cooling down in hot conditions are both active processes. The metabolic rate of endotherms is around five times higher than ectotherms, so they need to consume more food to meet their metabolic needs than ectotherms of a similar size.

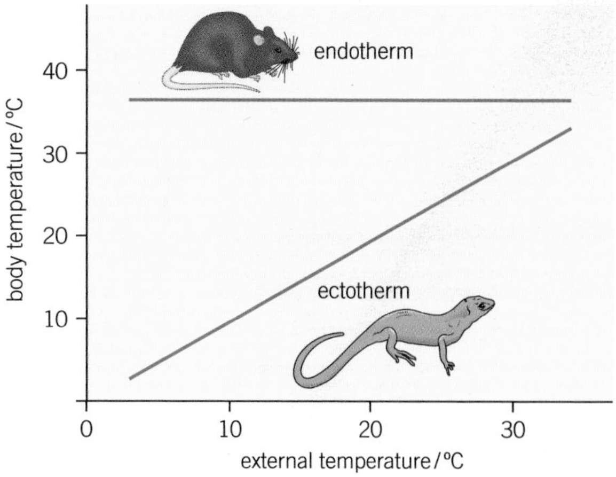

▲ Figure 2 *Graph to show simplified effect of changes in the internal and external temperature on ectotherms and endotherms*

Temperature regulation in ectotherms

Ectotherms cannot control their body temperature using their metabolism – however, they have evolved a range of behavioural responses that enable them to overcome the limitations imposed by the temperature of their surroundings.

Behavioural responses

Ectotherms display a number of behaviours which increase or reduce the radiation they absorb from the Sun. Sometimes they need to warm up to reach a temperature at which their metabolic reactions happen fast enough for them to be active. They may bask in the Sun, orientate their bodies so that the maximum surface area is exposed to the Sun, and even extend areas of their body to increase the surface area exposed to the Sun. For example, lizards often bask for long periods of time to get warm enough to move fast and hunt their prey, and insects such as locusts and butterflies orientate themselves for maximum exposure to the Sun and spread their wings to increase the available surface area to get warm enough to fly.

Ectotherms can increase their body temperature through conduction by pressing their bodies against the warm ground. They also get warmer as a result of exothermic metabolic reactions. Galapagos iguanas will contract their muscles and vibrate increasing cellular metabolism to raise their body temperature. Similarly, moths and butterflies may vibrate their wings to warm their muscles before they take flight.

Ectotherms sometimes need to cool down to prevent their core temperature reaching a point where enzymes begin to denature. To cool down, many of the warming processes are reversed. Ectotherms shelter from the sun by seeking shade, hiding in cracks in rocks, or even digging burrows. They will press their bodies against cool, shady earth or stones, or move into available water or mud. They orientate their bodies so that the minimum surface area is exposed to the sun, and minimise their movements to reduce the metabolic heat generated.

▲ Figure 3 *The black pigment of the marine iguanas observed by Darwin on the Galapagos Islands enables them to absorb enough heat to swim and feed in the relatively cold seawater surrounding the famous archipelago*

Physiological responses to warming

Much of the thermoregulation by ectotherms is the result of behavioural responses but some of them have physiological responses as well. Dark colours absorb more radiation than light colours. Lizards living in colder climates tend to be darker coloured than lizards living in hotter countries so that they get warmer. Some ectotherms also alter their heart rate to increase or decrease the metabolic rate and sometimes to affect the warming or cooling across the body surfaces.

Ectotherms are always more vulnerable to fluctuations in the environment than endotherms. However, by using a variety of behavioural and physiological strategies many of them can maintain relatively stable core temperatures. They need less food than endotherms as they use less energy regulating their temperatures, and so they can survive in some very difficult habitats where food is in short supply.

Summary questions

1 a What is an ectotherm?
(*2 marks*)
   b Give two examples of ectotherms. (*1 mark*)

2 Give an example of an ectotherm warming up or cooling down through interaction with the environment by:
   a radiation (*2 marks*)
   b conduction (*2 marks*)
   c convection (*2 marks*)
   d evaporation. (*2 marks*)

3 Galapagos marine iguanas are unique reptiles because they swim and feed in the sea. Read the following statements and discuss the observations in terms of thermoregulation.
   a Marine iguanas are black in colour. They spend a lot of time on the exposed rocks and alter their position and posture regularly. They need to have a body temperature of around 36 °C before they dive for food.
(*5 marks*)
   b Dives usually last a few minutes but can last up to 30 minutes. The length of the dive seems to be related to body size. The core temperature of the iguanas can drop by about 10 °C during a dive. The animals are slow and clumsy when they emerge from the sea.
(*6 marks*)

 ### The Namaqua chameleon – a highly adapted ectotherm

The Namaqua chameleon lives in the Namib desert, one of the most inhospitable hot and waterless environments on Earth. Several observations have been made on this rare and extremely well-adapted ectotherm:

- It is black in the morning. It may even appear black on the side exposed to the sun and pale grey on the other side of the body.
- It orientates its body sideways to the Sun.
- It has an increased heart rate early in the morning when basking.
- It inflates its body in the early morning.
- It presses its body to the desert sand in the morning.
- During the day the chameleon deflates its body.
- The animal becomes a very pale grey.
- It holds itself well away from the desert surface.
- The heart rate slows down.
- The chameleon opens its mouth and pants in the middle of the day.

▲ **Figure 4** *The Namaqua chameleon looks very different depending on whether it is trying to gain or lose heat*

Using each of the adaptations of the Namaqua chameleon described, explain how they help the animal to warm up or cool down.

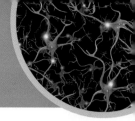

Endotherms can regulate their body temperature within a very narrow range in a wide variety of external conditions. Human beings, like all mammals, have a number of physiological responses that make this thermoregulation possible.

Detecting temperature changes

In any homeostatic system receptors are needed to detect a change in the internal environment. The peripheral temperature receptors are in the skin and detect changes in the surface temperature. Temperature receptors in the hypothalamus detect the temperature of the blood deep in the body. The temperature of the skin is much more likely to be affected by external conditions than the temperature of the hypothalamus. The combination of the two gives the body great sensitivity and allows it to respond not only to actual changes in the temperature of the blood but to pre-empt possible problems that might result from changes in the external environment.

The temperature receptors in the hypothalamus act as the thermostat of the body, controlling the responses that maintain the core temperature in a dynamic equilibrium to within about 1 °C of 37 °C.

Principles of thermoregulation in endotherms

Endotherms use their internal exothermic metabolic activities to keep them warm, and energy-requiring physiological responses to help them cool down. They also have passive ways of heating up and cooling down, to reduce the energy demands on their bodies. Like ectotherms, endotherms have a range of behavioural responses to temperature changes that include basking in the Sun, pressing themselves to warm surfaces, wallowing in water and mud to cool down, and digging burrows to keep warm or cool. Some animals even become dormant through the coldest weather (hibernation) or through the hottest weather (aestivation is a period of prolonged or deep sleep similar to hibernation but occurs in summer or during dry seasons to avoid heat stress rather than cold).

Humans have additional behavioural adaptations to help control body temperature – clothes are worn to stay warm, houses are built, and then heated up or cooled down to maintain the ideal temperature.

In spite of these behavioural responses, endotherms mainly rely on physiological adaptations to maintain a stable core body temperature, regardless of the environmental conditions or the amount of exercise being done. These adaptations include the peripheral temperature receptors, the thermoregulatory centres of the hypothalamus, the skin, and muscles.

Learning outcomes

Demonstrate knowledge, understanding, and application of:

→ the physiological and behavioural responses involved in temperature control in endotherms.

Synoptic link

You learnt about the areas of the brain in Topic 13.7, Structure and function of the brain.

▲ Figure 1 Thermoregulation makes it possible for endotherms to survive in many extreme environments

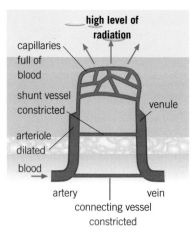

▲ **Figure 2** *Vasodilation leading to heat loss by radiation*

▲ **Figure 3** *Horses sweat all over when they are very hot. This horse is also panting, so he is cooling by evaporation both from his skin and his breathing passages. The hair on his body is also lying very flat to avoid trapping an insulating layer of air. The dilated arterioles and venules under the skin take blood to and from the capillary network*

Synoptic link

You learnt about surface area : volume ratio in Topic 7.1, Specialised exchange surfaces.

Cooling down

If the core body temperature increases it is important for an animal to cool down. There are a number of rapid responses to a rise in the core temperature that are common to all endotherms. These include:

Vasodilation

The arterioles near the surface of the skin dilate when the temperature rises. The vessels that provide a direct connection between the arterioles and the venules (the arteriovenous shunt vessels) constrict. This forces blood through the capillary networks close to the surface of the skin. The skin flushes, and cools as a result of increased radiation. If the skin is pressed against cool surfaces, then the cooling results from conduction.

Increased sweating

As the core temperature starts to increase, rates of sweating also increase. Sweat spreads out across the surface of the skin. In some mammals, including humans and horses, there are sweat glands all over the body. As the sweat evaporates from the surface of the skin, heat is lost, cooling the blood below the surface. In some animals, the sweat glands are restricted to the less hairy areas of the body such as the paws. These animals often open their mouths and pant when they get hot, again losing heat as the water evaporates. In human beings, around $1\,dm^3$ of sweat is lost by evaporation on a normal day. If the conditions are very hot and dry or the person is exercising very hard, up to $12\,dm^3$ of sweat a day can be lost. Kangaroos and cats often lick their front legs to keep cool in high temperatures.

Reducing the insulating effect of hair or feathers

As the body temperature begins to increase, the erector pili muscles (the hair erector muscles) in the skin relax – as a result, the hair or feathers of the animal lie flat to the skin. This avoids trapping an insulating layer of air. It has little effect in humans.

Endotherms that live in hot climates often have anatomical adaptations as well as the behavioural and physiological adaptations already described. These minimise the effect of high temperatures and maximise the ability of the animal to cool down through the surface area of the body. They include a relatively large surface area : volume (SA : V) ratio to maximise cooling (e.g., include large ears and wrinkly skin), and pale fur or feathers to reflect radiation.

Warming up

If the core temperature falls it is important for an animal to warm up and prevent further cooling. There are a number of rapid responses to a fall in the core temperature that are common to all endotherms.

Vasoconstriction

The arterioles near the surface of the skin constrict. The arteriovenous shunt vessels dilate, so very little blood flows through the capillary networks close to the surface of the skin. The skin looks pale, and very little radiation takes place. The warm blood is kept well below the surface.

Decreased sweating

As the core temperature falls, rates of sweating decrease and sweat production will stop entirely. This greatly reduces cooling by the evaporation of water from the surface of the skin, although some evaporation from the lungs still continues.

Raising the body hair or feathers

As the body temperature falls, the erector pili muscles in the skin contract, pulling the hair or feathers of the animal erect. This traps an insulating layer of air and so reduces cooling through the skin. The effect can be quite dramatic and it is a very effective way to reduce heat loss to the environment in many animals. In humans this has little effect although you can observe the hairs being pulled upright.

Shivering

As the core temperature falls the body may begin to shiver. This is the rapid, involuntary contracting and relaxing of the large voluntary muscles in the body. The metabolic heat from the exothermic reactions warm up the body instead of moving it and is an effective way of raising the core temperature.

Endotherms living in cold climates often have additional anatomical adaptations to help them keep warm. Many have adaptations that minimise their SA:V ratio to reduce cooling (e.g., small ears). Another common adaptation is a thick layer of insulating fat underneath the skin, for example, blubber in whales and seals. Some animals hibernate – they build up fat stores, build a well-insulated shelter, and lower their metabolic rate so they pass the worst of the cold weather in a deep sleep-like state.

Polar bears demonstrate many of the ways in which endotherms can survive in extremely cold conditions. They have small ears and fur on their feet to insulate them from the ice. The fur and skin of polar bears work together. The hairs are hollow so trap a permanent layer of insulating air. The skin underneath is black, so it absorbs warming radiation. They have a thick layer of fat under the skin. Polar bears are so well insulated that their external surfaces are similar in temperature to the snow and ice on which they live. Females dig dens in the snow and remain in them, warm and insulated, for months while they give birth to their cubs, only emerging when the cubs are large enough to survive the cold. Polar bears are so well adapted to life in temperatures down to −50 °C in the Arctic that they can overheat at temperatures over 10 °C.

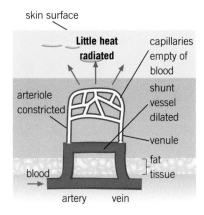

▲ Figure 4 *Vasoconstriction reduces heat loss by radiation*

▲ Figure 5 *The effect of the contraction of the erector pili muscles in robins*

Controlling thermoregulation

The physiological responses of endotherms to changes in the core temperature are the result of complex homeostatic mechanisms involving negative feedback control from the hypothalamus. There are two control centres:

The heat loss centre

This is activated when the temperature of the blood flowing through the hypothalamus increases. It sends impulses through autonomic motor neurones to effectors in the skin and muscles, triggering responses that act to lower the core temperature.

The heat gain centre

This is activated when the temperature of the blood flowing through the hypothalamus decreases. It sends impulses through the autonomic nervous system to effectors in the skin and the muscles, triggering responses that act to raise the core temperature.

The interaction of the sensory receptors, the autonomic nervous system, and the effectors in a sophisticated feedback system enables endotherms to maintain a very stable core body temperature regardless of environmental conditions or activity levels.

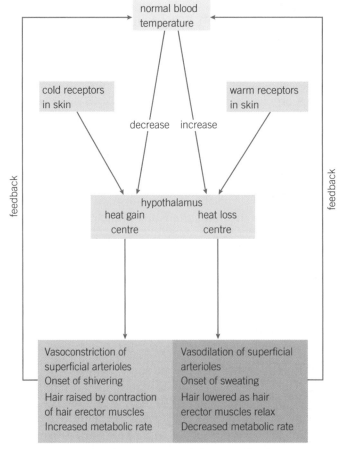

▲ **Figure 6** *Summary of the control of body temperature by the peripheral temperature receptors, the hypothalamus, and the autonomic nervous system*

Summary questions

1 Why is the control of the internal temperature so important to both ectotherms and endotherms? (*6 marks*)

2 Explain how the role of evaporation of water in thermoregulation differs between ectotherms and endotherms. (*6 marks*)

3 Explain the difference in the role of the peripheral temperature receptors and the temperature receptors in the hypothalamus in the regulation of the core body temperature in an endotherm. (*3 marks*)

4 Endotherms that live in very hot climates are often pale coloured.
   a Why is this? (*3 marks*)
   b Why might you expect endotherms that live in very cold environments to be dark coloured? (*2 marks*)
   c In fact, very few endotherms that live in very cold environments are dark coloured. Suggest reasons for this. (*4 marks*)

15.4 Excretion, homeostasis, and the liver

Specification reference: 5.1.2

Many of the chemical reactions of metabolism that take place in the cells of the body produce waste products that are toxic if they are allowed to build up. **Excretion** is the removal of the waste products of metabolism from the body.

Excretion in mammals

The main metabolic waste products in mammals are:

- Carbon dioxide – one of the waste products of cellular respiration which is excreted from the lungs.
- Bile pigments – formed from the breakdown of haemoglobin from old red blood cells in the liver. They are excreted in the bile from the liver into the small intestine via the gall bladder and bile duct. They colour the faeces.
- Nitrogenous waste products (urea) – formed from the breakdown of excess amino acids by the liver. All mammals produce **urea** as their nitrogenous waste. Fish produce ammonia while birds and insects produce uric acid. Urea is excreted by the kidneys in the urine.

The liver

The liver is one of the major body organs involved in homeostasis. It is a reddish-brown organ which makes up about 5% of the total body mass – the largest internal organ of the body. It lies just below the diaphragm and is made up of several lobes. The liver is very fast growing and damaged areas generally regenerate very quickly.

The liver has a very rich blood supply – about $1\,dm^3$ of blood flows through it every minute. Oxygenated blood is supplied to the liver by the hepatic artery and removed from the liver and returned to the heart in the hepatic vein. The liver is also supplied with blood by a second vessel, the **hepatic portal vein**. This carries blood loaded with the products of digestion straight from the intestines to the liver and this is the starting point for many metabolic activities of the liver (Figure 1). Up to 75% of the blood flowing through the liver comes via the hepatic portal vein.

The structure of the liver

The liver carries out many different complex functions but the cells are surprisingly simple and uniform in appearance. Liver cells or **hepatocytes** have large nuclei, prominent Golgi apparatus, and lots of mitochondria, indicating that they are metabolically active cells (Figure 1). They divide and replicate – even if around 65% of the liver is lost, it will regenerate in a matter of months.

The blood from the hepatic artery and the hepatic portal vein is mixed in spaces called sinusoids which are surrounded by hepatocytes. This

Learning outcomes

Demonstrate knowledge, understanding, and application of:

→ the term *excretion* and its importance in maintaining metabolism and homeostasis

→ the structure and mechanisms of action and functions of the mammalian liver

→ the examination and drawing of stained sections to show the histology of liver tissue.

Synoptic link

You learnt about the excretion of carbon dioxide from the body in Topic 7.2, Mammalian gaseous exchange system and Topic 8.4, Transport of oxygen and carbon dioxide in the blood.

Synoptic link

You learnt about macrophages in Topic 12.5, Non-specific animal defences against pathogens.

mixing increases the oxygen content of the blood from the hepatic portal vein, supplying the hepatocytes with enough oxygen for their needs. The sinusoids contain Kupffer cells, which act as the resident macrophages of the liver, ingesting foreign particles and helping to protect against disease. The hepatocytes secrete bile from the breakdown of the blood into spaces called canaliculi, and from these the bile drains into the bile ductules which take it to the gall bladder.

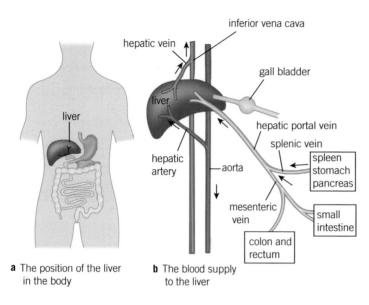

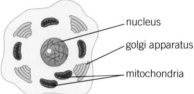

a The position of the liver in the body

b The blood supply to the liver

c The arrangement of the tissues in the liver, giving all of the hepatocytes close contact with the blood for both removing materials and adding substances.

d Single hepatocyte

▲ **Figure 1** *The structure of the liver is related to its variety of functions in the metabolism of carbohydrates, proteins, and fats and in excretion and homeostasis*

The functions of the liver

The liver has many functions – around 500 different metabolic pathways are linked to the liver. Several of these play a major role in homeostasis. These functions will now be explored in more detail.

Carbohydrate metabolism

Synoptic link

You learnt about the interconversion of glucose to glycogen and the storage of glycogen in the liver in Topic 14.3, Regulation of blood glucose concentration.

Hepatocytes are closely involved in the homeostatic control of glucose levels in the blood by their interaction with insulin and glucagon. When blood glucose levels rise, insulin levels rise and stimulate hepatocytes to convert glucose to the storage carbohydrate glycogen. About 100 g of glycogen is stored in the liver. Similarly, when blood sugar levels start to fall, the hepatocytes convert the glycogen back to glucose under the influence of the hormone glucagon.

Deamination of excess amino acids

The liver plays a vital role in protein metabolism where hepatocytes synthesise most of the plasma proteins. Hepatocytes also carry out *transamination* – the conversion of one amino acid into another. This is important because the diet does not always contain the required balance of amino acids but transamination can overcome the problems this might cause.

The most important role of the liver in protein metabolism is in **deamination** – the removal of an amine group from a molecule. The body cannot store either proteins or amino acids. Any excess ingested protein would be excreted and therefore wasted if it were not for the action of the hepatocytes. They deaminate the amino acids, removing the amino group, and converting it first into ammonia which is very toxic and then to urea. Urea is toxic in high concentrations but not in the concentrations normally found in the blood. Urea is excreted by the kidneys (you will learn about the role of the kidneys in excretion and water balance in Topics 15.5 and 15.6). The remainder of the amino acid can then be fed into cellular respiration or converted into lipids for storage.

The ammonia produced in the deamination of proteins is converted into urea in a set of enzyme-controlled reactions known as the **ornithine cycle**. Removing the amino group from amino acids and converting the highly toxic ammonia to the less toxic and more manageable compound urea involves some complex biochemistry. This is simplified and summarised in Figure 3.

Detoxification

The level of toxins in the body always tends to increase. Apart from urea, many other metabolic pathways produce potentially poisonous substances. We also take in a wide variety of toxins by choice such as alcohol and other drugs. The liver is the site where most of these substances are detoxified and made harmless.

One example is the breakdown of hydrogen peroxide, a by-product of various metabolic pathways in the body. Hepatocytes contain the enzyme **catalase**, one of the most active known enzymes, that splits the hydrogen peroxide into oxygen and water. Another example is the way in which liver detoxifies the ethanol – the active drug in alcoholic drinks. Hepatocytes contain the enzyme alcohol dehydrogenase that breaks down the ethanol to ethanal. Ethanal is then converted to ethanoate which may be used to build up fatty acids or used in cellular respiration.

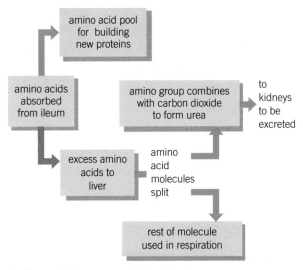

▲ **Figure 2** *The process of deamination*

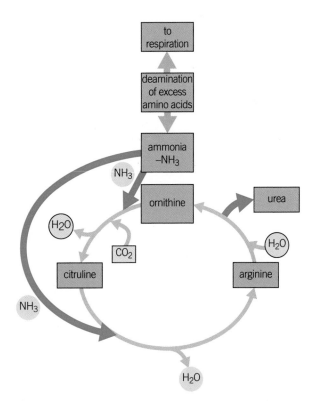

▲ **Figure 3** *The ornithine cycle where ammonia resulting from the breakdown of excess amino acids in the body is converted into useful products and urea*

Synoptic link

You learnt about catalase as an enzyme in Topic 3.7, Types of proteins and Topic 4.2, Factors affecting enzyme activity.

Looking at liver cells

When liver tissue is stained for use under the light microscope, different stains can be used to show up different things, for example glycogen stored in the hepatocytes.

Using low magnification with a light microscope enables you to see the arrangement of the liver cells, the blood vessels, and the sinusoids.

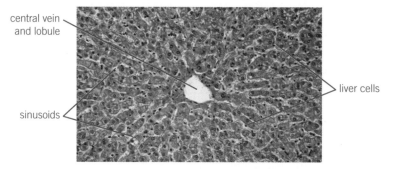

central vein and lobule

liver cells

sinusoids

▲ **Figure 4** *Light micrograph through a section of human liver ×34 magnification*

Using a high magnification gives more detail of the individual hepatocytes.

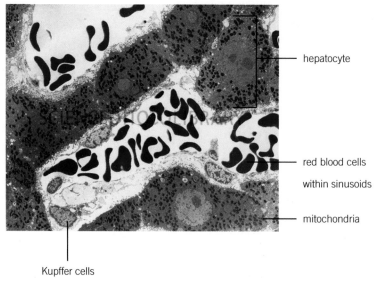

hepatocyte

red blood cells within sinusoids

mitochondria

Kupffer cells

▲ **Figure 5** *High magnification light micrograph of liver cells ×3 500 magnification*

Make a scientific drawing of some of the cells seen in Figure 5. Annotate your drawings to indicate the role of the cells in the liver.

Synoptic link

You learnt about the light microscope in Topic 2.1, Microscopy.

Cirrhosis of the liver

Cirrhosis is a disease where the normal liver tissue is replaced by fibrous scar tissue. There are lots of different causes including genetic conditions and hepatitis C, however, in the UK the most common cause is drinking excessive amounts of alcohol.

There are three stages of alcoholic liver disease – alcoholic fatty liver disease, alcoholic hepatitis, and liver cirrhosis. In fatty liver, big fat-filled vesicles displace the nuclei of the hepatocytes and the liver gets larger. In alcoholic hepatitis, the patient has fatty liver along with damaged hepatocytes and the sinusoids and hepatic veins become narrowed. In alcoholic cirrhosis the liver tissue is irreversibly damaged. Many hepatocytes die and are replaced with fibrous tissue. The hepatocytes can no longer divide and replace themselves so the liver shrinks and its ability to deal with toxins in the body decreases.

1 In many cases, if an affected person stops drinking alcohol the liver may recover. How is this possible?
2 Many people cannot stop drinking – explain why not and discuss ways of helping those affected.
3 Suggest possible effects on the body when the alcoholic liver damage becomes irreversible.
4 Discuss the pros and cons of giving someone with alcoholic cirrhosis of the liver a liver transplant.

Summary questions

1 Many people use the term excretion to describe defecation. Why is this not entirely accurate? *(6 marks)*

2 Hepatocytes make up approximately 70–85% of the liver's mass.
  a What is a hepatocyte? *(2 marks)*
  b Hepatocytes have large nuclei, lots of Golgi apparatus, and many mitochondria. What does this tell you about the cells? *(3 marks)*

3 a Draw a labelled diagram to show the structure of the liver. *(6 marks)*
  b Explain how the structure of the liver is adapted for its functions in the body *(6 marks)*

4 a Why do you think the liver is particularly affected by excess drinking? *(6 marks)*
  b Why is a build-up of fatty tissue a common symptom of excess drinking? *(4 marks)*

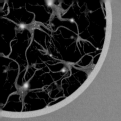

15.5 The structure and function of the mammalian kidney

Specification reference: 5.1.2

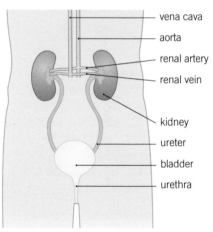

▲ Figure 1 *The human urinary system*

Human kidneys are typical of all mammalian kidneys. They are a pair of reddish-brown organs attached to the back of the abdominal cavity. They are usually surrounded by a thick, protective layer of fat and a layer of fibrous connective tissue. The kidneys play two important homeostatic roles in the body – they are involved in excretion and **osmoregulation**. They filter nitrogenous waste products out of the blood, especially urea. They also help to maintain the water balance and pH of the blood, and hence the tissue fluid that surrounds all the cells.

The anatomy of the kidneys

If you put your hands on your hips, the place where your thumbs are is the approximate position of your kidneys. The kidneys are supplied with blood at arterial pressure by the renal arteries that branch off from the abdominal aorta. Blood that has circulated through the kidneys is removed by the renal vein that drains into the inferior vena cava. About $90–120\,cm^3$ of blood passes through the kidneys every minute. All of the blood in the body passes through the kidneys about once an hour. The kidneys filter $180\,dm^3$ of blood a day, producing $1–2\,dm^3$ of urine. The final volume depends on many different factors. You will learn more about the factors that determine the amount of urine produced in Topic 15.6, The kidney and osmoregulation.

The kidneys are made up of millions of small structures called **nephrons** that act as filtering units. The sterile liquid produced by the kidney tubules is called urine. The urine passes out of the kidneys down tubes called **ureters**. It is collected in the bladder, a muscular sac that can store around $400–600\,cm^3$ of urine. When the bladder is getting full, the sphincter at the exit to the bladder opens and the urine passes out of the body down the **urethra**.

Kidney structure

If you slice open a kidney, you will see three main areas – the **cortex**, the **medulla**, and the **pelvis** (Figure 2).

- The cortex is the dark outer layer. This is where the filtering of the blood takes place and it has a very dense capillary network carrying the blood from the renal artery to the nephrons.

- The medulla is lighter in colour – it contains the tubules of the nephrons that form the pyramids of the kidney and the collecting ducts.

- The pelvis (which is Latin for *basin*) of the kidney is the central chamber where the urine collects before passing out down the ureter.

Nephrons – the functional units of the kidney

In the nephrons the blood is filtered and then the majority of the filtered material is returned to the blood, removing nitrogenous wastes and balancing the mineral ions and water. Each nephron is around 3 cm long and there are around 1.5 million nephrons in each kidney. This provides the body with several kilometres of tubules for the reabsorption of water, glucose, salts, and other substances back into the blood.

Structure of the nephron

The main structures and functions of the nephron are as follows:

- **Bowman's capsule** – cup-shaped structure that contains the glomerulus, a tangle of capillaries. More blood goes into the glomerulus than leaves it due to the ultrafiltration processes that take place.

- **Proximal convoluted tubule** – the first, coiled region of the tubule after the Bowman's capsule, found in the cortex of the kidney. This is where many of the substances needed by the body are reabsorbed into the blood

- **Loop of Henle** – a long loop of tubule that creates a region with a very high solute concentration in the tissue fluid deep in the kidney medulla. The descending loop runs down from the cortex through the medulla to a hairpin bend at the bottom of the loop. The ascending limb travels back up through the medulla to the cortex.

- **Distal convoluted tubule** – a second twisted tubule where the fine-tuning of the water balance of the body takes place. The permeability of the walls to water varies in response to the levels of the **antidiuretic hormone (ADH)** in the blood (ADH is explored in more detail in Topic 15.6, The kidney and osmoregulation). Further regulation of the ion balance and pH of the blood also takes place in this tubule.

- **Collecting duct** – the urine passes down the collecting duct through the medulla to the pelvis. More fine-tuning of the water balance takes place – the walls of this part of the tubule are also sensitive to ADH.

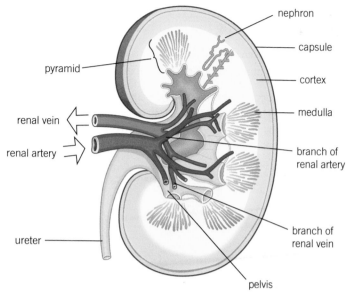

▲ **Figure 2** *Internal structure of a kidney*

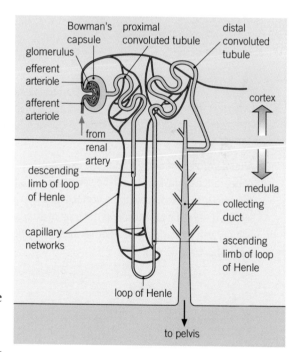

▲ **Figure 3** *The structure of a nephron and its blood supply*

The nephron has a network of capillaries around it which finally lead into a venule and then to the renal vein. The blood that leaves the kidney has greatly reduced levels of urea, but the levels of glucose and other substances such as amino acids needed by the body are almost the same as when the blood entered the kidneys (may be slightly less as some glucose will have been used for selective reabsorption). The mineral ion concentration in the blood has also been restored to ideal levels.

Synoptic link

You learnt about arteries, arterioles, capillaries, venules, and veins in Topic 8.2, The blood vessels, and about the formation of tissue fluid in Topic 8.3, Blood, tissue fluid, and lymph.

Investigating the kidneys

Lamb, pig, or even beef kidneys can be used to look at the external and internal structures of these fascinating organs.

1 The protective layer of fat around kidneys (renal capsule) is impressive – these vital organs are well-cushioned from physical damage in the body (Figure 4).

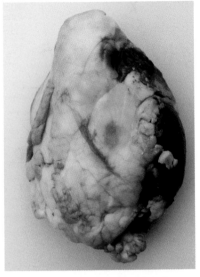

▲ **Figure 4** *Here the kidney is still surrounded by perirenal fat (adipose tissue)*

2 After carefully removing the fat, the external appearance of the kidney can be seen. You can see the colour, the fibrous coat, the ureter, and if you are skilful, you will be able to identify the renal artery and vein as well (depending on whether these have been left on by the butcher).

▲ **Figure 5** *External appearance of the kidney once perirenal fat has been removed*

3 Slice the kidney open carefully using a scalpel or a very sharp knife by making small cuts. Look carefully and identify as many of the internal features as you can (coloured pins are useful for this).

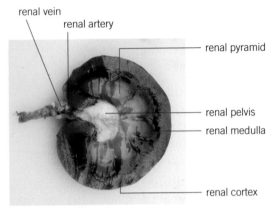

▲ **Figure 6** *Labelled photo of a dissected kidney*

4 Dissecting a kidney cannot show us the histology of the kidneys. To observe the structure of the individual nephrons requires a hand-lens or stained sections of kidney tissue and a microscope. It is very difficult to see an entire tubule but different parts of the tubules can be identified (Figures 7 and 8). To make the nephrons more visible a drop of hydrogen peroxide can be applied onto the cut surface of the kidney (wearing safety glasses and gloves). There will be rapid effervescence (foaming), after this is wiped off, the renal tubules, collecting duct, and loops of Henle should be a little clearer to see as shown by strings of bubbles.

In a section through the cortex you will see Bowman's capsules and glomeruli, as well as sections through proximal and distal convoluted tubules. The lumen of the distal tubules tends to be bigger and more open than those of proximal tubules which can be helpful in identifying them.

In a section through the medulla you will see mainly loops of Henle and collecting duct. In a transverse section you will see the lumens of the tubules – the collecting ducts are larger than the thick ascending loops of Henle while the thin-walled descending limbs are only visible at very high magnifications. In a longitudinal section you will see the parallel tubes – low magnifications give an overall impression whilst higher magnifications enable you to see individual tubules.

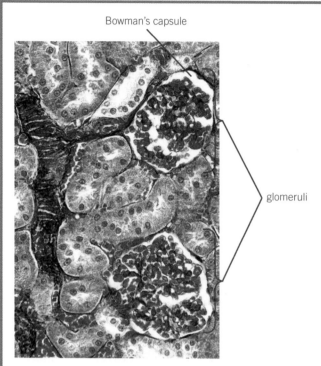

Bowman's capsule

glomeruli

▲ Figure 7 *Light micrograph of glomeruli in kidney. These are a tightly coiled network of capillaries surrounded by the lumen of the Bowman's capsule. Surrounding this are the tubules (rounded) where reabsorption takes place, × 230 magnification*

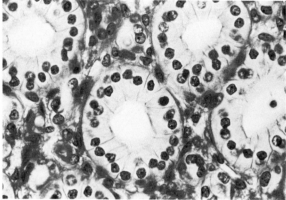

▲ Figure 8 *Light micrograph showing a section through a normal human kidney, ×275 magnification*

1 Compare the section of a kidney in Figure 6 with the stylised diagram in Figure 2. How does the dissection differ from the diagram? Compare their value in helping to understand the structure of the organ.

2 Which micrograph do you find most useful in developing an understanding of the structure of the kidney. Why?

The functions of the nephrons

Ultrafiltration

The first stage in the removal of nitrogenous waste and osmoregulation of the blood is **ultrafiltration**. Ultrafiltration in the kidney tubules is a specialised form of the process that results in the formation of tissue fluid in the capillary beds of the body and it is the result of the structure of the glomerulus and the cells lining the Bowman's capsule (Figure 9 and 10). The glomerulus is supplied with blood by a relatively wide afferent (incoming) arteriole from the renal artery. The blood leaves through a narrower efferent (outward) arteriole and as a result there is considerable pressure in the capillaries of the glomerulus. This forces the blood out through capillary wall – it acts rather like a sieve. Then the fluid passes through the basement membrane – scientists are increasingly recognising the basement membrane as an important factor in the filtration process. The basement membrane is made up of a network of collagen fibres and other proteins that make up a second 'sieve'. Most of the plasma contents can pass through the basement membrane but the blood cells and many proteins are retained in the capillary because of their size.

Synoptic link

You learnt about collagen and its properties in Topic 3.7, Types of proteins.

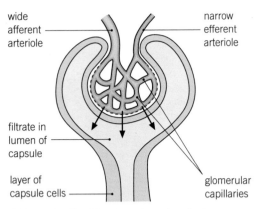

wide afferent arteriole

narrow efferent arteriole

filtrate in lumen of capsule

layer of capsule cells

glomerular capillaries

▲ **Figure 9** *Ultrafiltration takes place in the Bowman's capsule as a result of the high blood pressure in the glomerulus*

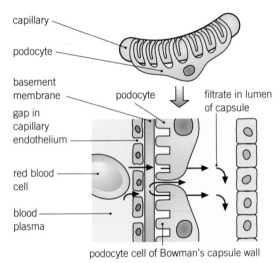

capillary

podocyte

basement membrane

gap in capillary endothelium

podocyte

filtrate in lumen of capsule

red blood cell

blood plasma

podocyte cell of Bowman's capsule wall

▲ **Figure 10** *The microstructure of the cells in the capillaries of the glomerulus and the podocytes of the Bowman's capsule is important in the ultrafiltration of the blood*

▼ **Table 1** *Comparison of the mean composition of human blood plasma and the ultrafiltrate as it enters the nephron*

Molecule or ion	Approximate concentration / $g\,dm^{-3}$	
	plasma	filtrate
water	900.0	900.0
protein	80.0	0.0
glucose	1.0	1.0
amino acids	0.5	0.5
urea	0.3	0.3
inorganic ions	7.2	7.2

The wall of the Bowman's capsule also involves special cells called *podocytes* that act as an additional filter. They have extensions called *pedicels* that wrap around the capillaries, forming slits that make sure any cells, platelets, or large plasma proteins that have managed to get through the epithelial cells and the basement membrane do not get through into the tubule itself. The filtrate which enters the capsule contains glucose, salt, urea, and many other substances in the same concentrations as they are in the blood plasma (Figure 11 and Table 1). The process is so efficient that up to 20% of the water and solutes are removed from the plasma as it passes through the glomerulus. The volume of blood that is filtered through the kidneys in a given time is known as the glomerular filtration rate.

Reabsorption

Ultrafiltration removes urea, the waste product of protein breakdown, from the blood but it also removes a lot of water along with the glucose, salt, and other substances which are present in the plasma. Many of these substances are needed by the body – for example, glucose is used for cellular respiration and is never normally excreted. The ultrafiltrate is also hypotonic to (less concentrated than) the blood plasma. The main function of the nephron after the Bowman's capsule is to return most of the filtered substances back to the blood.

The proximal convoluted tubule

In the proximal convoluted tubule all of the glucose, amino acids, vitamins, and hormones are moved from the filtrate back into the blood by active transport. Around 85% of the sodium chloride and water is reabsorbed as well – the sodium ions are moved by active transport while the chloride ions and water follow passively down concentration gradients. The cells lining the proximal convoluted tubule have clear adaptations:

- they are covered with microvilli, greatly increasing the surface area over which substances can be reabsorbed
- they have many mitochondria to provide the ATP needed in active transport systems.

Once the substances have been removed from the nephron, they diffuse into the extensive capillary network which surrounds the tubules down steep concentration gradients. These are maintained by the constant flow of blood through the capillaries. The filtrate reaching the loop of Henle at the end of the proximal convoluted tubule is isotonic (at same concentration) with the tissue fluid surrounding the tubule and isotonic with the blood. At this stage over 80% of the glomerular filtrate has been reabsorbed back into the blood. This remains the same regardless of the conditions in the body.

The loop of Henle

The loop of Henle is the section of the kidney tubule that enables mammals to produce urine more concentrated than their own blood. Different areas of the loop have different permeabilities to water and this is central to the way the loop of Henle functions. It acts as a countercurrent multiplier, using energy to produce concentration gradients that result in the movement of substances such as water from one area to another. Cells use ATP to transport ions using active transport and this produces a diffusion gradient in the medulla.

The changes that take place in the descending limb of the loop of Henle depend on the high concentrations of sodium and chloride ions in the tissue fluid of the medulla that are the result of events in the ascending limb of the loop.

- The descending limb leads from the proximal convoluted tubule. This is the region where water moves out of the filtrate down a concentration gradient. The upper part is impermeable to water but the lower part of the descending limb is permeable to water and runs down into the medulla. The concentration of sodium and chloride ions in the tissue fluid of the medulla gets higher and higher moving through from the cortex to the pyramids, as a result of the activity of the ascending limb of the loop of Henle.

The filtrate entering the descending limb of the loop of Henle is isotonic with the blood. As it travels down the limb, water passes out of the loop into the tissue fluid by osmosis down a concentration gradient. It then moves down a concentration gradient into the blood of the surrounding capillaries (the vasa recta).

The descending limb is not permeable to sodium and chloride ions, and no active transport takes place in the descending limb. The fluid that reaches the hairpin bend is very concentrated and hypertonic to the blood in the capillaries.

- The first section of the ascending limb of the loop of Henle is very permeable to sodium and chloride ions and they move out of the concentrated solution by diffusion down a concentration gradient. In the second section of the ascending limb, sodium and chloride ions are actively pumped out into the medulla tissue fluid against a concentration gradient. This produces very high sodium and chloride ion concentrations in the medulla tissue. Importantly, the ascending limb of the loop of Henle is impermeable to water, so water cannot follow the chloride and sodium ions down a concentration gradient. This means the fluid left in the ascending limb becomes increasingly dilute, while the tissue fluid of the medulla develops the very high concentration

Synoptic link

You learnt about the process of active transport in Topic 5.4, Active transport.

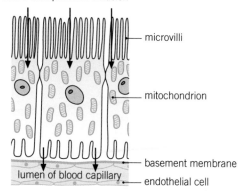

reabsorption of useful materials by active transport and diffusion

- microvilli
- mitochondrion
- basement membrane
- lumen of blood capillary
- endothelial cell

▲ **Figure 11** *The cells of the proximal convoluted tubule show a high level of adaptation to their function*

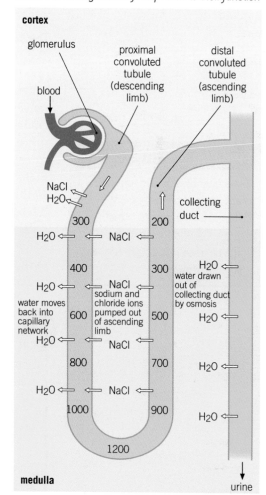

▲ **Figure 12** *The countercurrent multiplier system of the loop of Henle depends on a combination of active transport, diffusion, and osmosis. The higher the number the lower the water potential*

of ions that is essential for the kidney to produce urine that is more concentrated than the blood. This is a key part of the countercurrent multiplier system.

By the time the dilute fluid reaches the top of the ascending limb it is hypotonic to the blood again, and it then enters the distal convoluted tubule and collecting duct.

Distal convoluted tubule

Balancing the water needs of the body takes place in the distal convoluted tubule and the collecting duct. These are the areas where the permeability of the walls of the tubules varies with the levels of ADH. The cells lining the distal convoluted tubule also have many mitochondria so they are adapted to carry out active transport.

If the body lacks salt, sodium ions will be actively pumped out of the distal convoluted tubule with chloride ions following down an electrochemical gradient. Water can also leave the distal tubule, concentrating the urine, if the walls of the tubule are permeable in response to ADH. The distal convoluted tubule also plays a role in balancing the pH of the blood.

The collecting duct

The collecting duct passes down through the concentrated tissue fluid of the renal medulla (Figure 3 and 6). This is the main site where the concentration and volume of the urine produced is determined. Water moves out of the collecting duct by diffusion down a concentration gradient as it passes through the renal medulla. As a result the urine becomes more concentrated. The level of sodium ions in the surrounding fluid increases through the medulla from the cortex to the pelvis. This means water can be removed from the collecting duct all the way along its length, producing very hypertonic urine when the body needs to conserve water. The permeability of the collecting duct to water is controlled by the level of ADH, which determines how much or little water is reabsorbed. You will learn more about the role of the collecting duct in maintaining the water potential of the blood in Topic 15.6, The kidney and osmoregulation.

> **Study tip**
>
> Remember that the amount of reabsorption that occurs in the proximal tubule is always the same – the fine-tuning of the water balance takes place further along the nephron.

How long is your loop of Henle?

The ability of animals to produce very concentrated urine depends on several factors, one is the length of the loop of Henle. As you have seen the loop of Henle develops the concentration gradient across the kidney medulla, meaning water can leave the collecting duct all the way through, concentrating the urine as it goes. Fish have no loop of Henle and cannot produce urine that is more concentrated than their blood. Desert animals tend to have lots of nephrons that have very long loops of Henle that travel deep into the medulla.

▼ Table 2 *Urine concentrations and urine/plasma ratios in mammal species from different habitats*

Mammal	Urine concentration/mOsmol l^{-1}	Urine : plasma ratio
rat	2900	9
kangaroo rat	5500	16
beaver	520	1.7
human	1400	4–5
porpoise	1800	5
camel	2800	8

1 Demonstrate how this data could be presented in a different way.
2 How would you expect the loops of Henle of a camel to compare with those of a beaver? Explain your answer.
3 Why do you think a porpoise has a similar urine : plasma ratio as a human being?
4 The kangaroo rat has long loops of Henle but they are not comparatively as long as some animals such as the camel. The cells lining the loops in the kangaroo rat, however, have a very large number of mitochondria that have a very large number of cristae. Suggest why these observations are important in the development of the urine : plasma ratio seen in the kangaroo rat.

Summary questions

1 Give three examples of how the kidneys are well adapted for their functions in the body (excluding cellular adaptations). *(6 marks)*

2 a How much blood is filtered over 24 hours through the kidneys? *(1 mark)*
 b What percentage of this filtrate is lost to the body as urine? *(4 marks)*

3 a What is ultrafiltration? *(4 marks)*
 b What would you expect to see in normal glomerular filtrate? *(4 marks)*
 c Kidney infections can damage the lining of the Bowman's capsule. How might this result in protein appearing in the urine? *(6 marks)*

4 The normal concentration of urea in the blood is around 2.5–7.1 mmol / litre of blood. The concentration of urea in the urine varies considerably but most people pass 0.43–0.72 moles of urea in the urine over 24 hours.
 a Calculate the mean concentration of urea in the urine over a 24-hour period. *(3 marks)*
 b Approximately how much more urea is there in urine than in blood? *(1 mark)*
 c i Suggest three possible factors that might affect the amount of urea in the urine. *(3 marks)*
   ii Why do these factors not affect the concentration of urea in the blood? *(1 mark)*

5 Explain the main stages of how the kidney tubules produce urine that is more concentrated than the blood in these different regions
 a The proximal tubule *(4 marks)*
 b The descending limb of the loop of Henle *(5 marks)*
 c The ascending limb of the loop of Henle *(5 marks)*
 d The distal convoluted tubule *(6 marks)*
 e The collecting duct. *(6 marks)*

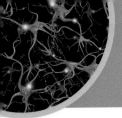
The mammalian kidney has a vital excretory function – it removes urea, the nitrogenous waste product of metabolism from the body. The kidney, however, also plays another important homeostatic role in the body – it is the main organ of *osmoregulation*. This involves controlling the water potential of the blood within very narrow boundaries, regardless of the activities of the body. Eating a salty meal, drinking large volumes of liquid, exercising hard, running a fever, or visiting a very hot climate can all put osmotic stresses on the body. It is very important to keep the water potential of the tissue fluid as stable as possible, because if water moves into or out of the cells by osmosis it can cause damage and even death.

Osmoregulation

Every day the body has to deal with many unpredictable events. The water potential of the blood has to be maintained regardless of the water and solutes taken in as you eat and drink, and the water and mineral salts lost by sweating, in defaecation, and in the urine. Changing the concentration of the urine is crucial in this dynamic equilibrium. The amount of water lost in the urine is controlled by ADH in a negative feedback system. ADH is produced by the hypothalamus and secreted into the posterior pituitary gland, where it is stored. ADH *increases* the permeability of the distal convoluted tubule and, most importantly, the collecting duct to water. You will be concentrating on the effect of ADH on the collecting duct walls.

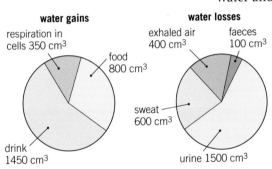

▲ Figure 1 *Typical water gains and losses over a 24-hour period in an adult human being*

The mechanism of ADH action

ADH is released from the pituitary gland and carried in the blood to the cells of the collecting duct where it has its effect. The hormone does not cross the membrane of the tubule cells – it binds to receptors on the cell membrane and triggers the formation of cyclic AMP (cAMP) as a second messenger inside the cell. A second messenger is a molecule which relays signals received at cell surface receptors to molecules inside the cell. The cAMP causes a cascade of events:

- Vesicles in the cells lining the collecting duct fuse with the cell surface membranes on the side of the cell in contact with the tissue fluid of the medulla.

- The membranes of these vesicles contain protein-based water channels (aquaporins) and when they are inserted into the cell surface membrane, they make it permeable to water.

- This provides a route for water to move out of the tubule cells into the tissue fluid of the medulla and the blood capillaries by osmosis.

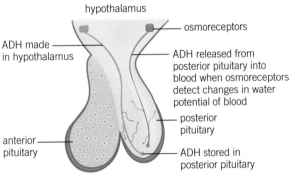

▲ Figure 2 *ADH is made in the hypothalamus but stored and released from the posterior pituitary gland when it is stimulated by neurones from the osmoreceptors in the hypothalamus*

The more ADH that is released, the more water channels are inserted into the membranes of the tubule cells. This makes it easy for more water to leave the tubules by diffusion, resulting in the formation of a small amount of very concentrated urine. Water is returned to the capillaries, maintaining the water potential of the blood and therefore the tissue fluid of the body.

When ADH levels fall, the reverse happens. Levels of cAMP fall, then the water channels are removed from the tubule cell membranes and enclosed in vesicles again. The collecting duct becomes impermeable to water once more, so no water can leave. This results in the production of large amounts of very dilute urine, and maintains the water potential of the blood and the tissue fluid.

Negative feedback control and ADH

The permeability of the collecting ducts is controlled to match the water requirements of the body very closely. This is brought about by a complex negative feedback system that involves **osmoreceptors** in the hypothalamus of the brain. These osmoreceptors are sensitive to the concentration of inorganic ions in the blood and are linked to the release of ADH.

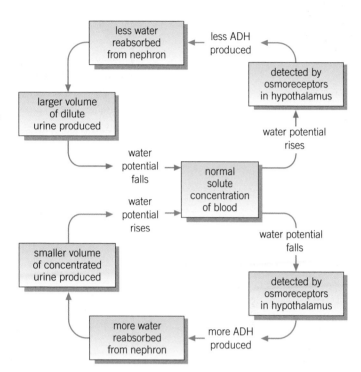

▲ Figure 3 *The negative feedback loop that controls the water potential of the blood*

When water is in short supply

When water is in short supply in the body, the concentration of inorganic ions in the blood rises and the water potential of the blood and tissue fluid becomes more negative. This is detected by the osmoreceptors in the hypothalamus. They send nerve impulses to the posterior pituitary which in turn releases stored ADH into the blood. The ADH is picked up by receptors in the cells of the collecting duct and increases the permeability of the tubules to water. Water leaves the filtrate in the tubules and passes into the blood in the surrounding capillary network. A small volume of concentrated urine is produced (Figure 3).

An excess of water

When large amounts of liquid are taken in, the blood becomes more dilute and its water potential becomes less negative. Again, the change is detected by the osmoreceptors of the hypothalamus. Nerve impulses to the posterior pituitary are reduced or stopped and so the release of ADH by the pituitary is inhibited. Very little reabsorption of water can take place because the walls of the collecting duct remain impermeable to water. In this way the concentration of the blood is maintained – and large amounts of dilute urine are produced (Figure 3).

ADH, water balance, and blood pressure

The osmoreceptors in the hypothalamus are not the only sensory receptors that exert control over the release of ADH. It is also stimulated or inhibited by changes in the blood pressure, detected by baroreceptors in the aortic and carotid arteries. These baroreceptors are also involved in the control of the heart rate.

A rise in blood pressure can often be caused by a rise in blood volume. The increase in pressure is detected by the baroreceptors and in turn they prevent the release of ADH. This increases the volume of water lost in the urine, reducing the blood volume and so the blood pressure falls.

If the blood pressure falls it can be a signal that the blood volume has fallen. If the baroreceptors detect a fall in blood pressure there is an increase in the release of ADH from the pituitary, so the kidneys respond to reduce water loss from the body. Water is returned to the blood and a small amount of concentrated urine is produced.

People with severe diarrhoea or more than 20% blood loss often produce little or no urine. In each case explain why.

Summary questions

1 In cool weather you may produce more urine than on a similar day in hot weather. Suggest a reason for this. *(2 marks)*

2 Draw a diagram or a flow chart to explain how ADH increases the permeability of the cells lining the distal convoluted tubules and the collecting duct to water. *(6 marks)*

3 There is a rare condition called diabetes insipidus where the body does not make ADH (or very rarely, the kidneys do not respond to ADH).
   a What would you expect the symptoms of diabetes insipidus to be? *(4 marks)*

   b Explain what is happening in the kidney tubules in a patient with diabetes insipidus. *(6 marks)*

   c Suggest how you might treat mild diabetes insipidus (when some ADH is made) and severe diabetes insipidus (where no ADH is made). *(3 marks)*

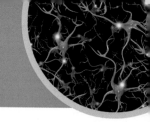

15.7 Urine and diagnosis

Specification reference: 5.1.2

For many centuries people have used urine to try and diagnose diseases by looking at the colour, the smell, and even the taste of it. Urine is still a useful diagnostic tool, although our methods of analysis are more sophisticated.

Urine samples and diagnostic tests

Urine contains water, urea, mineral salts – and much more. It contains the breakdown products of a whole range of chemicals, including hormones and any toxins taken into the body. If you are affected by one of a number of different diseases, new substances will show up in your urine. The presence of glucose in the urine is a well-known symptom of type 1 and type 2 diabetes. If you have muscle damage, large amounts of creatinine will show up in your urine.

Urine and pregnancy testing

The human embryo implants in the uterus, around six days after conception. The site of the developing placenta then begins to produce a chemical called human chorionic gonadotrophin (hCG). Some of this hormone is found in the blood and the urine of the mother. Until the 1960s, the most reliable available pregnancy test was to inject the urine from a pregnant woman into an African clawed toad (*Xenopus laevis*). If she was pregnant, the hCG triggered egg production in the toad within 8–12 hours of the injection. It could not be used until the woman was several weeks pregnant.

Modern pregnancy tests still test for hCG in the urine, but they rely on **monoclonal antibodies**. Some are so sensitive that pregnancy can be detected within hours of implantation.

Making monoclonal antibodies

Monoclonal antibodies are antibodies from a single clone of cells that are produced to target particular cells or chemicals in the body.

A mouse is injected with hCG so it makes the appropriate antibody. The B-cells that make the required antibody are then removed from the spleen of the mouse and fused with a myeloma, a type of cancer cell which divides very rapidly. This new fused cell is known as a hybridoma. Each hybridoma reproduces rapidly, resulting in a clone of millions of 'living factories' making the desired antibody. These monoclonal antibodies are collected, purified and used in a variety of ways.

The main stages in a pregnancy test are as follows:

- the wick is soaked in the first urine passed in the morning – this will have the highest levels of hCG.

- The test contains mobile monoclonal antibodies that have very small coloured beads attached to them. They will only bind to hCG. If the woman is pregnant the hCG in her urine binds to the mobile monoclonal antibodies and forms a *hCG/antibody complex* (complete with coloured bead).

Learning outcomes

Demonstrate knowledge, understanding, and application of:

→ how excretory products can be used in medical diagnosis.

▲ **Figure 1** *The mixture of urine and a toad may not sound very scientific but the combination made a relatively accurate and available pregnancy test for many years*

Synoptic link

You learnt about the specific immune system and antibodies in Topic 12.6, The specific immune system.

▲ **Figure 2** *Pregnancy test*

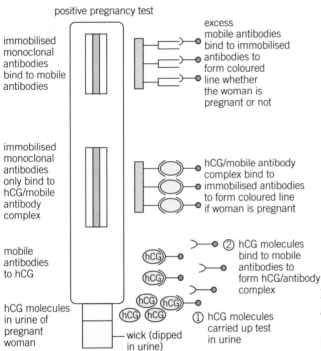

positive pregnancy test

immobilised monoclonal antibodies bind to mobile antibodies

excess mobile antibodies bind to immobilised antibodies to form coloured line whether the woman is pregnant or not

immobilised monoclonal antibodies only bind to hCG/mobile antibody complex

hCG/mobile antibody complex bind to immobilised antibodies to form coloured line if woman is pregnant

mobile antibodies to hCG

② hCG molecules bind to mobile antibodies to form hCG/antibody complex

hCG molecules in urine of pregnant woman

wick (dipped in urine)

① hCG molecules carried up test in urine

▲ Figure 3 *The presence of hCG in the urine can be picked up using monoclonal antibodies to produce an easy-to-use and very accurate pregnancy test*

- The urine carries on along the test structure until it reaches a window.
- Here there are immobilised monoclonal antibodies arranged in a line or a pattern such as a positive (+) sign that only bind to the hCG/antibody complex. If the woman is pregnant, a coloured line or pattern appears in the first window.
- The urine continues up through the test to a second window.
- Here there is usually a line of immobilised monoclonal antibodies that bind only to the mobile antibodies, regardless of whether they are bound to hCG or not. This coloured line forms regardless of whether the woman is pregnant – it simply indicates that the test is working.

If the woman is pregnant, two coloured patterns appear. If she is not pregnant, only one appears.

Urine and anabolic steroids

Athletes and body builders may try to cheat by using anabolic steroids. Anabolic steroids are drugs that mimic the action of the male sex hormone testosterone and they stimulate the growth of muscles. They are, however, excreted in the urine. By testing the urine using gas chromatography and mass spectrometry, scientists can show that an individual has been using these drugs, which are banned in all sports. The urine sample is vaporised with a known solvent and passed along a tube. The lining of the tube absorbs the gases and is analysed to give a chromatogram that can be read to show the presence of the drugs.

Urine and drug testing

Urine is tested for the presence of many different drugs, including alcohol. Because drugs or metabolites – the breakdown products of drugs – are filtered through the kidneys and stored in the bladder, it is possible to find drug traces in the urine some time after a drug has been used.

If someone is suspected of having taken an illegal drug, they may be asked to provide a urine sample and this will be divided into two. The first sample may be tested by an immunoassay, using monoclonal antibodies to bind to the drug or its breakdown product. If this shows positive, the second sample may be run through a gas chromatograph/mass spectrometer to confirm the presence of the drug. A wide range of drugs can be detected in the urine in this way as illustrated in Table 1.

▼ Table 1

Substance	Time it persists in the urine
ethanol (alcohol)	6–24 hours
amphetamines	1–3 days
cannabis	22 hours to 30 days depending on use
cocaine	2–5 days

Summary questions

1 Why is urine so useful for diagnostic tests?   (3 marks)

2 The professional bodies of different sports carry out random urine testing both during training and competition. The winning athletes always have their urine tested.
  a Urine samples are divided into two and only one of them is tested initially. Explain why.   (4 marks)
  b Why do you think urine tests are carried out at random during training as well as at competitions?   (3 marks)

3 a Compare the process of pregnancy tests with the process of testing for illegal drugs.   (3 marks)
  b Some pregnancy tests come with two tests. Explain how a pregnancy test might show a false-negative result.   (3 marks)

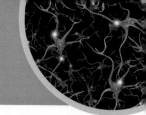

15.8 Kidney failure

Specification reference: 5.1.2

The kidneys play a vital role in homeostasis. If they are damaged and become less efficient or stop working, the effects on the body are serious and may be fatal if they are not treated.

Causes and effects of kidney failure

There are a number of reasons why the kidneys may fail. They include kidney infections, where the structure of the podocytes and the tubules themselves may be damaged or destroyed; raised blood pressure that can damage the structure of the epithelial cells and basement membrane of the Bowman's capsule; and genetic conditions such as polycystic kidney disease where the healthy kidney tissue is replaced by fluid-filled cysts or damaged by pressure from cysts.

If the kidneys are infected or affected by high blood pressure this may cause:

- Protein in the urine – if the basement membrane or podocytes of the Bowman's capsule are damaged, they no longer act as filters and large plasma proteins can pass into the filtrate and are passed out in the urine.

- Blood in the urine – another symptom that the filtering process is no longer working.

If the kidneys fail completely, the concentrations of urea and mineral ions build up in the body. The effects include:

- Loss of electrolyte balance – if the kidneys fail, the body cannot excrete excess sodium, potassium, and chloride ions. This causes osmotic imbalances in the tissues and eventual death.

- Build-up of toxic urea in the blood – if the kidneys fail, the body cannot get rid of urea and it can poison the cells.

- High blood pressure – the kidneys play an important role in controlling the blood pressure by maintaining the water balance of the blood. If the kidneys fail, the blood pressure increases and this can cause a range of health problems including heart problems and strokes.

- Weakened bones as the calcium/phosphorus balance in the blood is lost.

- Pain and stiffness in joints as abnormal proteins build up in the blood.

- Anaemia – the kidneys are involved in the production of a hormone called erythropoietin that stimulates the formation of red blood cells. When the kidneys fail it can reduce the production of red blood cells causing tiredness and lethargy.

Measuring glomerular filtration rate

Kidney problems almost always affect the rate at which blood is filtered in the Bowman's capsules of the nephrons. The glomerular

Learning outcomes

Demonstrate knowledge, understanding, and application of:

→ the effects of kidney failure and its potential treatments.

▼ Table 1 *To show average GFR in healthy adults*

Age (years)	Average eGFR
20–29	116
30–39	107
40–49	99
50–59	93
60–69	85
70+	75

filtration rate (GFR) is widely used as a measure to indicate kidney disease. The rate of filtration is not measured directly – a simple blood test measures the level of creatinine in the blood. Creatinine is a breakdown product of muscles and it is used to give an estimated glomerular filtration rate (eGFR). The units are cm^3/min. If the levels of creatinine in the blood go up, it is a signal that the kidneys are not working properly. Certain factors need to be taken into account in the calculations to work out GFR. For example, GFR decreases steadily with age even if you are healthy, and men usually have more muscle mass and therefore more creatinine than women.

Treating kidney failure with dialysis

As you can see in Table 1, normal GFRs do not fall below 70 even in very elderly people. A GFR of below 60 for more than three months is taken to indicate moderate to severe chronic kidney disease – and if it falls below 15, that is kidney failure. The kidneys are filtering so little blood they are virtually ineffective.

There are two main ways in which kidney failure is treated. In **renal dialysis**, the function of the kidneys is carried out artificially. In a transplant, a new healthy kidney is put into the body to replace the functions of the failed kidneys – an animal can function perfectly well with just one healthy kidney.

There are two main types of dialysis – haemodialysis and peritoneal dialysis.

Haemodialysis

This involves the use of a dialysis machine. It is usually carried out in hospital although sometimes patients will have a machine in their own home. Blood leaves the patient's body from an artery and flows into the dialysis machine, where it flows between partially permeable dialysis membranes. These membranes mimic the basement membrane of the Bowman's capsule. On the other side of the membranes is the dialysis fluid. During dialysis it is vital that patients lose the excess urea and mineral ions that have built up in the blood. It is equally important that they do not lose useful substances such as glucose and some mineral ions.

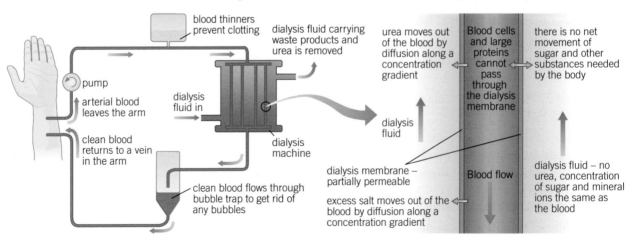

▲ Figure 1 *A dialysis machine relies on simple diffusion to remove waste products from the body*

The loss of these substances is prevented by careful control of the dialysis fluid. It contains normal plasma levels of glucose to ensure there is no net movement of glucose out of the blood. The dialysis fluid also contains normal plasma levels of mineral ions, so any excess mineral ions in the blood move out by diffusion down a concentration gradient into the dialysis fluid, thus restoring the correct electrolyte balance of the blood. The dialysis fluid contains no urea meaning there is a very steep concentration gradient from the blood to the fluid, and as a result, much of the urea leaves the blood. The blood and the dialysis fluid flow in opposite directions to maintain a countercurrent exchange system and maximise the exchange that takes place.

The whole process of dialysis depends on diffusion down concentration gradients – there is no active transport. Dialysis takes about eight hours and has to be repeated regularly. Patients with kidney failure who rely on haemodialysis have to remain attached to a dialysis machine several times a week for many hours. They also need to manage their diets carefully, eating relatively little protein and salt and monitoring their fluid intake to keep their blood chemistry as stable as possible. The only time they can eat and drink what they like is at the beginning of the dialysis process.

Peritoneal dialysis

Peritoneal dialysis is done inside the body – it makes use of the natural dialysis membranes formed by the lining of the abdomen, that is, the peritoneum. It is usually done at home and the patient can carry on with their normal life while it takes place. The dialysis fluid is introduced into the abdomen using a catheter. It is left for several hours for dialysis to take place across the peritoneal membranes, so that urea and excess mineral ions pass out of the blood capillaries, into the tissue fluid, and out across the peritoneal membrane into the dialysis fluid. The fluid is then drained off and discarded, leaving the blood balanced again and the urea and excess minerals removed.

Treating kidney failure by transplant

Long-term dialysis has some serious side effects. The best solution for the patient is a kidney transplant, where a single healthy kidney from a donor is placed within the body. The blood vessels are joined and the ureter of the new kidney is inserted into the bladder. If the transplant is successful, the kidney will function normally for many years.

The main problem with transplanted organs is the risk of rejection. The antigens on the donor organ differ from the antigens on the cells of the recipient and the immune system is likely to recognise this. This can result in rejection and the destruction of the new kidney.

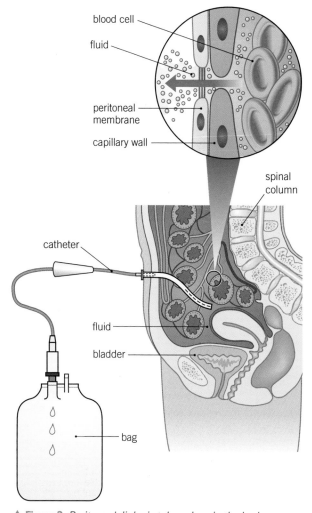

▲ **Figure 2** *Peritoneal dialysis takes place in the body cavity across the peritoneal membranes*

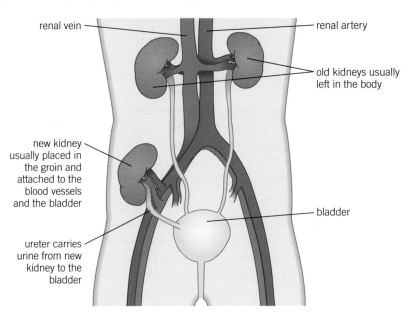

renal vein — — renal artery

old kidneys usually left in the body

new kidney usually placed in the groin and attached to the blood vessels and the bladder

bladder

ureter carries urine from new kidney to the bladder

▲ **Figure 3** *A donor kidney takes over the functions of the failed kidneys, which are usually left in the body*

There are a number of ways of reducing the risk of rejection. The match between the antigens of the donor and the recipient is made as close as possible. For example, a donor kidney can be used with a 'tissue type' very similar to the recipient (from people with the same blood group).

The recipient is given drugs to suppress their immune response (immunosuppressant drugs) for the rest of their lives. This helps to prevent the rejection of their new organ. Immunosuppressant drugs are improving all the time and the need for a really close tissue match is becoming less important.

The disadvantage of taking immunosuppressant drugs is that they prevent the patients from responding effectively to infectious diseases. They have to take great care if they become ill in any way. However, most people feel this is a small price to pay for a new, functioning kidney.

Transplanted organs don't last forever with the average transplanted kidney functioning for around 9–10 years, although some have continued working for around 50 years. Once the organ starts to fail the patient has to return to dialysis and wait until another suitable kidney is found.

Dialysis or transplant?

Dialysis is much more readily available than donor organs, so it is there whenever kidneys fail. It enables the patient to lead a relatively normal life. However, patients have to monitor their diet carefully and need regular sessions on the machine. Long-term dialysis is much more expensive than a transplant and can eventually cause damage to the body.

If a patient receives a kidney transplant they are free from the restrictions which come with regular dialysis sessions and dietary monitoring. This is generally the ideal scenario for patients waiting for a transplant.

Synoptic link

You learnt about stem cells in Topic 6.5, Stem cells.

The main source of donor kidneys is from people who die suddenly, often from road accidents, strokes, and heart attacks. In the UK, organs can be taken from people if they carry an organ donor card or are on the online donor register. Alternatively, a relative of someone who has died suddenly can give his or her consent.

Unfortunately for people needing a transplant, there is a shortage of donor kidneys. Many people do not register as donors. In addition, as cars become safer, fewer people die suddenly in traffic accidents. Whilst good, this means there are fewer potential donors. At any one time there are thousands of people undergoing kidney dialysis. Most will not have the opportunity to have a kidney transplant. In 2013–14, 3257 people in the UK had kidney transplants. However, there were still around 6000 people on dialysis waiting for a kidney. There are increasing numbers of live donor transplants, where a family member of someone with a tissue match donates a kidney. In 2013–14, 2143 people received kidneys from donors who had died, and 1114 were from living donors.

In 2011, scientists grew functioning embryonic kidney tissue from stem cells. Going forward, the hope is that whole new kidneys can be grown – perhaps even without the antigens which trigger the immune reaction so that patients don't need to take immunosuppressant drugs.

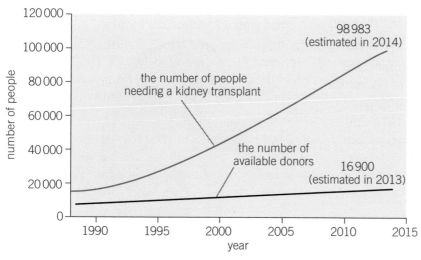

▲ **Figure 4** *This graph shows how the gap between people needing a kidney transplant and the available donors is widening in the US*

Summary questions

1 a Why is kidney failure such a threat to life? *(3 marks)*
 b On what process does dialysis depend? *(1 mark)*

2 Produce a flow chart to describe how a dialysis machine works. *(6 marks)*

3 Sometimes a live donor, usually a close family member, will donate a kidney. These transplants have a higher rate of success than normal transplants from dead, unrelated donors.
 a Suggest two reasons why live transplants from a close family member have a higher success rate than normal transplants. *(2 marks)*
 b Why do you think that live donor transplants are still the minority? *(2 marks)*

4 a Explain the importance of dialysis fluid containing no urea and normal plasma levels of salt, glucose, and minerals. *(4 marks)*
 b Both blood and dialysis fluid are constantly circulated through the dialysis machine. Explain why it is important that the blood and dialysis fluid flow in opposite directions and that there is a constant circulation of dialysis fluid. *(3 marks)*
 c Why can patients with kidney failure eat and drink what they like during the first few hours of dialysis? *(2 marks)*

Practice questions

1   The Cori cycle is the name given to the metabolic pathway that recycles the lactate produced during anaerobic respiration.

The diagram summarises the production of lactate and the Cori cycle.

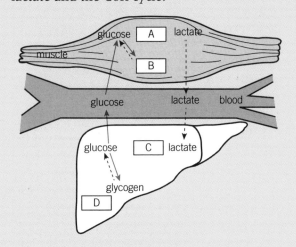

a   (i)   Name the processes occurring at A and C in the diagram   (*2 marks*)

   (ii)   Name the molecule B.   (*1 mark*)

   (iii)   Name the organ D.   (*1 mark*)

b   Suggest why the Cori cycle is important for homeostasis.   (*4 marks*)

It was believed for many years that lactate was a toxic waste product. In the 1970's this began to be questioned and the phrase '*lactic acid, friend or foe*' was coined.

c   Discuss the phrase 'lactic acid, friend or foe'.   (*5 marks*)

2   The light micrograph shows a section through a human liver.

a   Name the structures shown by A and B   (*2 marks*)

b   The diagram outlines the formation of urea in the liver.

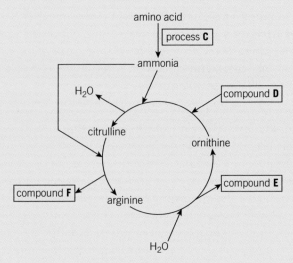

Using the diagram identify:

Process **C**

Compound **D**

Compound **E**

Compound **F**   (*4 marks*)

OCR F214 2010 (apart from a)

3   The diagram represent a vertical section through a mammalian kidney.

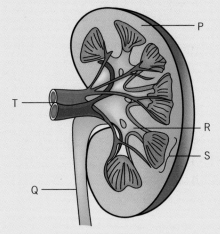

a   Name: region P, structure Q, structure R, structure S, and structure T   (*5 marks*)

b   (i)   Draw a labelled diagram of a glomerulus and Bowman's capsule.   (*4 marks*)

   (ii)   Outline the roles of the structures you have drawn in the process of ultrafiltration.   (*4 marks*)

c   Nephritis is a condition in which the tissue of the glomerulus and proximal convoluted tubule becomes inflamed and damaged.

Suggest and explain two differences in the composition of urine of a person with nephritis when compared to the urine of a person with healthy kidneys. (*4 marks*)

d Caffeine is a mild diuretic. Caffeine prevents the introduction of additional aquaporins into the wall of the collecting duct of the nephron and therefore additional water is not removed from the urine.

Aquaporins are channels in the cell surface membrane that allow water molecules to pass through.

The diagram represents an aquaporin.

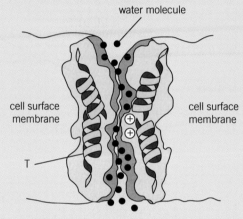

(i) Identify the type of molecule labelled **T** (*1 mark*)

(ii) The aquaporin allows water to travel from the collecting duct into the surrounding tissues but prevents the passage of ions such as sodium ions and potassium ions.

With reference to the diagram, suggest two ways in which the structure of this aquaporin prevents the passage of ions.

(*2 marks*)

OCR F214 2011 (apart from a and b)

4 The graph shows the metabolic rate changes as ambient temperature (temperature of the surroundings) changes.

Non-shivering thermogenesis is stimulated by cold temperatures and leads to an increase in metabolic activity that is not associated with muscle contraction. The increased metabolic activity results in the production of heat.

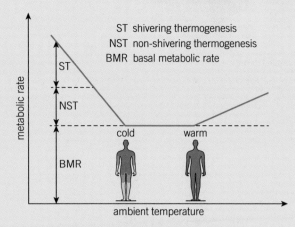

a Describe the change in metabolic rate shown in the graph. (*3 marks*)

b State which part of the involuntary nervous system stimulates thermogenesis. (*1 mark*)

c Explain why an increased metabolic rate leads to an increase in the body temperature. (*3 marks*)

d Explain the change in metabolic rate as the external temperature increases. (*3 marks*)

Recent research has shown that the metabolic activity associated NST occurs mainly in a specialised type of fat tissue called brown adipose tissue (BAT).

A typical brown adipose cell and white adipose cell are shown in the diagram.

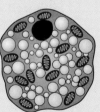

brown adipose cell   white adipose cell

e (i) Describe the visible differences between the two cells in the diagrams. (*3 marks*)

(ii) Suggest how the brown adipose cell is better adapted for the increased heating of the body. (*3 marks*)

f The mitochondria in brown adipose cells contain a protein called uncoupling protein 1 (UCP1) that inhibits the synthesis of ATP.

Explain how this would result in increased warming OR a greater temperature rise. (*2 marks*)

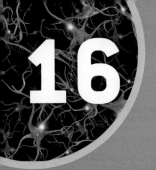

16 PLANT RESPONSES

16.1 Plant hormones and growth in plants

Specification reference: 5.1.5

People often regard plants as passive objects in the environment that simply grow and sometimes flower. However, plants are dynamic systems, not only photosynthesising and producing food but also responding to their environment in many different ways. They have evolved to cope with abiotic stresses such as lack of water, and they have a range of adaptations to protect them against the attentions of herbivores. They also show directional growth in response to environmental cues such as light and gravity – these are known as **tropisms**.

Chemical coordination

Plants are multicellular organisms living in a complex and ever-changing environment. The key limitations on plants are that they are rooted – they are not mobile, and they do not have a rapidly responding nervous system. They are, however, coordinated organisms that show clear responses to their environment, communication between cells, and even communication between different plants. The timescales of most plant responses are slower than animal responses, but they still respond as a result of complex chemical interactions. Plants have evolved a system of hormones – chemicals that are produced in one region of the plant and transported both through the transport tissues and from cell to cell and have an effect in another part of the plant. Important plant hormones include **auxins**, **gibberellins**, abscisic acid (ABA), and ethene. These chemicals have a wide range of functions within the plant.

▼ Table 1 *Some of the roles of plant hormones*

Hormone	Some of their known roles in plants
auxins	control cell elongation, prevent leaf fall (abscission), maintain apical dominance, involved in tropisms, stimulate the release of ethene, involved in fruit ripening
gibberellin	cause stem elongation, trigger the mobilisation of food stores in a seed at germination, stimulate pollen tube growth in fertilisation
ethene	causes fruit ripening, promotes abscission in deciduous trees
ABA (abscisic acid)	maintains dormancy of seeds and buds, stimulates cold protective responses, for example, antifreeze production, stimulates stomatal closing

Plants produce chemicals which signal to other species – for example, to protect themselves from attack by insect pests – and may communicate with other plants. They also produce chemical defences against herbivores. In plant responses, chemicals are essential.

The growth of plants, from the germination of a seed to the long-term growth of a tree, is controlled by plant hormones. You will look at the different chemicals and their roles in isolation, but in fact the growth and form of a plant are the result of the interaction of many different hormonal and environmental factors.

Scientists are still unsure about the details of many plant responses. There are a number of reasons for this. Plant hormones work at very low concentrations, so isolating them and measuring changes in concentrations is not easy. The multiple interactions between the different chemical control systems also make it very difficult for researchers to isolate the role of a single chemical in a specific response.

Outlined are a number of key aspects of plant growth with the role of hormones highlighted.

Plant hormones and seed germination

For a plant to start growing, the seed must germinate.

- When the seed absorbs water, the embryo is activated and begins to produce gibberellins. They in turn stimulate the production of enzymes that break down the food stores found in the seed. The food store is in the cotyledons in dicot seeds and the endosperm in monocot seeds. The embryo plant uses these food stores to produce ATP for building materials so it can grow and break out through the seed coat. Evidence suggests that gibberellins switch on genes which code for amylases and proteases – the digestive enzymes required for germination. There is also evidence suggesting that another plant hormone, ABA, acts as an antagonist to gibberellins (interferes with the action of gibberellin), and that it is the relative levels of both hormones which determine when a seed will germinate.

Experimental evidence

Experimental evidence supporting the role of gibberellins in the germination of seeds includes:

- Mutant varieties of seeds have been bred which lack the gene that enables them to make gibberellins. These seeds do not germinate. If gibberellins are applied to the seeds externally, they then germinate normally.
- If gibberellin biosynthesis inhibitors are applied to seeds, they do not germinate as they cannot make the gibberellins needed for them to break dormancy. If the inhibition is removed, or gibberellins are applied, the seeds germinate.

> **Synoptic link**
>
> You learnt about plant transport systems in Chapter 9, Transport in plants.

▲ **Figure 1** *Plants' responses enable them to grow and flower in almost every land environment*

> **Synoptic link**
>
> You learnt about the role of cotyledons as food stores in the seeds of dicotyledenous plants in Topic 9.1, Transport systems in dicotyledonous plants.

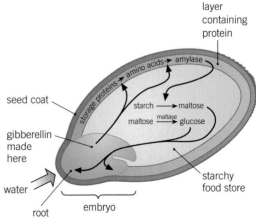

▲ **Figure 2** *The role of gibberellins in germination*

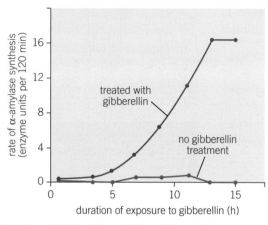

▲ **Figure 3** *The effect of gibberellins on the synthesis of amylase, a starch-digesting enzyme, in isolated tissue from barley seeds*

Plant hormones, growth, and apical dominance

The growth of a plant shoot after a seed has germinated is controlled by a number of plant hormones.

Auxins

Auxins such as indoleacetic acid (IAA) are growth stimulants produced in plants. Small quantities can have powerful effects. They are made in cells at the tip of the roots and shoots, and in the meristems. Auxins can move down the stem and up the root both in the transport tissue and from cell to cell. The effect of the auxin depends on its concentration and any interactions it has with other hormones. Auxins have a number of major effects on plant growth.

- They stimulate the growth of the main, apical shoot. Evidence suggests that auxins affect the plasticity of the cell wall – the presence of auxins means the cell wall stretches more easily. Auxin molecules bind to specific receptor sites in the plant cell membrane, causing a fall in the pH to about 5. This is the optimum pH for the enzymes needed to keep the walls very flexible and plastic. As the cells mature, auxin is destroyed. As the hormone levels fall, the pH rises so the enzymes maintaining plasticity become inactive. As a result, the wall becomes rigid and more fixed in shape and size and the cells can no longer expand and grow.

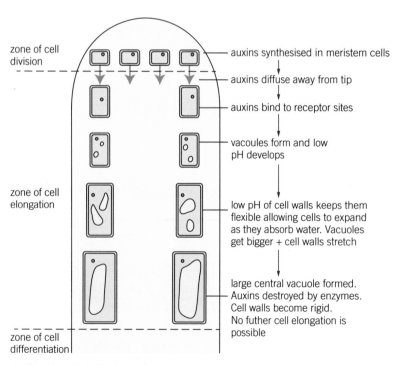

▲ **Figure 4** *The effect of auxin on apical shoot growth*

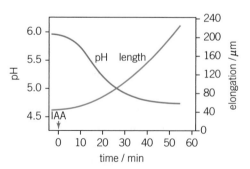

▲ **Figure 5** *Graph to show the effect of an external application of auxin on pH levels in the cell walls and on shoot growth*

High concentrations of auxins suppress the growth of lateral shoots. This results in apical dominance. Growth in the main shoot is stimulated by the auxin produced at the tip so it grows quickly. The lateral shoots are inhibited by the hormone that moves back down the stem, so they do not grow very well. Further down the stem, the auxin concentration is lower and so the lateral shoots grow more strongly. There is a lot of experimental evidence for the role of auxins in apical dominance. For example, if the apical shoot is removed, the auxin-producing cells are removed and so there is no auxin. As a result, the lateral shoots, freed from the dominance of the apical shoot, grow faster. If auxin is applied artificially to the cut apical shoot, apical dominance is reasserted and lateral shoot growth is suppressed.

Low concentrations of auxins promote root growth. Up to a given concentration, the more auxin that reaches the roots, the more they grow. Auxin is produced by the root tips and auxin also reaches the roots in low concentrations from the growing shoots. If the apical shoot is removed, then the amount of auxin reaching the roots is greatly reduced and root growth slows and stops. Replacing the auxin artificially at the cut apical shoot restores the growth of the roots. High auxin concentrations inhibit root growth.

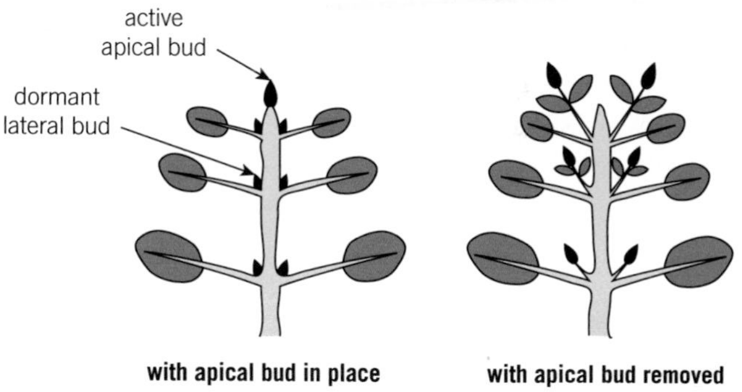

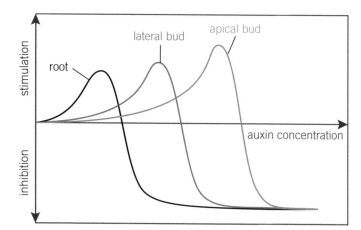

▲ Figure 6 *Apical dominance affects the form of a plant and can clearly be seen in conifers*

Gibberellins

As you have learnt, gibberellins are involved in the germination of seeds. They are also important in the elongation of plant stems during growth. Gibberellins affect the length of the internodes – the regions between the leaves on a stem. Gibberellins were discovered because they are produced by a fungus from the genus *Gibberella* that affects rice. The infected seedlings grew extremely tall and thin. Scientists investigated the rice and isolated chemicals – gibberellins – which produce the same spindly growth in the plants. It was then discovered that plants themselves produce the same compounds. Plants that have short stems produce few or no gibberellins. There are well over a hundred different naturally produced gibberellins. Scientists have bred many dwarf varieties of plants where the gibberellin synthesis pathway is interrupted. Without gibberellins the plant stems are much shorter. This reduces waste and also makes the plants less vulnerable to damage by weather and harvesting.

▲ Figure 7 *Different concentrations of hormones affect different tissues in different ways – the graph shows the effect of auxin on roots, lateral buds, and apical buds*

Synoptic link

You learnt about the use of standard deviation to measure the spread of data in Topic 10.6, Representing variation graphically.

Investigating the effect of hormones on plant growth 🧪

There are many different ways to investigate the effect of plant hormones on the growth of the shoots, the roots, and the germination of seeds. These include growing seedlings hydroponically (in nutrient solution rather than soil) in serial dilutions of different hormones, or applying different concentrations of hormones to the cut ends of stems or roots and observing the effects.

In most experiments, it is important to make serial dilutions to observe the effects of different concentrations of the hormones, as they can have different effects on growth at different concentrations.

Experiments investigating the effect of hormones on plant growth usually involve large numbers of plants. When you have completed your measurements, the spread of data from each experimental group should be measured using standard deviation.

1 Suggest an advantage and a disadvantage of using serial dilutions of hormones in nutrient solution to investigate the effect of a hormone on plant growth.

2 An experiment was conducted to investigate using plant hormones to increase the growth of cuttings. It was found that in the first experiment there was a significant increase ($p \leq 0.05$) in fresh mass with potting medium 2, and leaf length with potting medium 1, both at 100 mg/l IAA. Explain the meaning of the term **significant increase** ($p \leq 0.05$) and explain how it can be calculated.

Synoptic link

You learnt about serial dilutions in Topic 4.2, Factors affecting enzyme activity.

Synergism and antagonism

Most plant hormones do not work on their own but by interacting with other substances. In doing so, very fine control over the responses of the plant can be achieved. If different hormones work together, complementing each other and giving a greater response than they would on their own, the interaction is known as synergism. If the substances have opposite effects, for example one promoting growth and one inhibiting it, the balance between them will determine the response of the plant. This is known as antagonism. Our knowledge of plant hormones and the mechanisms by which they have an effect is still far from complete – this is an active and important area of research.

Summary questions

1 Why are chemicals so important in coordinating the growth of plants? *(3 marks)*

2 a Give three examples of plant hormones and for each give one function in the plant. *(3 marks)*
   b Why are the chemicals you have listed in answer 2a described as plant hormones? *(3 marks)*

3 Plant hormones have very different effects on different plant tissues.
   a Give an example with experimental evidence. *(6 marks)*
   b Explain the importance of these multifunction hormones in a plant. *(2 marks)*

4 Explain how the data in Figure 5 appear to confirm our current model of auxin action. *(6 marks)*

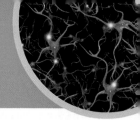

When the environmental conditions around plants change, they have to cope or die. Abiotic stresses include changes in day length, cold and heat, lack of water, excess water, high winds, and changes in salinity. Plants need to be able to cope with these changes. As you have already learnt, plant responses involve both physical and physiological adaptations. They may have very thick cuticles, hairy leaves, sunken stomata or a wilting response in hot, dry or extremely windy conditions, or develop aerenchyma if they grow in an aquatic environment.

Leaf loss in deciduous plants as a response to abiotic stress

Plants that grow in temperate climates experience great environmental changes during the year. For example, the range of daylight hours in parts of northern Scotland ranges from about 6.5 hours midwinter to just under 18.5 hours midsummer. Temperatures vary as well – in England the mean temperature is 3–6°C in winter and 16–21°C in summer. As light and temperature affect the rate of photosynthesis, seasonal changes have a big impact on the amount of photosynthesis possible. The point comes when the amount of glucose required for respiration to maintain the leaves, and to produce chemicals from chlorophyll that might protect them against freezing is greater than the amount of glucose produced by photosynthesis. In addition, a tree that is in leaf is more likely to be damaged or blown over by winter gales. This means deciduous trees in temperate climates lose all of their leaves in winter and remain dormant until the days lengthen and temperatures rise again in spring.

Daylength sensitivity

Scientists have discovered that plants are sensitive to a lack of light in their environment. This is known as photoperiodism. For many years it was assumed that plants responded to the length of daylight, but more recent evidence suggests that it is lack of light that is the trigger for change. Many different plant responses are affected by the photoperiod including the breaking of the dormancy of the leaf buds so they open up, the timing of flowering in a plant and when tubers are formed in preparation for overwintering.

The sensitivity of plants to day length (or dark length) results from a light-sensitive pigment called phytochrome. This exists in two forms – P_r and P_{fr}. Each absorbs a different type of light and the ratio of P_r to P_{fr} changes depending on the levels of light.

Learning outcomes

Demonstrate knowledge, understanding, and application of:

→ the types of plant responses

→ the roles of plant hormones.

Synoptic link

You learnt about the way plants respond to very dry and to watery conditions in Topic 9.5, Plant adaptations to water availability.

Synoptic link

You will learn more about photosynthesis in Chapter 17, Energy for biological processes.

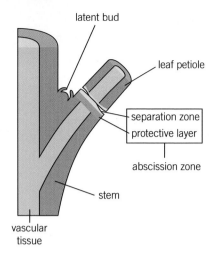

▲ **Figure 1** *The abscission layer in a leaf stalk*

▲ **Figure 2** *The scar where a leaf has fallen from this horse chestnut twig is clear to see, including the dark dots which are the sealed ends of the xylem and phloem vessels*

Synoptic link

You learnt about the mechanism of opening and closing the stomata as a result of turgor changes in Topic 9.3, Transpiration.

Abcission or leaf fall

After a summer of long days, short nights, and warm temperatures, the nights lengthen, days shorten, and temperatures fall as autumn develops. The lengthening of the dark period triggers a number of changes, including abscission or leaf fall and a period of dormancy during the winter months.

The falling light levels result in falling concentrations of auxin. The leaves respond to the falling auxin concentrations by producing the gaseous plant hormone ethene. At the base of the leaf stalk is a region called the abscission zone, made up of two layers of cells sensitive to ethene. Ethene seems to initiate gene switching in these cells resulting in the production of new enzymes. These digest and weaken the cell walls in the outer layer of the abscission zone, known as the separation layer.

The vascular bundles which carry materials into and out of the leaf are sealed off. At the same time fatty material is deposited in the cells on the stem side of the separation layer. This layer forms a protective scar when the leaf falls, preventing the entry of pathogens. Cells deep in the separation zone respond to hormonal cues by retaining water and swelling, putting more strain on the already weakened outer layer. Then further abiotic factors such as low temperatures or strong autumn winds finish the process – the strain is too much and the leaf separates from the plant. A neat, waterproof scar is left behind.

Preventing freezing

Another major abiotic factor which affects plants is a decrease in temperature. If cells freeze, their membranes are disrupted and they will die. Many plants, however, have evolved mechanisms that protect their cells in freezing conditions. The cytoplasm of the plant cells and the sap in the vacuoles contain solutes which lower the freezing point. Some plants produce sugars, polysaccharides, amino acids, and even proteins which act as antifreeze to prevent the cytoplasm from freezing, or protect the cells from damage even if they do freeze.

Most species only produce the chemicals which make them hardy and frost resistant during the winter months. It appears that different genes are suppressed and activated in response to a sustained fall in temperatures along with a reduction in day length, effectively preparing the plants to withstand frosty conditions. A sustained spell of warm weather along with extended day length reverses these changes in the spring.

Stomatal control

As you have already learnt, heat and water availability are major abiotic stresses for plants. One of the major ways in which plants can respond to these stresses is to open the stomata to cool the plant as water evaporates from the cells in the leaves in transpiration, or to close the stomata to conserve water.

The opening and closing of the stomata in response to abiotic stresses is largely under the control of the hormone ABA. The leaf cells appear to release ABA under abiotic stress, causing stomatal closure. However, scientists now think that the roots also provide an early warning of water stresses through ABA. So, for example, when the levels of soil water fall and transpiration is under threat, plant roots produce ABA which is transported to the leaves where it binds to receptors on the plasma membrane of the stomatal guard cells. ABA activates changes in the ionic concentration of the guard cells, reducing the water potential and therefore turgor of the cells. As a result of reduced turgor, the guard cells close the stomata and water loss by transpiration is greatly reduced.

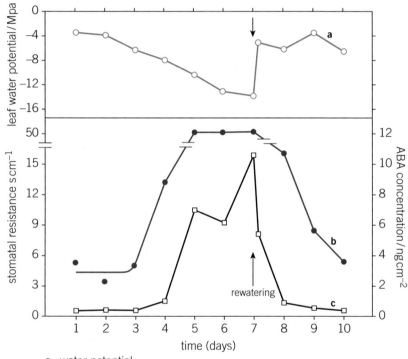

a   water potential
b   stomatal resistance
c   ABA content in corn

▲ **Figure 3** *Changes in water potential, stomatal resistance, and ABA content in corn in response to water stress – stomatal resistance increases as stomata close. All measurements of stomatal resistance in excess of $20\,s\,cm^{-1}$ are shown as $50\,s\,cm^{-1}$*

1 Why is it so important for plants to be able to respond to their surroundings? (*2 marks*)

2 Why do many trees in temperate climates lose all of their leaves in winter? (*6 marks*)

3 Produce a flow diagram to explain the process of abscission. (*6 marks*)

4 a Explain how plant hormones are involved in protecting the plant cells from damage in freezing conditions. (*5 marks*)
   b Give two more adaptations by which plant cells may be protected against damage by freezing during the winter. (*3 marks*)
   c If there is a sudden spell of freezing weather early in the autumn, many plants which can normally survive the winter may be killed. Similarly, if there is a late frost after several weeks of warm spring weather, plants that have already survived a harsh winter can die. Explain the difference in the response of the plants in these circumstances compared with the winter months. (*6 marks*)

5 Explain how the experimental evidence shown in Figure 3 supports the idea of ABA from the roots affecting stomatal opening in times of water stress. (*6 marks*)

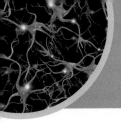

Synoptic link

You learnt about the responses of plants to attack by pathogens and the defensive chemicals they make in Topic 12.4, Plant defences against pathogens.

▲ **Figure 1** *These thorns act to deter herbivores. The fruit, which have evolved to be eaten to spread the seeds, are held on thorn-free stalks*

▲ **Figure 2** *A stinging nettle magnified 20×, these hollow stinging hairs act like hypodermic needles and inject histamines and other irritant chemicals into animals that try to eat the leaves*

Herbivores are animals that eat plants. They range from tiny insects to enormous animals such as elephants and rhinos. Herbivory is the process by which herbivores eat plants.

Responses to herbivory

Plants cannot escape animals which want to eat them, so they have evolved a wide range of defences to prevent attack by herbivores or minimise the damage they do.

Physical defences

Common physical defences include thorns, barbs, spikes, spiny leaves, fibrous and inedible tissue, hairy leaves, and even stings to protect themselves and discourage herbivores from eating them.

Chemical defences

Plants have also evolved a wide range of chemical responses to herbivory – the stinging nettle manages to include both physical and chemical defences in its vicious trichomes (stinging hairs), but many other plants produce a cocktail of unpleasant chemicals too. These include:

- **Tannins** – part of a group of compounds called phenols produced by many plants. Tannins can make up to 50% of the dry weight of the leaves. They have a very bitter taste which puts animals off eating the leaves. They are toxic to insects – they bind to the digestive enzymes produced in the saliva and inactivate them. Tea and red wine are both rich in plant tannins.

- **Alkaloids** – a large group of very bitter tasting, nitrogenous compounds found in many plants. Many of them act as drugs, affecting the metabolism of animals that take them in and sometimes poisoning them. Alkaloids include caffeine, nicotine, morphine, and cocaine. Caffeine is toxic to fungi and insects, and the caffeine produced by coffee bush seedlings spreads through the soil and prevents the germination of the seeds of other plants – so caffeine protects the plant both against herbivores and against plant rivals. Nicotine is a toxin produced in the roots of tobacco plants, transported to the leaves and stored in vacuoles to be released when the leaf is eaten.

- **Terpenoids** – a large group of compounds produced by plants which often form essential oils but also often act as toxins to insects and fungi that might attack the plant. Pyrethrin, produced by chrysanthemums, acts as an insect neurotoxin, interfering with the nervous system. Some terpenoids act as insect repellents, for example, citronella produced by lemon grass repels insects.

Pheromones

A pheromone is a chemical made by an organism which affects the social behaviour of other members of the same species. Because plants do not behave socially, they do not rely a lot on pheromones. There are a few instances where they could be regarded as using pheromones to defend themselves:

- If a maple tree is attacked by insects, it releases a pheromone which is absorbed by leaves on other branches. These leaves then make chemicals such as callose to help protect them if they are attacked. Scientists have observed that leaves on the branches of nearby trees also prepare for attack in response to these chemical signals.

- There is some evidence that plants communicate by chemicals produced in the root systems and one plant can 'tell' a neighbour if it is under water stress.

▲ Figure 3 *Human uses for some of the compounds produced by plants*

However, plants do produce chemicals called volatile organic compounds (VOCs) which act rather like pheromones between themselves and other organisms, particularly insects. They diffuse through the air in and around the plant. Plants use these chemical signals to defend themselves in some amazing ways. They are usually only made when the plant detects attack by an insect pest through chemicals in the saliva of the insect. This may elicit gene switching. For example:

- When cabbages are attacked by the caterpillars of the cabbage white butterfly, they produce a chemical signal which attracts the parasitic wasp *Cotesia glomerata*. This insect lays its eggs in the caterpillars which are then eaten alive, protecting the plant. The signal from the plant also deters any other female cabbage white butterflies from laying their eggs. Scientists estimate up to 90% of cabbage white caterpillars are affected by the parasite. If the cabbage is attacked by the mealy cabbage greenfly, it sends out a different signal which attracts the parasitic wasp *Diaretiella rapae* which only attacks greenfly.

- When apple trees are attacked by spider mites, they produce VOCs which attract predatory mites that come and destroy the apple tree pests.

- Some types of wheat seedling produce VOCs when they have been attacked by aphids and these repel other aphids from the plant.

▲ Figure 4 *Foxgloves produce a chemical, digitalis, which is toxic to mammals – it slows the heart rate. It also contains chemicals which cause vomiting, so after eating a foxglove leaf a small mammal will feel very ill but then vomit, removing the toxins. Digitalis can kill people too, but drugs based on the chemical are also used to treat human heart problems and save lives*

Sometimes a VOC produced by a plant that has been attacked will not only attract predators of the pest organism – it may also act as a 'pheromone' so that neighbouring plants begin to produce the VOC before they are actually attacked.

Folding in response to touch

Most plants, with the exception of a few insectivorous plants such as the Venus fly trap, move so slowly you cannot follow the movement with the naked eye. It is revealed over hours or days, or by time-lapse photography. There are, however, some exceptions – the sensitive plant *Mimosa pudica* is one of a small number of plants which move at a speed you can see. This plant uses conventional defences against herbivores – it contains a toxic alkaloid and the stem has sharp prickles, but if the leaves are touched, they fold down and collapse. Scientists think this frightens off larger herbivores, and dislodges small insects which have landed on the leaves. The leaf falls in a few seconds, and recovers over 10–12 minutes as a result of potassium ion movement into specific cells, followed by osmotic water movement. The exact causes of the dramatic change in the leaves are still being researched.

Mimosa pudica – nerves and muscles in plants

The dramatic leaf-folding in response to touch seen in *Mimosa pudica* seems to be the result of chemical changes in some large, fairly thin-walled cells found at the bases of the leaves and the individual leaflets. These special cells have relatively elastic walls and they surround a vascular bundle to form a thickened region called a pulvinus, which acts rather like a joint (Figure 5). The cells at the top are the flexor region and the cells at the bottom are the extensor region.

When the leaf is touched, scientists think there is an electrochemical change in the cells which results in something rather like the action potential in nerve cells which causes the active movement of potassium ions into the cells on the upper, flexor side of the pulvinus, while potassium ions are similarly moved out of cells on the lower, extensor side. Water follows the potassium ions by osmosis, so turgor in the top cells increases and in the lower cells decreases. There are elastic tissues in the cells that increase this effect. They include actin, one of the proteins found in mammalian muscle cells. As a result the leaflet, or whole leaf, bends down. When the plant recovers the situation is reversed – potassium ions return to their resting levels down concentration and electrochemical gradients and water follows by osmosis. (Figure 6).

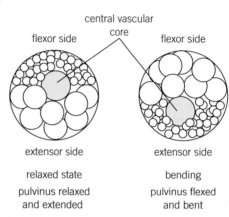

▲ **Figure 5** *The structure of a pulvinus when extended and flexed. The best current model for the cellular basis of the rapid folding response of* Mimosa pudica *to touch*

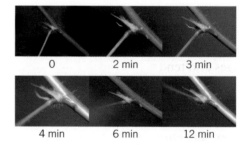

▲ **Figure 6** *The change in angle of pulvinus during recovery*

The initial leaf folding of *Mimosa pudica* takes seconds but recovery takes 10–12 minutes. Both involve similar changes in potassium levels. Suggest why such different rates of change may have evolved.

Summary questions

1 What is herbivory? (*1 mark*)

2 Describe two examples of chemical defences against herbivory by animals, explaining how they protect the plant and how they are used by people. (*6 marks*)

3 a Why are the chemicals sometimes known as plant pheromones not strictly pheromones? (*2 marks*)
  b Why is it so important that these chemicals are volatile? (*2 marks*)
  c Give two examples of how these chemicals may be used to protect a plant against herbivory, including a discussion as to whether your examples are pheromones or not. (*6 marks*)

16.4 Tropisms in plants

Specification reference: 5.1.5

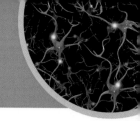

Plant growth responses to stimuli from one direction are known as tropisms. The growth of plants in response to light which comes from one direction only is called **phototropism**; the response to gravity is called **geotropism**; the response to chemicals chemotropism; and to touch thigmotropism. Tropisms involve differential growth of plant cells triggered by chemical messages produced in response to a particular stimulus.

Tropisms as a response to environmental cues

To be able to make the maximum use of the environmental conditions, plants must grow and respond to variations in those conditions. For example, once a seed begins to germinate in the soil, the shoot and root must keep growing in the right direction if the developing plant is to survive. The shoot must grow up towards the light source for photosynthesis to take place. The roots must grow downwards into the soil which will provide support, minerals, and water for the plant. The movements of the root and shoot take place in direct response to environmental stimuli. The direction of the response is related to the direction from which the stimulus comes. These responses are examples of tropisms.

Much of the research on tropisms uses germinating seeds and very young seedlings. They are easy to work with and manipulate and as they are growing and responding rapidly, any changes show up quickly. Changes also tend to affect the whole organism rather than a small part (as with a mature plant) and this makes any tropisms much easier to observe and measure. The seedlings of monocotyledonous plants – usually cereals such as oats and wheat – are most commonly used as the shoot that emerges is a single spike with no apparent leaves known as a coleoptile. It is easier to manipulate and observe than a dicotyledonous shoot. However, coleoptiles are relatively simple plant systems, so it is important to remember that the control of the responses to light in an intact adult plant may be more complex.

Phototropism

The basic model of the way plants respond to light as they grow was based on experiments where shoots were kept entirely in the dark or in full illumination. However, this is rarely the case in real life. Phototropisms are the result of the movement of auxins across the shoot or root if it is exposed to light that is stronger on one side than the other.

If plants are grown in bright, all-round light in normal conditions of gravity they grow more or less straight upwards. In even but low light they will also grow straight upwards – in fact in these conditions they will grow faster and taller than in bright light. If plants, however, are exposed to light which is brighter on one side than another, or

Learning outcomes

Demonstrate knowledge, understanding, and application of:

→ the types of plant responses

→ practical investigations into phototropism and geotropism.

Synoptic link

You learnt about dicotyledonous plants in Topic 9.1, Transport systems in dicotyledonous plants.

to unilateral light that only shines from one side, then the shoots of the plant will grow towards that light and the roots, if exposed, will grow away. Shoots are said to be positively phototropic and roots are negatively phototropic. This response has an obvious survival value for a plant. It helps to ensure that the shoots receive as much all-round light as possible, allowing the maximum amount of photosynthesis to take place. Also, if the roots should emerge from the soil – as they might do after particularly heavy rain, for example – they will rapidly turn back to the soil.

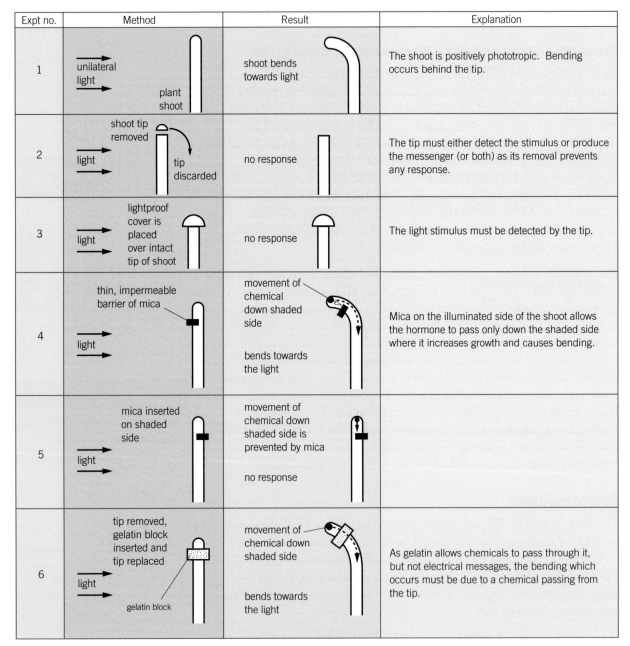

Expt no.	Method	Result	Explanation
1	unilateral light, plant shoot	shoot bends towards light	The shoot is positively phototropic. Bending occurs behind the tip.
2	shoot tip removed, light, tip discarded	no response	The tip must either detect the stimulus or produce the messenger (or both) as its removal prevents any response.
3	lightproof cover is placed over intact tip of shoot, light	no response	The light stimulus must be detected by the tip.
4	thin, impermeable barrier of mica, light	movement of chemical down shaded side; bends towards the light	Mica on the illuminated side of the shoot allows the hormone to pass only down the shaded side where it increases growth and causes bending.
5	mica inserted on shaded side, light	movement of chemical down shaded side is prevented by mica; no response	
6	tip removed, gelatin block inserted and tip replaced, light, gelatin block	movement of chemical down shaded side; bends towards the light	As gelatin allows chemicals to pass through it, but not electrical messages, the bending which occurs must be due to a chemical passing from the tip.

▲ Figure 1 *Experimental observations such as these by Darwin and Boysen-Jensen are the basis of our understanding of phototropisms. Darwin's experiments helped to show that it is the tips of shoots which are the source of the phototropic response, meanwhile Boysen-Jensen helped show the nature of the 'messenger' in the phototropic response*

The effect of unilateral light

Examples of the response of plants to unilateral light can be seen in any garden or woodland. Where plants are partially shaded the shoots grow towards the light and then grow on straight towards it. This response appears to be the result of the way auxin moves within the plant under the influence of light.

Figure 2 shows that the side of a shoot exposed to light contains less auxin than the side which is not illuminated. It appears that light causes the auxin to move laterally across the shoot, so there is a greater concentration on the unilluminated side. This in turn stimulates cell elongation and growth on the dark side, resulting in observed growth towards the light. Once the shoot is growing directly towards the light, the unilateral stimulus is removed. The transport of auxin stops and the shoot then grows straight towards the light. The original theory was that light destroyed the auxin, but this has been disproved by experiments showing that the levels of auxin in shoots are much the same regardless of whether they have been kept in the dark or under unilateral illumination.

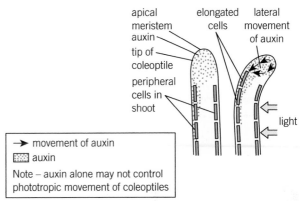

apical meristem | elongated cells | lateral movement of auxin
auxin
tip of coleoptile
peripheral cells in shoot
light

→ movement of auxin
▓ auxin
Note – auxin alone may not control phototropic movement of coleoptiles

▲ **Figure 2** *In unilateral light, auxin moves laterally across the tip of the shoot away from the light. This stimulates growth on the shady side, so the shoot grows asymmetrically towards the light*

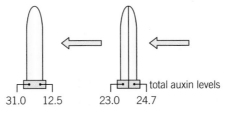

intact shoot | split shoot

Shoots kept in the dark – total auxin produced approximately the same whether shoot split or not.

25.5 | 24.1 total auxin levels

Shoot in unilateral light – total auxin produced approximately the same whether shoot split or not.

26.2 | 23.4 total auxin levels

Shoot in unilateral light but undivided – auxin accumulates on the dark side and lower on the lit side when the shoot is intact, but when it is divided the auxin concentration is approximately the same both sides. This suggests that normally auxin is transported across the shoot in unilateral light from the lit side to the dark side.

31.0 12.5 | 23.0 24.7 total auxin levels

▲ **Figure 3** *Experiments such as these with maize coleoptiles help us to determine what happens to auxin levels within a shoot illuminated by unilateral light*

Practical investigations into phototropisms

There are many different ways to investigate phototropisms. Some of them are listed here.

- Germinate and grow seedlings in different conditions of dark, all-round light, and unilateral light. Observe, measure, and record the patterns of growth. Time-lapse photography can give a good record of the changes as they take place.

- Germinate and grow seedlings in unilateral light with different colour filters to see which wavelengths of light trigger the phototropic response.

- Repeat some of the classic experiments (Figure 1) – cover the tips of coleoptiles with foil, remove the tips of some coleoptiles, place auxin-impregnated agar jelly blocks or lanolin on decapitated coleoptiles, place auxin-impregnated agar blocks on one side only of decapitated coleoptiles.

▲ **Figure 4** *Experiments on phototropisms can be done on dicot seedlings as well as coleoptiles – these cress seedlings have been grown in all-round light, the dark, and unilateral light from left to right respectively*

Discuss potential advantages and disadvantages of using new technology such as smart phones and tablets in recording and displaying data in practical investigations into phototropisms.

Growing in the dark

The fact that plants grow more rapidly in the dark than when they are illuminated can at first seem illogical. If a plant, however, is in the dark the biological imperative is to grow upwards rapidly to reach the light to be able to photosynthesise. The seedlings that break through the soil first will not have to compete with other seedlings for light. Evidence suggests that it is gibberellins that are responsible for the extreme elongation of the internodes when a plant is grown in the dark. Once a plant is exposed to the light, a slowing of upwards growth is valuable. Resources can be used for synthesising leaves, strengthening stems, and overall growth. Scientists have demonstrated that levels of gibberellin fall once the stem is exposed to light.

Gardeners sometimes use this response to 'force' growth in plants – early rhubarb is famously grown in dark sheds in Yorkshire. The rapid upward growth which takes place in a plant grown in the dark is known as etiolation. Etiolated plants are thin and pale – because the plant is deprived of light little chlorophyll develops in the leaves.

Geotropisms

Light is not the only thing to which plants are sensitive. Plants are also sensitive to gravity, and the different responses of the roots and shoots are very important in the control of plant growth.

In normal conditions, plants always receive a unilateral gravitational stimulus – gravity always acts downwards. The response of plants to gravity can be seen in the laboratory using seedlings placed on their sides either in all-round light or in the dark. Shoots are usually negatively geotropic (grow away from gravitational pull) and roots are positively geotropic (grow towards gravitational pull). This adaptation ensures that the roots grow down into the soil and the shoots grow up to the light. Geotropisms are also known as gravitropisms.

Practical investigations into geotropisms

Figure 5 shows two ways in which geotropic responses can be demonstrated. These basic techniques can be adapted to investigate different aspects of the geotropic response.

- The geotropic response can be investigated in shoots and roots using a rotating drum known as a clinostat. The plants can be grown on a slowly rotating clinostat (about four revolutions per hour) so the gravitational stimulus is applied evenly to all sides of the plant – and the root and (in the dark) shoot grow straight.

- Alternatively, the seeds can be placed in petri dishes stuck to the wall of the lab, and the dishes rotated 90° at intervals as the seedlings grow. A geotropic response in the roots can be seen within about two hours.

Investigations into geotropisms are usually carried out with the plants exposed to all round light or in the dark. Suggest a reason for these conditions.

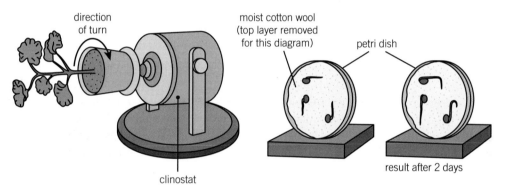

▲ **Figure 5** *Simple investigations can demonstrate geotropisms in shoots (left) and roots (right)*

Summary questions

1 a What is a tropism?   (*1 mark*)
   b Explain what is meant by phototropisms and geotropisms. (*2 marks*)
   c Describe the different phototropic and geotropic responses
      in shoots and roots.   (*2 marks*)

2 Produce a flow diagram to explain the events which bring
   about a phototropic response in a shoot.   (*6 marks*)

3 If a block of butter is used in Figure 1 instead of the gelatin block,
   there is no response in the decapitated shoot. Explain how
   this informs scientists that the message is water soluble. (*5 marks*)

4 Originally scientists thought geotropisms were the result of auxin
   movements in response to gravity. Investigate current models
   of how geotropisms occur and write a brief report.   (*6 marks*)

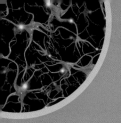

16.5 The commercial use of plant hormones

Specification reference: 5.1.5

Plant hormones are involved in the control of many different aspects of plant life, from germination of the seeds and growth of the stems to ripening of the fruit and the fall of the leaves. As scientists have unravelled more and more about the role of these fascinating chemicals in the life of plants, people have developed a number of ways of using plant hormones commercially in agriculture and horticulture.

Control of ripening

The gaseous plant hormone ethene is involved in the ripening of climacteric fruits. These are fruits that continue to ripen after they have been harvested. Their ripening is linked to a peak of ethene production triggering a series of chemical reactions including a greatly increased respiration rate. Climacteric fruits include bananas, tomatoes, mangoes, and avocados. Non-climacteric fruit (such as oranges, strawberries, and watermelon) do not produce large amounts of ethene and do not ripen much after picking.

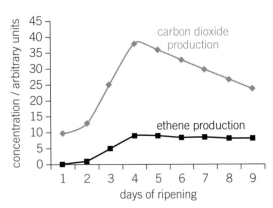

▲ Figure 1 Graph to show the effect of ethene on respiration rate of ripening fruit

The effect of ethene on climacteric fruit can easily be seen if part of a bunch of green bananas is put in a bag with a single ripe banana. The bunch with the ripe banana will ripen faster than the rest of the bunch, even if the temperature is exactly the same in both cases. Ethene from the ripe banana stimulates the rapid ripening of the green ones.

Ethene is widely used commercially in the production of perfectly ripe climacteric fruit for greengrocers and supermarkets. These fruit are harvested when they are fully formed but long before they are ripe, and then cooled, stored, and transported. The unripe fruit is hard and much less easily damaged during transport around the world than the ripe versions. When the fruit are needed for sale, they are exposed to ethene gas under controlled conditions. This ensures that each batch of fruit ripens at the same rate and are all at the same stage to be put on the shelves for sale to the public. This careful control of ripening prevents a lot of wastage of fruit during transport, and increases the time available for them to be sold.

Hormone rooting powders and micropropagation

Auxin affects the growth of both shoots and roots. Scientists have discovered that the application of auxin to cut shoots stimulates the production of roots. This makes it much easier to propagate new plants from plant cuttings. A cutting is a small piece of the stem of a plant, usually with some leaves on. If this is placed in compost or soil – or even water – roots may eventually appear and a new plant forms. Dipping the cut stem into hormone rooting powder increases the chances of roots forming, and of successful propagation taking place. This has made it much easier for horticulturists to develop cuttings to sell and for individuals taking their own cuttings.

In both horticulture and agriculture, many plants are now propagated on a large scale by micropropagation, when thousands of new plants are grown from a few cells of the original plant. Plant hormones are essential in this process – they control the production of the mass of new cells and then the differentiation of the clones into tiny new plants.

Hormonal weedkillers

As you have seen, the interactions between the different plant hormones are finely balanced to enable the plant to grow. If this balance is lost it can interrupt the metabolism of the whole plant and may lead to plant death. Sometimes, this is exactly what we want to achieve, and plant hormones can help us. Weeds are plants that grow where they are not wanted. Commercial food crops are vital globally for producing the food people need to eat. Weeds interfere with crop plants, competing for light, space, water, and minerals.

Scientists have developed synthetic auxins which act as very effective weedkillers. Many of the main staple foods around the world are narrow-leaved monocot plants such as rice, maize, and wheat. Most of the weeds are broad-leaved dicots. If synthetic dicot auxins are applied as weedkiller, they are absorbed by the broad-leaved plants and affect their metabolism. The growth rate increases and becomes unsustainable, so they die. The narrow-leaved crop plants are not affected and continue to grow normally, freed from competition. The synthetic auxins used by farmers and gardeners are simple and cheap to produce, have a very low toxicity to mammals, and are selective.

Other uses of plant hormones

There are many different ways in which plant hormones are used commercially as well as those explored here. Examples include:

- Auxins can be used in the production of seedless fruit.

- Ethene is used to promote fruit dropping in plants such as cotton, walnuts, and cherries.

- Cytokinins are used to prevent ageing of ripened fruit and products such as lettuces, and in micropropagation to control tissue development.

- Gibberellins can be used to delay ripening and ageing in fruit, to improve the size and shape of fruits, and in beer brewing to speed up the malting process.

Synoptic link

You will learn more about the use of plant hormones in taking cuttings in Topic 22.1, Natural cloning in plants and the use of plant hormones in micropropagation in Topic 22.2, Artificial cloning in plants.

▲ Figure 2 *Using rooting powder on cuttings is very simple and effective*

Summary questions

1 How are plant hormones used to control the ripening of fruit? (*4 marks*)

2 Why is it commercially important to be able to control fruit ripening? (*6 marks*)

3 Produce a table to summarise as many commercial uses as you can find for four named plant hormones. (*6 marks*)

4 Look at the graph in Figure 1.
   a Describe the changes in ethene production and carbon dioxide production in the tomato as it ripens. (*6 marks*)
   b Suggest what is happening. (*6 marks*)
   c Why do cool conditions slow the rate of ripening even if ethene is present? (*3 marks*)

Practice questions

1 a Plant responses to environmental changes are co-ordinated by plant growth substances (plant hormones).

Explain why plants need to be able to respond to their environment. *(2 marks)*

b The following investigation was carried out into the effects of plant growth substances on germination:

● A large number of lettuce seeds was divided into eight equal batches

● Each batch of seeds was placed on moist filter paper in a Petri dish and given a different treatment.

The different treatments are shown in the table. Each tick represents one of the eight batches of seeds.

treatment	concentration of gibberellin (mol dm^{-3})			
	0.00	0.05	0.50	5.00
A	water	✔	✔	✔	✔
B	Abscisic acid	✔	✔	✔	✔

The batches of seeds were left to germinate at 25°C in identical conditions and the percentage germination was calculated. The graph shows the results of this investigation.

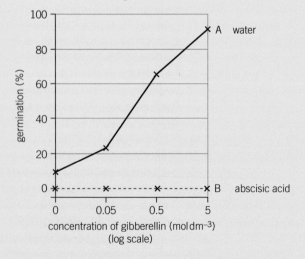

(i) Describe, with reference to the graph, the effects of the plant growth substances on the germination of lettuce seeds. *(4 marks)*

(ii) Explain how the plant hormones have these effects. *(4 marks)*

(iii) Explain why all the lettuce seeds were kept at 25°C. *(2 marks)*

(iv) State **three** variables, **other than temperature**, that needed to be controlled in the investigation. *(3 marks)*

c State two commercial uses of plant growth substances. *(2 marks)*

OCR F215 2010 (apart from 1b(ii))

2 Plants are able to respond to changes in their environment.

a Describe two ways in which hormones may alter a plants growth in response to overcrowding by other plants. *(4 marks)*

OCR F215 2012

b Suggest how hormones alter the growth and morphology, or growth and development of a plant *(4 marks)*

3 The growth and development of a fruit tree is controlled by plant growth regulators. The table shows the stages that occur as the tree grows and develops and indicates the stages at which giberellin is involved (green shading).

	Germination	Growth to maturity	Flowering	Fruit development	Abscission	Seed dormancy
Gibberellin						
Auxin						
Cytokinins						
Ethylene						
ABA						

a In the past plant chemicals such as auxins and gibberellins were referred to as plant hormones. At one stage this was changed to plant growth regulators. Now they are again generally referred to in university plant biology departments as plant hormones.

Explain why plant hormone is a more accurate term than plant growth regulator. *(5 marks)*

b Copy and complete the table and indicate which hormone(s) is/are involved at each stage, using crosses. Gibberellins have already been completed as an example. *(4 marks)*

c Compare and contrast the activity of auxins and cytokinins. *(6 marks)*

4 In an investigation into the effects of water stress, cowpea seeds were sown and the seedlings were thinned to one per pot. The plants were watered normally until mature then watering was completely stopped for 14 days to induce water stress.

The number of leaves and total leaf surface area were measured daily throughout this period.

The results are shown in the graph.

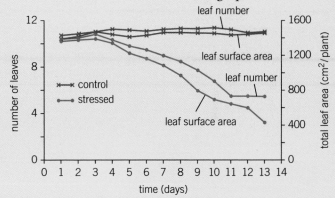

a (i) Describe the changes in leaf number and surface area over the 14 day period shown in the graph. *(3 marks)*

(ii) Describe and explain the effects of water stress on the growth and development of plants. *(3 marks)*

b Outline the process of leaf abscission including the roles of plant hormones. *(5 marks)*

c Compare and contrast the response of cowpeas to water stress to the loss of leaves by trees in the UK in autumn. *(5 marks)*

17 ENERGY FOR BIOLOGICAL PROCESSES

17.1 Energy cycles

Specification reference: 5.2.1 and 5.2.2

Synoptic link

You learnt about active transport in Topic 5.4, Active transport.

Synoptic link

You will learn more about the transfer of energy through ecosystems in Topic 23.2, Biomass transfer through an ecosystem.

Synoptic link

You learnt about glucose as an energy store in Topic 3.3, Carbohydrates, and about ATP as a molecule of energy in Topic 3.11, ATP.

Study tip

A glucose molecule contains more energy than the single metabolic reactions needed to break it down. The energy contained within a glucose molecule is used to synthesise many molecules of ATP, and it is molecules of ATP that drive metabolic reactions.

The need for energy

Living organisms have to be active to survive. Organisms grow, respond to changes in their environment, and deal with threats from other organisms. They have to find or make food, and reproduce. All of this activity depends on metabolic reactions and processes continually taking place in individual cells.

A few examples of these metabolic activities amongst many include:

- active transport, which is essential for the uptake of nitrates by root hair cells, loading sucrose into sieve tube cells, the selective reabsorption of glucose and amino acids in the kidney, and the conduction of nerve impulses
- anabolic reactions, such as the building of polymers like proteins, polysaccharides, and nucleic acids essential for growth and repair
- movement brought about by cilia, flagella, or the contractile filaments in muscle cells.

All of these metabolic activities require energy.

Energy flow through living organisms

The total amount of energy in the universe hasn't changed from the time of the Big Bang, when the universe began, until now. The universe is gradually cooling as it expands because this energy is being dispersed over a larger area. This is inevitable and irreversible. Energy cannot be created or destroyed.

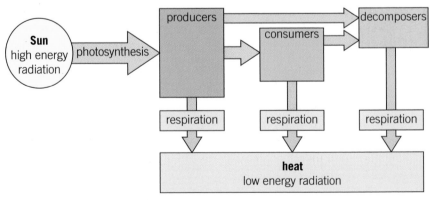

▲ Figure 1 Transfer of energy through ecosystems

Radiation from the Sun is used to fuel the metabolic reactions and processes necessary to keep organisms alive before, eventually, being transferred back to the atmosphere as heat (Figure 1).

Photosynthesis

Organisms make use of the energy in the bonds of organic molecules, such as glucose. These bonds are formed during **photosynthesis** in plants and other photosynthetic organisms (Topic 17.3, Photosynthesis).

Light is trapped by chlorophyll molecules. This energy is used to drive the synthesis of glucose from carbon dioxide and water.

Respiration

All organisms need to respire. **Respiration** is the process by which organic molecules, such as glucose, are broken down into smaller inorganic molecules, like carbon dioxide and water. The energy stored within the bonds of the organic molecules is used to synthesise adenosine triphosphate (ATP).

Two of the most important chemical reactions in the living world are photosynthesis and respiration. Photosynthesis is the reaction behind the production of most of the biomass on the earth. Respiration is the process by which organisms break down biomass to provide the ATP needed to drive the metabolic reactions that take place in cells.

The two reactions are intimately linked (Table 1). The raw materials for one are the products of the other so they are interrelated throughout the living world. The overall reactions for the two are as follows:

Photosynthesis

$$6CO_2 + 6H_2O \leftrightarrow C_6H_{12}O_6 + 6O_2$$

Respiration

$$C_6H_{12}O_6 + 6O_2 \rightarrow 6CO_2 + 6H_2O$$

The importance of carbon–hydrogen bonds

A general rule in biochemistry is that energy is used to break bonds, and energy is released when bonds are formed. The same quantity of energy is involved whether a particular bond is being broken or formed. This is called the bond energy. Whether an overall reaction is **exothermic** (releases energy) or **endothermic** (takes in energy) depends on the total number and strength of bonds that are broken or formed during the reaction.

The atoms in small inorganic molecules like water and carbon dioxide are joined by strong bonds that release a lot of energy when they form but require a lot of energy to break. Organic molecules like glucose and amino acids contain many more bonds than small inorganic molecules. These are weaker bonds compared with those in inorganic molecules and, therefore, release less energy when they form and require less energy to be broken.

In respiration large organic molecules are broken down forming small inorganic molecules. The total energy required to break all the bonds in a complex organic molecule is less than the total energy released in the formation of all the bonds in the smaller inorganic products. The excess energy released by the formation of the bonds is used to synthesise ATP.

> ## Synoptic link
>
> You will learn more about respiration in Chapter 18, Respiration.

▼ Table 1 *Comparison of respiration and photosynthesis*

	Respiration	Photosynthesis
Reactants	glucose and oxygen	water and carbon dioxide
Products	water and carbon dioxide	glucose and oxygen
Purpose	release energy	trap energy

> ## Study tip
>
> It is important to remember that energy is not produced, created, made, or lost. Say instead that energy is released or absorbed, or transferred as heat.

Organic molecules contain large numbers of carbon–hydrogen bonds, particularly lipids. Carbon and hydrogen share the electrons almost equally in bonds that form between them. This results in a non-polar bond which does not require a lot of energy to break. The carbon and hydrogen released then form strong bonds with oxygen atoms, forming carbon dioxide and water, resulting in the release of large quantities of energy. The reverse happens in photosynthesis when organic molecules are made from small inorganic molecules. The energy required to build these molecules comes from the Sun.

Energy transfer

When thinking about energy transfer:
Bond energy = energy required to break bond = energy required to form bond.

The overall reaction that takes place in photosynthesis is:

$$6CO_2 + 6H_2O \rightarrow C_6H_{12}O_6 + 6O_2$$

The overall reaction that takes place in respiration is the reverse of this reaction.

The structures of the molecules involved are shown in Figure 2.

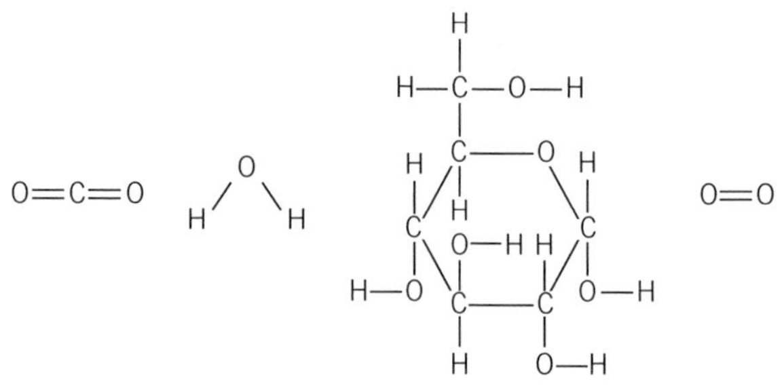

▲ Figure 2

1 Copy and complete the table.

Bond	Number of bonds involved in reaction	Bond energy (kJ/mol)	Total (kJ/mol)
C=C		803	
O—H		464	
C—H		414	
C—C		347	
C—O		358	
O=O		495	

2 Using information from the completed table, calculate how much energy is required to make one mole of glucose from six moles of carbon dioxide and six moles of water.

3 State how much energy would be released when one mole of glucose is respired.

4 The breakdown of one glucose molecule results in the production of 38 molecules of ATP. The formation of ATP requires 30.6 kJ/mol. Calculate the percentage of energy released when one mole of glucose is respired.

Summary questions

1 Explain why it is incorrect to say that energy is produced.   (2 marks)

2 Explain why ATP is not a good energy storage molecule but why organic molecules, like lipids or carbohydrates, are.   (4 marks)

3 Explain the interrelationship between respiration and photosynthesis in organisms.   (5 marks)

17.2 ATP synthesis

Specification reference: 5.2.2

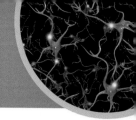

ATP production

As you have already learnt, ATP is the universal energy currency in cells. The bond energy in the ATP molecule is used to drive essential metabolic processes. The production of ATP is therefore fundamental to all forms of life.

In photosynthesis, light provides the energy needed to build organic molecules like glucose. This energy is used to form chemical bonds in ATP, which are then broken to release the energy needed to make bonds as glucose is formed.

In respiration organic molecules, such as glucose, are broken down and the energy released is used to synthesise ATP. This ATP is then used to supply the energy needed to break bonds in the metabolic reactions of the cell.

Synthesis of ATP is therefore a crucial step in both respiration and photosynthesis.

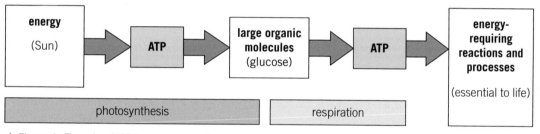

▲ **Figure 1** *The role of ATP in energy transfer*

Chemiosmosis

The ATP produced in both photosynthesis and respiration is synthesised primarily by a process called **chemiosmosis**. Chemiosmosis involves the diffusion of protons from a region of high concentration to a region of low concentration through a partially permeable membrane. The movement of the protons as they flow down their concentration gradient releases energy that is used in the attachment of an inorganic phosphate (P_i) to ADP, forming ATP.

Chemiosmosis depends on the creation of a proton concentration gradient. The energy to do this comes from high-energy electrons (excited electrons).

Excited electrons

Electrons are raised to higher energy levels, or excited, in two ways:

- electrons present in pigment molecules (e.g., **chlorophyll**) are excited by absorbing light from the Sun

- high energy electrons are released when chemical bonds are broken in respiratory substrate molecules (e.g., glucose).

The excited electrons pass into an electron transport chain and are used to generate a proton gradient.

> ### Synoptic link
>
> You learnt about ATP–ADP in Topic 3.11, ATP.

Electron transport chain

An electron transport chain is made up of a series of **electron carriers**, each with progressively lower energy levels. As high energy electrons move from one carrier in the chain to another, energy is released. This is used to pump protons across a membrane, creating a concentration difference across the membrane and therefore a proton gradient. The proton gradient is maintained as a result of the impermeability of the membrane to hydrogen ions.

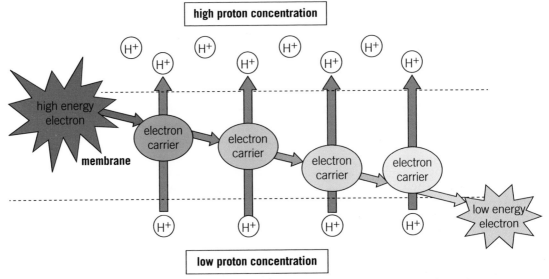

▲ Figure 2 *The flow of electrons along an electron transport chain releases energy which is used to pump protons across a membrane, resulting in the formation of a proton gradient*

The only way the protons can move back through the membrane down their concentration gradient is through hydrophilic membrane channels linked to the enzyme ATP synthase (catalyses the formation of ATP). The flow of protons through these channels provides the energy used to synthesise ATP (from ADP and P_i).

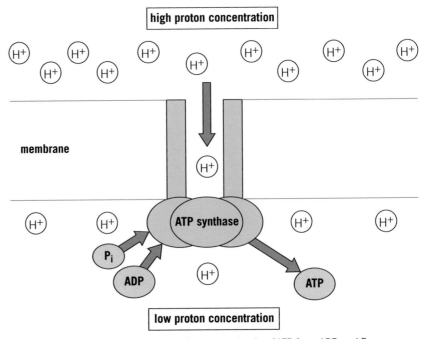

▲ Figure 3 *The role of ATP synthase in the synthesis of ATP from ADP and P_i*

In photosynthesis, ATP is used to synthesise glucose and other organic molecules. The ATP produced in respiration provides the energy needed by metabolic processes and reactions essential to life.

A simple way of modelling this process is to think of the flow of water through a hydroelectric power station causing turbines to spin, generating electricity. Both chemiosmosis and hydroelectric power generation result in energy in a very useable form.

The processes of oxidative phosphorylation (in respiration) and photophosphorylation (in photosynthesis) are also vital in chemiosmosis. You will learn about these in more detail in Topic 18.4, Oxidative phosphorylation and Topic 17.3, Photosynthesis, respectively.

Synoptic link

You learnt about diffusion and active transport across cell membranes in Chapter 5, Plasma membranes.

Summary questions

1   Explain the importance of ATP to living organisms.   (*3 marks*)

2   Describe the properties of cell membranes necessary for the formation of a proton gradient.   (*5 marks*)

3   Name the type of diffusion which enables protons to move through ATP synthase and explain the role of ATP synthase in the production of ATP.   (*4 marks*)

4   The synthesis and breakdown of ATP is an example of a reversible reaction:

$$ADP + P_i \rightleftharpoons ATP$$

ATPase is often the name given to the enzyme which hydrolyses ATP, producing ADP and P_i. ATPase and ATP synthase are, in fact, the same enzyme. Explain how this is possible.   (*5 marks*)

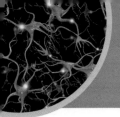

17.3 Photosynthesis

Specification reference: 5.2.1

Synoptic link

You will learn more about how respiration releases energy in Chapter 18, Respiration.

Synoptic link

You learnt about endosymbiosis in Topic 2.6, Prokaryotic and eukaryotic cells.

Photosynthesis is the process by which energy, in the form of light from the Sun, is used to build complex organic molecules, such as glucose. Light energy is transformed into chemical energy trapped in the bonds of the complex organic molecules produced. Organisms that can photosynthesise, like plants and algae, are said to be **autotrophic**.

Heterotrophic organisms, like animals, obtain complex organic molecules by eating other (heterotrophic and/or autotrophic) organisms.

Both autotrophic and heterotrophic organisms then break down complex organic molecules during the process of respiration to release the energy they need to drive metabolic processes.

Photosynthesis can be summarised by the equation:

$$6CO_2 \quad + \quad 6H_2O \quad \rightarrow \quad C_6H_{12}O_6 \quad + \quad 6O_2$$

carbon dioxide + water → glucose + oxygen

Structure and function of chloroplasts

As you have learnt, photosynthesis takes place in chloroplasts. The network of membranes present within chloroplasts provides a large surface area to maximise the absorption of light essential in the first step of photosynthesis. The membranes form flattened sacs called thylakoids which are stacked to form grana (singular granum) (Figure 1). The grana are joined by membranous channels called lamellae.

Light is absorbed by complexes of pigments, such as chlorophyll, which are embedded within the thylakoid membranes.

The fluid enclosed in the chloroplast is called the stroma and is the site of the many chemical reactions resulting in the formation of complex organic molecules.

Chlorophyll

Pigment molecules absorb specific wavelengths (colours) of light and reflect others. Different pigments absorb and reflect different wavelengths and this is why they have different colours. The primary pigment in photosynthesis is chlorophyll. Chlorophyll absorbs mainly red and blue light and reflects green light. The presence of large quantities of chlorophyll is the reason for the familiar green colour of plants.

Although there are a number of different pigments that absorb light, the primary pigment is chlorophyll a. Other pigments, like chlorophyll b, xanthophylls, and carotenoids absorb different wavelengths of light than those absorbed by chlorophyll a. Different combinations of pigments are the reason for the different shades and colours of leaves.

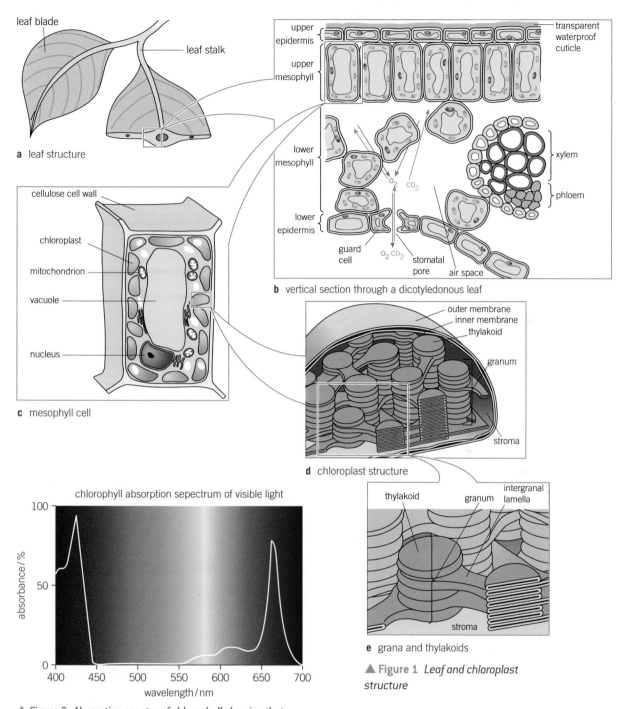

a leaf structure

b vertical section through a dicotyledonous leaf

c mesophyll cell

d chloroplast structure

e grana and thylakoids

▲ Figure 1 *Leaf and chloroplast structure*

chlorophyll absorption sepectrum of visible light

▲ Figure 2 *Absorption spectra of chlorophyll showing that red and blue light is absorbed and green light is reflected*

Chlorophyll b, xanthophylls, and carotenoids are embedded in the thylakoid membrane of the chloroplast. These and other proteins and pigments form a light harvesting system (also known as an antennae complex). The role of the system is to absorb, or harvest, light energy of different wavelengths and transfer this energy quickly and efficiently to the reaction centre. Chlorophyll a is located in the reaction centre, which is where the reactions involved in photosynthesis take place.

The light harvesting system and reaction centre are collectively known as a photosystem.

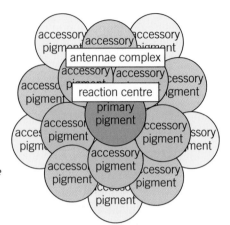

▶ **Figure 3** *The reaction centre is surrounded by an antennae complex, maximising the absorption of light*

Plants use sunscreen as well

Plants need light to photosynthesise but if sunlight is too intense chlorophyll is destroyed. Chlorophyll has to be continuously synthesised during the summer to maintain the level needed to photosynthesise at the required rate. Chlorophyll is not produced when there is little or no sunlight – this is why areas of grass that have been covered turn yellow.

Carotenoids are accessory pigments responsible for the yellow/orange colours seen in plant leaves. Orange carotene and yellow xanthophyll are two examples. These colours are not normally seen because they are masked by the green colour of chlorophyll. Carotenoids are not broken down, unlike chlorophyll, in strong sunlight and are present throughout the growing season.

The shorter days and cooler nights of autumn cause changes in the pigment composition in leaves. Chlorophyll a is no longer produced and leaves turn yellow/orange as we see the carotenoids.

Anthocyanin is a red/purple pigment formed from a reaction between sugars and proteins present in cell sap. It is produced when the concentration of sugars is high. High light intensity also promotes the production of anthocyanins.

Anthocyanin produces the red skin of apples and the purple of black grapes. The colour of the anthocyanin

pigments is pH-dependent, leading to a range of different colours from red to purple.

Anthocyanins act as a sunscreen by absorbing blue-green and ultraviolet light, thereby inhibiting the destruction of chlorophyll.

In their role as pigments they help trees maximise production towards the end of the growing season as the weather changes in autumn.

The red/purple coloration of leaves is also thought to camouflage leaves from herbivores blind to red wavelengths.

1 Suggest explanations for the following observations:
   a Apples are often red on one side and green on the other.
   b Leaves with more vibrant red colours are seen during years when there has been lots of sunlight and dry weather. When it has been raining and overcast there will not be as much red foliage present.
2 Suggest why the production of anthocyanins is temperature-dependent.

Investigating photosynthetic pigments

Chromatography can be used to separate the different pigments in a plant extract. The mobile phase would be the solution containing a mixture of pigments and the stationary phase a thin layer of silica gel applied to glass.

The different solubilities of the pigments in the mobile phase, and their differing interactions with the stationary phase, lead to them moving at different rates. This results in the pigments being separated as they move through the silica gel.

The retention value (R_f) for each pigment can be calculated using the formula:

$$R_f = \frac{\text{distance travelled by component}}{\text{distance travelled by solvent}}$$

Synoptic link

You also learnt about chromatography and how it is used to separate amino acids in solution in Topic 3.6, Structure of proteins.

1 List the apparatus you would use and outline the procedure you would follow to separate a mixture of plant pigments.

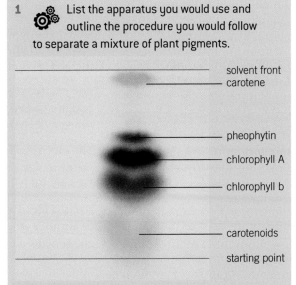

solvent front
carotene
pheophytin
chlorophyll A
chlorophyll b
carotenoids
starting point

▲ **Figure 4** *Thin layer chromatogram (TLC) of an extract of thylakoid membranes from the leaf of annual meadow grass. A drop of extract was laid at the bottom of the sheet. The sheet was then placed in a beaker of solvent separating out the pigments. Five bands can be seen.*

2 Calculate the R_f value for each pigment.

The two stages of photosynthesis

There are two stages in photosynthesis:

- Light-dependent stage – energy from sunlight is absorbed and used to form ATP (Figure 5). Hydrogen from water is used to reduce coenzyme **NADP** to reduced NADP.

- Light-independent stage – hydrogen from reduced NADP and carbon dioxide is used to build organic molecules, such as glucose. ATP supplies the required energy (Figure 6).

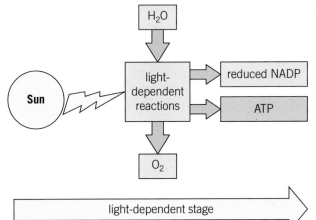

▲ **Figure 5** *Summary of the light-dependent stage of photosynthesis which occurs within and across the thylakoid membranes*

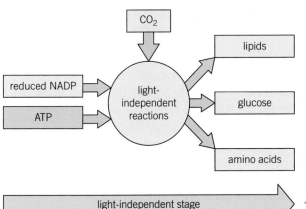

◀ **Figure 6** *Summary of the light-independent stage of photosynthesis which occurs in the stroma*

Synoptic link

You learnt about coenzymes in Topic 4.4, Cofactors, coenzymes, and prosthetic groups.

Study tip

The coenzyme used in photosynthesis is NADP, this is not the same coenzyme as NAD, used in respiration.

The light-dependent stage of photosynthesis

Non-cyclic photophosphorylation

Two photosystems are involved in **non-cyclic photophosphorylation**, photosystem II (PSII) followed by photosystem I (PSI). The reaction centre of PSI absorbs light at a higher wavelength (700 nm) than PSII (680 nm). The light absorbed excites electrons at the reaction centres of the photosystems.

The excited electrons are released from the reaction centre of PSII and are passed to an electron transport chain. ATP is produced by the process of chemiosmosis (Topic 17.2, ATP synthesis).

The electrons lost from the reaction centre at PSII are replaced from water molecules broken down using energy from the Sun (Topic 17.4, Factors affecting photosynthesis).

Excited electrons are released from the reaction centre at PSI, passed to another electron transport chain, and ATP is again produced by chemiosmosis. The electrons lost from this reaction centre are replaced by electrons that have just travelled along the first electron transport chain after being released from PSII.

The electrons leaving the electron transport chain following PSI are accepted, along with a hydrogen ion, by the coenzyme NADP, forming reduced NADP. Reduced NADP provides the hydrogen or reducing power in the production of organic molecules, such as glucose, in the light-independent stage which follows.

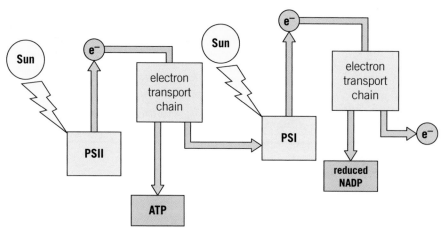

▲ **Figure 7** *Diagram summarising the two electron transport chains involved in cyclic and non-cyclic photophosphorylation, this is often referred to as the Z-scheme*

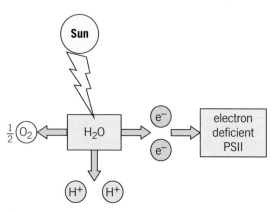

▲ **Figure 8** *Photolysis which is catalysed by the oxygen-evolving complex present in PSII*

Photolysis

Water molecules are split into hydrogen ions, electrons, and oxygen molecules using energy from the Sun in a process called photolysis. The electrons released replace the electrons lost from the reaction centre of PSII. This is why water, along with light and carbon dioxide, is a raw material of photosynthesis.

The oxygen-evolving complex which forms part of PSII is an enzyme that catalyses the breakdown of water. Here water molecules are split into hydrogen ions, electrons, and oxygen molecules using energy from the Sun in a process called photolysis. The electrons released replace the electrons lost from the reaction centre of PSII. This is why water, along with light and carbon dioxide, is a raw material of photosynthesis. The photolysis reaction is summarised as:

$$H_2O \longrightarrow 2H^+ + 2e^- + \frac{1}{2}O_2$$

Oxygen gas is released as a by-product. The protons are released into the lumen of the thylakoids, increasing the proton concentration across the membrane. As they move back through the membrane down a concentration and electrochemical gradient, they drive the formation of more ATP. Once the hydrogen ions are returned to the stroma, they combine with NADP and an electron from PSI to form reduced NADP. This is used in the light-independent reactions of photosynthesis. This process removes hydrogen ions from the stroma so it helps to maintain the proton gradient across the thylakoid membranes.

Cyclic photophosphorylation

The electrons leaving the electron transport chain after PSI can be returned to PSI, instead of being used to form reduced NADP, leading to **cyclic photophosphorylation**. This means PSI can still lead to the production of ATP without any electrons being supplied from PSII. Reduced NADP is not produced when this happens.

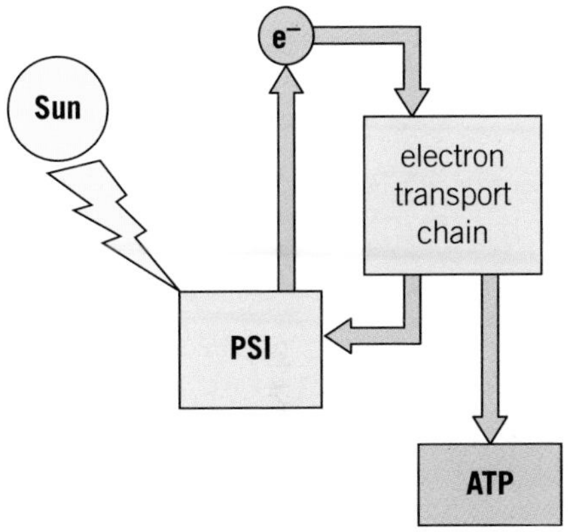

The light-independent stage of photosynthesis

The light-independent stage of photosynthesis takes place in the stroma of chloroplasts and uses carbon dioxide as a raw material. The products from the light-dependent stage – ATP and reduced NADP, are also required. Organic molecules, like glucose, are produced in a series of reactions collectively known as the Calvin cycle.

▲ Figure 9 *Cyclic photophosphorylation*

Calvin cycle

Carbon dioxide enters the intercellular spaces within the spongy mesophyll of leaves by diffusion from the atmosphere through stomata. It diffuses into cells and into the stroma of chloroplasts where it is combined with a five-carbon molecule called **ribulose bisphosphate (RuBP)**. The carbon in carbon dioxide is therefore **fixed**, meaning that it is incorporated into an organic molecule.

The enzyme **ribulose bisphosphate carboxylase (RuBisCO)** catalyses the reaction and an unstable six-carbon intermediate is produced. RuBisCO is the key enzyme in photosynthesis. It is a very inefficient enzyme as it is competitively inhibited by oxygen (see the Photorespiration application box) so a lot of it is needed to carry out photosynthesis successfully. Biologists estimate that it is probably the most abundant enzyme in the world.

The unstable six-carbon compound formed immediately breaks down, forming two three-carbon **glycerate 3-phosphate (GP)** molecules.

Each GP molecule is converted to another three-carbon molecule, **triose phosphate (TP)**, using a hydrogen atom from reduced NADP and energy supplied by ATP, both supplied from the light-dependent reactions of photosynthesis.

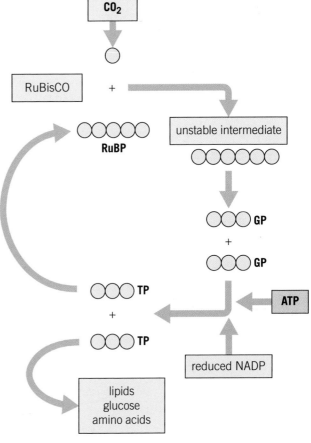

▲ Figure 10 *The Calvin cycle*

Triose phosphate is a carbohydrate, a three-carbon sugar, the majority of which is recycled to regenerate RuBP so that the Calvin cycle can continue. It is the starting point for the synthesis of many complex biological molecules, including other carbohydrates, lipids, proteins, and nucleic acids.

The Calvin cycle can be summarised in three steps:

- *Fixation* – carbon dioxide is fixed (incorporated into an organic molecule) in the first step.
- *Reduction* – GP is reduced to TP by the addition of hydrogen from reduced NADP using energy supplied by ATP.
- *Regeneration* – RuBP is regenerated from the recycled TP.

Regeneration of RuBP

For one glucose molecule to be produced six carbon dioxide molecules have to enter the Calvin cycle, resulting in six full turns of the cycle. This will result in the production of 12 TP molecules, two of which will be removed to make the glucose molecule.

This means that 10 TP molecules are recycled to regenerate six RuBP molecules (used in the six turns of the cycle).

10 × three-carbon TP = 30 carbons 'shuffled' gives 6 × five-carbon RuBP = 30 carbons

Energy is supplied by ATP for the reactions involved in the regeneration of RuBP.

Study tip

Reduced NADP must supply a hydrogen atom to GP in the Calvin cycle, not a hydrogen ion or proton. If there is no electron present a bond will not be formed with carbon.

Synoptic link

You learnt about transpiration in Topic 9.3, Transpiration.

Photorespiration

Stomata need to be open for plants to obtain carbon dioxide for photosynthesis. Water vapour leaves plants through open stomata by the process of transpiration. When the temperature is high and humidity of the atmosphere is low plants can lose too much water. To prevent excess water loss, stomata close.

This prevents the entry of carbon dioxide into the leaves of the plant. The plant will still be photosynthesising and so the carbon dioxide levels fall and the oxygen levels increase.

Oxygen is a competitive inhibitor of the enzyme RuBisCO, leading to the production of phosphoglycolate and reducing the production of GP. This only happens when the concentration of carbon dioxide becomes very low.

Phosphoglycolate is a toxic two-carbon molecule that needs to be removed. It is converted by the plant into other organic molecules and energy from ATP is needed for the conversion.

RuBisCO has a higher affinity for carbon dioxide than oxygen and approximately 25% of the products of the Calvin cycle are lost in photorespiration, reducing the efficiency of photosynthesis.

1 Explain why photorespiration is not something commercial producers would want to encourage.

2 Suggest why plants evolved with such an important enzyme as RuBisCO being inhibited by such a common molecule as oxygen.

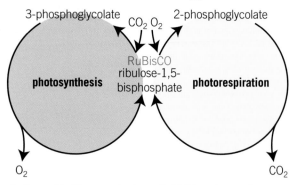

▲ Figure 11 *Photorespiration – RuBisCO and two substrates*

Summary of photosynthesis

The light-dependent stage, including photolysis, and the light-independent stage are summarised in Figure 12 showing how energy from the Sun in the form of light is used to build the chemical bonds in complex biological molecules.

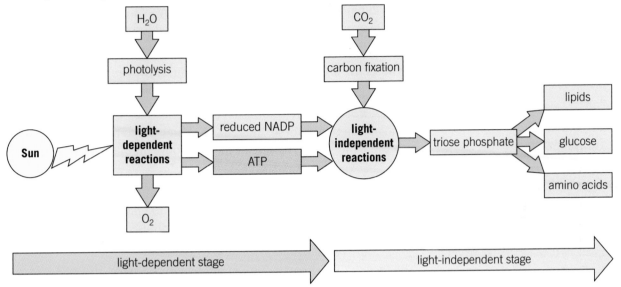

▲ **Figure 12** *Overview of the light-dependent and light-independent stages*

Summary questions

1 Explain the meaning of the term photophosphorylation. *(2 marks)*

2 Look at Figure 13. Explain why the absorption spectra of the pigments present in the thylakoid membranes of chloroplasts (top graph) follows the same pattern as the action spectrum (the rate of photosynthesis at different wavelengths of light, (bottom graph)). *(4 marks)*

3 Explain why photosynthesis stops when plants are exposed to green light only. *(3 marks)*

4 Explain what is meant by the term fixation. *(2 marks)*

5 a The Calvin cycle used to be called the 'dark reaction'. This term is now rarely used.

   Explain why this term is incorrect. *(1 mark)*

   b Explain why the alternate name of the Calvin cycle, the light-independent stage, is also not completely accurate. *(3 marks)*

6 Suggest the possible benefits of cyclic photophosphorylation. *(4 marks)*

7 Describe how RuBP is regenerated from TP in the Calvin cycle. *(4 marks)*

8 ⚙ The overall reaction for photosynthesis is summarised by the chemical equation:

$$6CO_2 + 6H_2O \rightarrow C_6H_{12}O_6 + 6O_2$$

Outline why this is an oversimplification. *(6 marks)*

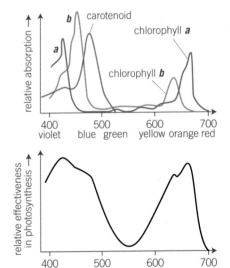

▲ **Figure 13**

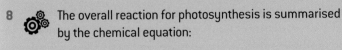

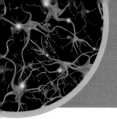

Photosynthesis is a multi-step reaction pathway that takes place in plants. Like any other chemical reaction, it is affected by various environmental factors. Plants are affected by changes in temperature and the availability of raw materials required for photosynthesis.

Plants obtain carbon dioxide through open stomata but this also involves the loss of water by transpiration. However, the loss of water vapour results in the transport of important ions and minerals from the soil to the leaves.

A balance between these different processes has to be maintained and this can be upset by changes in the environment of the plant.

Limiting factors

When one of the factors needed for a plant to photosynthesise is in short supply, it reduces the rate of photosynthesis, and is therefore a **limiting factor**.

The factors that affect the rate of photosynthesis are:

- *Light intensity* – light is needed as an energy source. As light intensity increases, ATP and reduced NADP are produced at a higher rate.

- *Carbon dioxide concentration* – carbon dioxide is needed as a source of carbon, so if all other conditions are met, increasing the carbon dioxide concentration increases the rate of carbon fixation in the Calvin cycle and, therefore, the rate of TP production.

- *Temperature* – affects the rate of enzyme-controlled reactions. As temperature increases, the rate of enzyme activity increases until the point at which the proteins denature. An increase in temperature increases the rates of the enzyme-controlled reactions in photosynthesis, such as carbon fixation. The rate of photorespiration, however, also increases above 25°C meaning higher photosynthetic rates may not be seen at higher temperatures even if enzymes are not actually denatured.

Stomata on plant leaves and other surfaces will close to avoid water loss by transpiration during dry spells when plants undergo water stress. The closure of stomata stops the diffusion of carbon dioxide into the plant, reducing the rate of the light-independent reaction, and eventually stopping photosynthesis.

Although water is required for photosynthesis it is never considered a limiting factor because for water potential to have become low enough to limit the rate of photosynthesis the plant will already have closed its stomata and ceased photosynthesis. Plants, except those with

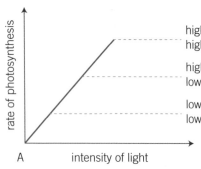

▲ Figure 1 *Following the blue line from A, rate of photosynthesis increases with increasing light intensity meaning light is the limiting factor. This happens until carbon dioxide concentration becomes limiting. Increasing the light intensity will then no longer have an effect until carbon dioxide concentration is increased*

adaptations to tolerate drought conditions, are unlikely to survive these conditions.

The law of limiting factors states that the rate of a physiological process will be limited by the factor which is in shortest supply.

Investigating the factors that affect the rate of photosynthesis

Data loggers are electronic devices that record data over time using sensors. Physical properties are recorded such as light intensity, temperature, pressure, pH (which can be used as a measure of carbon dioxide concentration), and humidity. Readings can be displayed in graphical form or on a spreadsheet.

They are usually equipped with a microprocessor (which inputs digital data) and internal memory for data storage. Data loggers can usually interface with a computer using specialised software.

Readings are taken with high degrees of accuracy and can be taken over long periods of time. They can be set to take many readings in a short period of time or used when there is a risk involved, for example, extreme cold or heat.

The factors affecting rate of photosynthesis can be investigated using a live pond weed, such as *Elodea*. The rate of photosynthesis can be estimated by calculating the rate of oxygen produced, carbon dioxide used, or increase in dry mass of a plant.

Apparatus could be set up as shown in Figure 2. Sodium hydrogen carbonate would be used to provide carbon dioxide. The pond weed should be kept illuminated before use. The apparatus should be left to equilibrate for 10 minutes or so before readings are taken. The oxygen sensor may also need to be calibrated using the oxygen concentration of air (21%).

The software can be set to take readings at desired intervals for the required length of time.

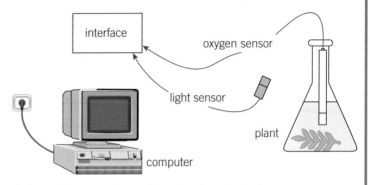

▲ **Figure 2** *The plant is supplied with carbon dioxide from sodium hydrogen carbonate in the solution containing the pond weed*

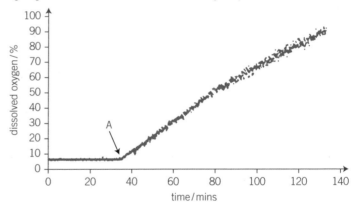

▲ **Figure 3** *Shows an example of a graph produced using the apparatus described*

1 Use the graph to calculate the rate of oxygen production after point A.
2 Suggest what happened at point A.
3 Outline how you could use the apparatus to investigate the effects of changing light intensity, temperature, and carbon dioxide concentration on the rate of photosynthesis.
4 Elodea will release bubbles of oxygen when photosynthesising which can be counted and used as an estimate of the rate of photosynthesis.
Evaluate the advantages and disadvantages of this method and of using data loggers in estimating the rate of photosynthesis.

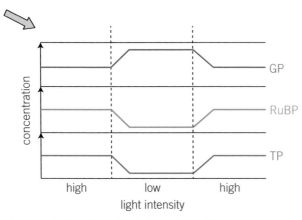

▲ Figure 4 *The effect of light intensity on GP, TP, and RuBP concentrations*

The effect of reducing light intensity on the Calvin cycle

Reducing light intensity will reduce the rate of the light-dependent stage of photosynthesis. This will reduce the quantity of ATP and reduced NADP produced. ATP and reduced NADP are needed to convert GP to TP. The concentration of GP will therefore increase and the concentration of TP will decrease. As there will be less TP to regenerate RuBP, the concentration of RuBP will also decrease. The reverse will happen when the light intensity is increased (Figure 4).

The effect of carbon dioxide concentration and temperature on the Calvin cycle

All the reactions making up the Calvin cycle are catalysed by enzymes, for example, RuBisCO in carbon fixation. At lower temperatures enzyme and substrate molecules have less kinetic energy resulting in fewer successful collisions and a reduced rate of reaction. This means decreasing temperature results in lower concentrations of GP, TP, and RuBP.

The same effect will be seen at high temperatures as enzymes will be denatured – this is irreversible.

As carbon dioxide is an essential substrate of the Calvin cycle, low concentrations will lead to reduced concentrations of GP (as there is less carbon dioxide to be fixed) and TP. The concentration of RuBP will increase as it is still being formed from TP but not being used to fix carbon dioxide.

 Artificial photosynthesis, a win-win solution

The burning of fossil fuels, and respiration, is continually releasing huge quantities of carbon dioxide into the atmosphere. The overall concentration of carbon dioxide, a greenhouse gas, is increasingly leading to more heat from the Sun being trapped in the atmosphere. This is enhanced global warming and is causing the polar ice caps to melt, increasing sea levels, and changing the climate around the world.

Fossil fuels have a limited supply and will eventually run out, leading to fuel shortages, and many parts of the world already suffer from food shortages.

Therefore, we have a surplus of carbon dioxide, which needs removing, and a shortage of fuel and food, both of which are forms of biomass produced by plants using carbon dioxide.

Photosynthesis would appear to offer a solution. It uses carbon dioxide and energy from the Sun to produce carbohydrates. Carbohydrates can be used as both food and fuel.

It is said that 'more energy hits Earth from the Sun in one hour than mankind uses in an entire year', so there is no shortage of energy. We already collect and use energy from the Sun in the form of solar power but the Sun doesn't always shine and at the moment there are no practical ways to store energy for a 'rainy day'. However, the carbohydrate fuel produced by plants can be stored for long periods.

The problem is that the process of photosynthesis, which has taken millions of years to evolve, is still not particularly efficient, so relying on plants is not the answer.

Artificial photosynthesis is seen as a possible solution. By improving on the natural process of photosynthesis carried out by plants, more carbon dioxide could be removed from the atmosphere and more carbohydrate products could be produced which could help with fuel and food shortages.

Suggest what you think would be the basic components of an artificial photosynthetic process.

Different types of photosynthesis

Most plants use the form of photosynthesis that you have learnt about in this chapter. This is referred to as C3 photosynthesis, and it is most efficient in cool, wet climates with average sunshine values.

Plants that live in hot, arid climates like the desert which are exposed to intense sunlight use different types of photosynthesis. Plants which use C4 and crassulacean acid metabolism (CAM) types of photosynthesis use water more efficiently and can photosynthesise at faster rates at higher temperatures and light intensities. Plants are adapted to their different environments to photosynthesise as efficiently as possible.

C4 photosynthesis

Plants that undergo C4 photosynthesis are adapted to high temperatures and limited water supply. They are able to fix carbon dioxide more efficiently and so do not need to have their stomata open for as long as C3 plants meaning there is less water lost by transpiration.

PEP carboxylase present in mesophyll cells, which first fixes carbon dioxide, is not inhibited by oxygen (like RuBisCO), increasing the efficiency of fixation. The four-carbon molecules produced are transported to bundle sheaths, formed from tightly packed cells,

deeper inside the plant. These molecules are then decarboxylated, and the carbon dioxide is then fixed by RuBisCO and enters the Calvin cycle. As RuBisCO is shielded from atmospheric oxygen, the waste of resources by photorespiration is reduced. Corn is an example of a C4 plant.

CAM photosynthesis

Other plants use CAM photosynthesis and open their stomata at night, usually closing them during the day, again reducing water loss by transpiration. Carbon dioxide is converted to an acid and stored during the night. During the day the acid is broken back down releasing carbon dioxide to RuBisCO. During very dry spells stomata can remain closed night and day. The carbon dioxide released from respiration is used in photosynthesis and the oxygen released by photosynthesis is used for respiration. Cacti are CAM plants.

1   Some plants drop their leaves and twigs and become dormant during dry spells. Describe the way in which a cactus survives dry spells and the advantages of this method.

2   Suggest why CAM plants can only keep their stomata closed night and day for short periods.

Summary questions

1   Describe what is meant by a limiting factor.   (*3 marks*)

2   Suggest why the rate of oxygen production is only an estimate of the rate of photosynthesis.   (*2 marks*)

3   Discuss, using what you have learnt in this chapter, how an understanding of the effect of limiting factors on the rate of photosynthesis is used to design more efficient glasshouses.   (*5 marks*)

Study tip

Always refer to light 'intensity' rather than 'level' or 'amount'. When referring to rate of photosynthesis it is often a good idea to refer to the rate of the Calvin cycle or carbon fixation as an example.

Practice questions

1 Photosynthesis is dependent on a supply of water, light energy, and carbon dioxide. Restricting the supply of any one of these factors reduces the rate of photosynthesis regardless of the availability of the other factors. A lower rate of photosynthesis results in a lower rate of growth and this has important implications for commercial plant growers.

a Outline the problems that commercial plant growers might have when using glasshouses. *(3 marks)*

The graph shows the changes in photosynthesis and respiration rates over a range of temperatures for tomato plants grown in a glasshouse. Tomatoes need to have reached a certain size and sweetness before they can be sold.

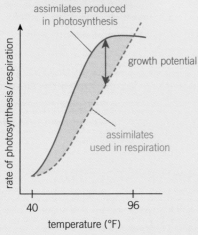

b (i) Describe, and explain, the changes in rates of photosynthesis and respiration shown in the diagram. *(5 marks)*

(ii) Discuss how these changes would affect the saleability of the tomatoes grown at different temperatures. *(4 marks)*

(iii) Comment on why there might be a difference between maximum rate of photosynthesis and optimum economic rate of photosynthesis. *(3 marks)*

2 It is known that C4-type plants carry out a much more efficient type of photosynthesis than C3 plants particularly under conditions of high light intensity and temperature.

The graph shows the changes in rate of photosynthesis for a C3 and C4 plant at different light intensities.

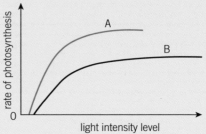

a State which curve in the graph is the C3 plant. *(1 mark)*

Plants are often adapted to grow in bright sunshine or shade, even leaves on the same tree can develop differently depending on the conditions they are exposed to.

The graph shows the changes in rate of photosynthesis in sun and shade leaves at different light intensities.

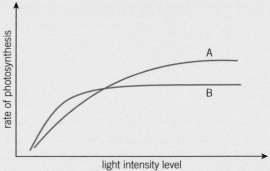

b (i) Suggest how leaves on the same plant can be exposed to different conditions that might affect leaf development. *(2 marks)*

(ii) State which curve in the diagram is most adapted to photosynthesis in direct sunlight. *(1 mark)*

(iii) Describe, making reference to cells, leaves and whole plants, how plants can be adapted to different light conditions. *(5 marks)*

3 The leaves of flowering plants have the ability to develop differently, depending on environmental conditions such as the amount of sun or shade a leaf receives.

A student carried out an investigation into sun and shade leaves from different parts of the same plant. Their observations and results are shown in the table.

Type of leaf	Number of leaves studied	Mean no. of stomata per mm² on lower surface	Mean thickness of leaf (μm)	Cuticle
sun	55	170	208	thick
shade	8	92	93	thin

a Calculate the percentage difference in the mean thickness of the sun leaves compared to the shade leaves.

Show your working. (*2 marks*)

b Suggest and explain one benefit of the greater mean number of stomata per mm² on the lower surface of the sun leaves. (*2 marks*)

c Describe two ways in which the student could improve her investigation. (*2 marks*)

OCR F214 2011

4 One way to determine the rate of photosynthesis is to measure the uptake of carbon dioxide.

a Discuss why measuring carbon dioxide uptake may or may not give a better indication of photosynthetic activity than measuring oxygen production. (*2 marks*)

b The graph shows the relationship between light intensity and the relative carbon dioxide uptake and production in a plant.

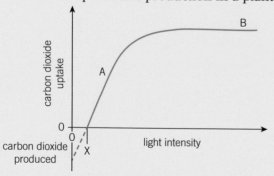

(i) State the factor limiting the rate of photosynthesis at **A** on the graph. (*1 mark*)

(ii) Suggest one factor that may limit the rate of photosynthesis at **B**. (*1 mark*)

(iii) Carbon dioxide is given off by the plant when the light intensity is lower than **X**.

Name the process that **produces** carbon dioxide in the plant. (*1 mark*)

(iv) With reference to the graph, explain the biochemical processes that are occurring in the plant:

- as light intensity increases from 0 (zero) to **X**.
- at light intensity **X**.
- at light intensities greater than **X**. (*3 marks*)

c (i) Name the products of the light-dependent stage of photosynthesis. (*3 marks*)

(ii) Paraquat is a weed killer. It binds with electrons in photosystem I.

Suggest how paraquat results in the death of a plant. (*2 marks*)

OCR F214 2012

Cellular respiration

Glucose is a hexose (six-carbon sugar) produced during photosynthesis. It is a complex molecule containing energy absorbed from sunlight 'trapped' within its carbon hydrogen bonds. Respiration is essentially the reverse of photosynthesis.

The carbon framework of glucose is broken down and the carbon-hydrogen bonds broken. The energy released is then used in the synthesis of ATP by chemiosmosis. ATP, the universal energy currency, is constantly synthesised and used in energy-requiring reactions and processes.

Respiration is a complex multi-step reaction pathway (Figure 1). You will be considering respiration in eukaryotic cells. A similar process takes place in prokaryotic cells but they do not have mitochondria so many of the reactions take place on cell membranes. To make the biochemistry of respiration clearer to understand you will look at it in stages, but it is important to remember that in the cell the process is continuous. The first stage of respiration is glycolysis.

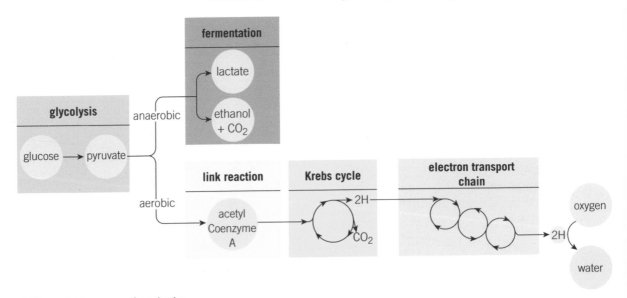

▲ Figure 1 Summary of respiration

Glycolysis

Glycolysis occurs in the cytoplasm of the cell. It does not require oxygen – it is an **anaerobic** process. Glucose, a six-carbon sugar, is split into two smaller, three-carbon pyruvate molecules. ATP and reduced nicotinamide adenine dinucleotide (NAD) are also produced. Glycolysis, summarised here, actually involves 10 reaction steps involving many enzymes.

The main steps in glycolysis are:

1   *Phosphorylation* – the first step of glycolysis requires two molecules of ATP. Two phosphates, released from the two ATP molecules, are attached to a glucose molecule forming **hexose bisphosphate.**

2   *Lysis* – this destabilises the molecule causing it to split into two **triose phosphate** molecules.

3   *Phosphorylation* – another phosphate group is added to each triose phosphate forming two triose bisphosphate molecules. These phosphate groups come from free inorganic phosphate (P_i) ions present in the cytoplasm.

4   *Dehydrogenation and formation of ATP* – the two triose bisphosphate molecules are then oxidised by the removal of hydrogen atoms (**dehydrogenation**) to form two **pyruvate** molecules. NAD coenzymes accept the removed hydrogens – they are reduced, forming two reduced NAD molecules.

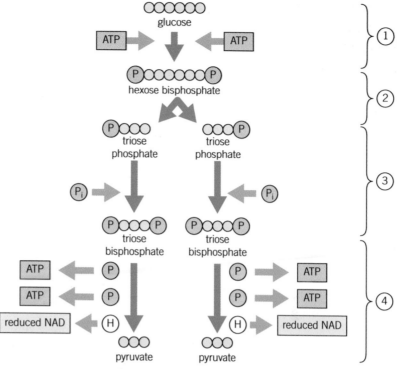

▲ **Figure 2** *Summary of glycolysis*

At the same time, four ATP molecules are produced using phosphates from the triose bisphosphate molecules.

This is an example of substrate level phosphorylation – the formation of ATP without the involvement of an electron transport chain. ATP is formed by the transfer of a phosphate group from a phosphorylated intermediate (in this case triose bisphosphate) to ADP.

Two ATP molecules are used to prime the process at the beginning, and four ATP molecules are produced, so the overall net ATP yield from glycolysis is two molecules of ATP.

The reduced NAD is used in a later stage to synthesise more ATP.

Synoptic link

It may be useful to look back to Topic 4.4, Cofactors, coenzymes, and prosthetic groups and Topic 3.11, ATP.

Study tip

Hexose bisphosphate, triose phosphate, and pyruvate are the only compounds that you have to be able to recall.

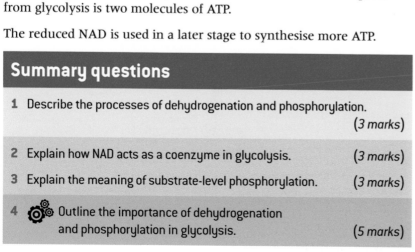

Summary questions

1   Describe the processes of dehydrogenation and phosphorylation.
(*3 marks*)

2   Explain how NAD acts as a coenzyme in glycolysis.   (*3 marks*)

3   Explain the meaning of substrate-level phosphorylation.   (*3 marks*)

4   Outline the importance of dehydrogenation and phosphorylation in glycolysis.   (*5 marks*)

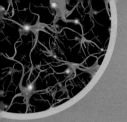

18.2 Linking glycolysis and the Krebs cycle

Specification reference: 5.2.2

Learning outcomes

Demonstrate knowledge, understanding, and application of:

→ the link reaction and its site in the cell

→ the structure of the mitochondrion.

Synoptic link

You learnt about mitochondria in Topic 2.4, Eukaryotic cell structure.

Study tip

Be careful to distinguish, and correctly use, the terms 'inner' (mitochondrial membrane) and 'inter' (membrane space).

As you have learnt, glycolysis takes place in the cytoplasm of the cell. In eukaryotic cells the remaining aerobic (oxygen-requiring) reactions of cellular respiration take place inside the mitochondria.

outer mitochondrial membrane separates the contents of the mitochondrion from the rest of the cell, creating a cellular compartment with ideal conditions for aerobic respiration

matrix contains enzymes for the Krebs cycle and the link reaction, also contains mitochondrial DNA

inner mitochondrial membrane contains electron transport chains and ATP synthase

intermembrane space Proteins are pumped into this space by the electron transport chain. The space is small so the concentration builds up quickly

cristae are projections of the inner membrane which increase the surface area available for oxidative phosphorylation

▲ Figure 1 *In the presence of oxygen, aerobic respiration occurs in the mitochondria of eukaryotic cells*

Oxidative decarboxylation (the link reaction)

The first step in aerobic respiration is oxidative decarboxylation. This is sometimes referred to as the link reaction, because it is the step that links anaerobic glycolysis, occurring in the cytoplasm, to the aerobic steps of respiration, occurring in the mitochondria.

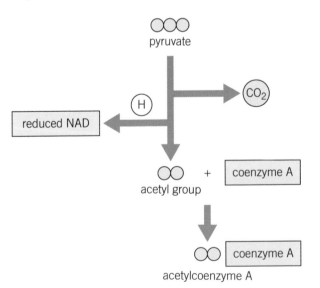

▲ Figure 2 *Summary of oxidative decarboxylation, linking the anaerobic and aerobic stages of cellular respiration in eukaryotic cells*

In eukaryotic cells, pyruvate enters the **mitochondrial matrix** by active transport via specific carrier proteins. Pyruvate then undergoes oxidative decarboxylation – carbon dioxide is removed (**decarboxylation**) along with hydrogen (oxidation). The hydrogen atoms removed are accepted by NAD. NAD is reduced to form **NADH** (reduced NAD). The resulting two-carbon acetyl group is bound by **coenzyme A** forming **acetylcoenzyme A (acetyl CoA)**. The International Union of Pure and Applied Chemistry (IUPAC) name for the acetyl group is the ethanoyl group, but the terms acetyl and acetylcoenzyme A are so widely known and used by biologists that the traditional names are retained.

Acetyl CoA delivers the acetyl group to the next stage of aerobic respiration, known as the Krebs cycle. The reduced NAD is used in oxidative phosphorylation to synthesise ATP (Topic 18.4, Oxidative phosphorylation).

Acetyl groups are now all that is left of the original glucose molecules. The carbon dioxide produced will either diffuse away and be removed from the organism as a metabolic waste or, in autotrophic organisms, it may be used as a raw material in photosynthesis.

Summary questions

1 Explain why the removal of carbon dioxide in the link reaction is called oxidative. *(2 marks)*

2 Name one organic compound and one inorganic compound produced in the link reaction. *(2 marks)*

3 Copy and complete the equation. *(3 marks)*

_____ + CoA + NAD ⟶ _____ + CO_2 + _____

4 Suggest why glycolysis occurs in the cytoplasm but not the mitochondrial matrix. *(4 marks)*

Synoptic link

You learnt about coenzymes in Topic 4.4, Cofactors, coenzymes, and prosthetic groups.

Study tip

Remember that each glucose molecule produces two pyruvate molecules which both go through the subsequent stages.

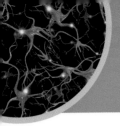

18.3 The Krebs cycle

Specification reference: 5.2.2

The Krebs cycle also takes place in the mitochondrial matrix and each complete cycle results in the breakdown of an acetyl group. Acetyl groups are all that remain of the glucose that entered glycolysis. It is another complex multi-step pathway, summarised here.

As in the previous stages, the Krebs cycle involves decarboxylation, dehydrogenation, and substrate-level phosphorylation. The hydrogen atoms released are picked up by the coenzymes NAD and flavin adenine dinucleotide (**FAD**). Carbon dioxide is a by-product of these reactions and the ATP produced is available for use by energy-requiring processes within the cell.

The reduced NAD and reduced FAD produced are used in the final, oxygen-requiring step of aerobic respiration (Topic 18.4, Oxidative phosphorylation) to produce large quantities of ATP by chemiosmosis.

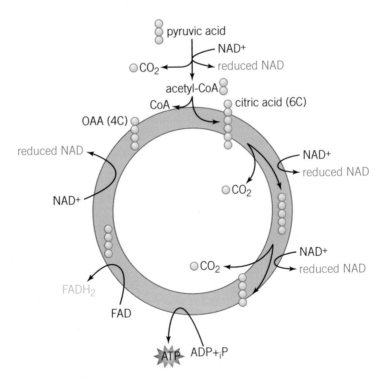

▲ **Figure 1** *The Krebs cycle*

1 Acetyl CoA delivers an acetyl group to the Krebs cycle. The two-carbon acetyl group combines with four-carbon **oxaloacetate** to form six-carbon **citrate**.

2 The citrate molecule undergoes decarboxylation and dehydrogenation producing one reduced NAD and carbon dioxide. A five-carbon compound is formed.

3 The five-carbon compound undergoes further decarboxylation and dehydrogenation reactions, eventually regenerating oxaloacetate,

and so the cycle continues. More carbon dioxide, two more reduced NADs, and one reduced FAD are produced. ATP is also produced by substrate-level phosphorylation.

The importance of coenzymes in respiration

Respiration is a complex multi-step reaction pathway. Coenzymes are required to transfer protons, electrons, and functional groups between many of these enzyme-catalysed reactions.

Redox reactions have an important role in respiration and without coenzymes transferring electrons and protons between these reactions many respiratory enzymes would be unable to function.

NAD and FAD are both coenzymes that accept protons and electrons released during the breakdown of glucose in respiration. The differences between these two enzymes are:

- NAD takes part in all stages of cellular respiration but FAD only accepts hydrogens in the Krebs cycle
- NAD accepts one hydrogen and FAD accepts two hydrogens
- reduced NAD is oxidised at the start of the electron transport chain releasing protons and electrons while reduced FAD is oxidised further along the chain
- reduced NAD results in the synthesis of three ATP molecules but reduced FAD results in the synthesis of only two ATP molecules.

You will see reduced NAD represented in a number of ways – for example, NADH, NADH + H^+, or $NADH_2^+$. The reason for this is that NAD is actually charged so is more accurately represented as NAD^+. When NAD^+ is reduced it accepts two protons and an electron pair (from a C–H bond) forming NADH + H^+. NADH, or reduced NAD, then transfers the proton and electron pair to a subsequent reaction.

Coenzymes are usually derived from vitamins. This is why, although coenzymes are mostly recycled, vitamins are an essential micronutrient.

> **Study tip**
>
> Oxaloacetate and citrate are the only names of Krebs cycle intermediate compounds that you need to remember.
>
> The number of ATP molecules produced as a result of reduced NAD and reduced FAD can vary.

Summary questions

1 Compare the structures of ATP and NAD.   (*3 marks*)

2 ⚙ ATP can be described as a coenzyme. Explain why.   (*2 marks*)

3 Draw a simple diagram summarising the breakdown of glucose to carbon dioxide and reduced coenzymes.   (*4 marks*)

4 Calculate the number of ATP molecules produced by substrate-level phosphorylation after two rounds of the Krebs cycle.   (*2 marks*)

5 The Krebs cycle does not use oxygen at any point. Suggest why the Krebs cycle is termed aerobic.   (*4 marks*)

6 ⚙ Suggest a reason for the involvement of FAD rather than NAD in only one specific step of the Krebs cycle.   (*6 marks*)

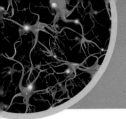

18.4 Oxidative phosphorylation

Specification reference: 5.2.2

Learning outcomes

Demonstrate knowledge, understanding, and application of:

→ the process and site of oxidative phosphorylation.

Synoptic link

You learnt about chemiosmosis in Topic 17.2, ATP synthesis.

Oxidative phosphorylation

The hydrogen atoms that have been collected by the coenzymes NAD and FAD are delivered to electron transport chains present in the membranes of the cristae of the mitochondria.

The hydrogen atoms dissociate into hydrogen ions and electrons. The high energy electrons are used in the synthesis of ATP by chemiosmosis. Energy is released during redox reactions as the electrons reduce and oxidise electron carriers as they flow along the electron transport chain. This energy is used to create a proton gradient leading to the diffusion of protons through ATP synthase resulting in the synthesis of ATP.

At the end of the electron transport chain the electrons combine with hydrogen ions and oxygen to form water. Oxygen is the final electron acceptor and the electron chain cannot operate unless oxygen is present. Respiration which involves the complete breakdown of glucose is therefore an aerobic process.

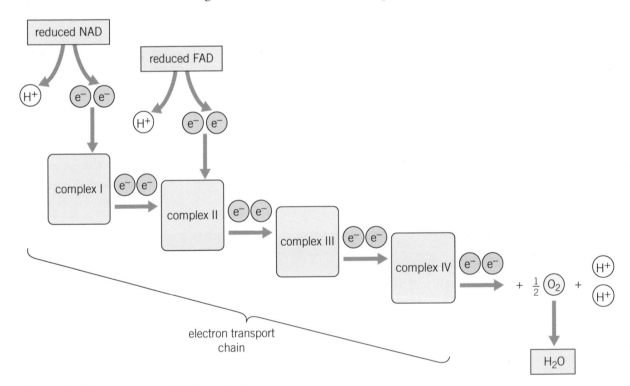

▲ Figure 1 *Electron transport chain in respiration*

Synoptic link

You learnt about coenzymes in Topic 4.4, Cofactors, coenzymes.

The phosphorylation of ADP to form ATP is dependent on electrons moving along electron transport chains. This requires the presence of oxygen and is known as oxidative phosphorylation.

The hydrogens released from NAD and FAD could combine directly with oxygen, releasing energy from the formation of bonds during the production of water. However, this energy could not be used to synthesise ATP. Heat released in the exothermic reaction would simply raise the temperature of the cell.

Substrate level phosphorylation

Substrate level phosphorylation is the production of ATP involving the transfer of a phosphate group from a short-lived, highly reactive intermediate such as creatine phosphate. This is different from oxidative phosphorylation which couples the flow of protons down the electrochemical gradient through ATP synthase to the phosphorylation of ADP to produce ATP.

Summary questions

1 Explain why hydrogens have to be actively pumped across the membrane from the matrix and return to the matrix by diffusion through ATP synthase. *(4 marks)*

2 Explain why the electrons released from reduced FAD lead to the synthesis of less ATP than the electrons released from reduced NAD. *(4 marks)*

3 Cyanide is a respiratory poison. It attaches to the iron in the haem group of cytochrome c oxidase in complex IV of the electron transport chain.
Suggest an explanation for the toxicity of cyanide. *(4 marks)*

4 Explain, with reasons, whether you agree with the following statements. *(6 marks)*
- ATP synthase is not actually part of the electron transport chain.
- Oxygen is required for the transfer of electrons along the electron transport chain.
- Hydrogen ions return to the matrix by facilitated diffusion.

Study tip

Oxidative phosphorylation is a process that occurs along the electron transport chain, which involves a series of membrane-bound enzyme complexes.

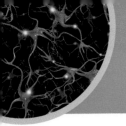

18.5 Anaerobic respiration

Specification reference: 5.2.2

Aerobic respiration was not possible when life began, as there was no oxygen present in the atmosphere of the Earth at that time. It is a relatively new process in evolutionary terms but as it is a far more efficient process than **anaerobic respiration** it was rapidly selected for. Aerobic respiration produces around 38 molecules of ATP per glucose molecule whereas fermentation (a form of anaerobic respiration) only produces two molecules of ATP (net).

Anaerobic respiration in eukaryotic organisms

Eukaryotic cells respire aerobically if enough oxygen is available. Anaerobic respiration, resulting in the synthesis of smaller quantities of ATP, occurs in the absence of oxygen and is also used when oxygen cannot be supplied fast enough to respiring cells. The use of this less efficient process to produce ATP is a temporary 'emergency' measure to keep vital processes functioning.

Organisms fall into different categories determined by their dependence on oxygen or not:

- **obligate anaerobes** – cannot survive in the presence of oxygen. Almost all obligate anaerobes are prokaryotes, for example, *Clostridium* (bacteria that cause food poisoning), although there are some fungi as well.

- **facultative anaerobes** – synthesise ATP by aerobic respiration if oxygen is present, but can switch to anaerobic respiration in the absence of oxygen, for example, yeast.

- **obligate aerobes** – can only synthesise ATP in the presence of oxygen, for example, mammals. The individual cells of some organisms, such as muscle cells in mammals, can be described as facultative anaerobes because they can supplement ATP supplies by employing anaerobic respiration in addition to aerobic respiration when the oxygen concentration is low. However, this is only for short periods and oxygen is eventually required. The shortfall of oxygen during the period of anaerobic respiration produces compounds that have to be broken down when oxygen becomes available again, so the organism as a whole is an obligate aerobe.

Fermentation

Fermentation (a form of anaerobic respiration) is the process by which complex organic compounds are broken down into simpler inorganic compounds without the use of oxygen or the involvement of an electron transport chain. The organic compounds, such as glucose, are not fully broken down so fermentation produces much less ATP than aerobic respiration. The small quantity of ATP produced is synthesised by substrate-level phosphorylation alone.

The end products of fermentation differ depending on the organism. **Alcoholic fermentation** occurs in yeast and some plant root cells.

Here the end products are ethanol (an alcohol) and carbon dioxide. **Lactate fermentation** results in the production of lactate and is carried out in animal cells.

When there is no oxygen to act as the final electron acceptor at the end of the electron transport chain in oxidative phosphorylation, the flow of electrons stops. This means the synthesis of ATP by chemiosmosis also stops.

As the flow of electrons along the electron transport chain has stopped, the reduced NAD and reduced FAD are no longer able to be oxidised because there is nowhere for the electrons to go. This means NAD and FAD cannot be regenerated and so the decarboxylation and oxidation of pyruvate and the Krebs cycle comes to a halt as there are no coenzymes available to accept the hydrogens being removed.

Glycolysis would also come to halt due to the lack of NAD if it were not for the process of fermentation.

Lactate fermentation in mammals

In mammals, pyruvate can act as a hydrogen acceptor taking the hydrogen from reduced NAD, catalysed by the enzyme **lactate dehydrogenase**. The pyruvate is converted to lactate (lactic acid) and NAD is regenerated. This can be used to keep glycolysis going so a small quantity of ATP is still synthesised. In mammals in particular, anaerobic respiration in the muscles is often supported by ATP from aerobic respiration, which is still being produced as fast as oxygen can be delivered in other parts of the body.

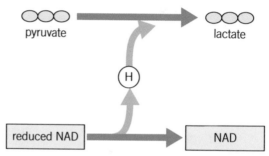
▲ Figure 1 *Summary of lactate fermentation in mammals*

Lactic acid is converted back to glucose in the liver but oxygen is needed to complete this process. This is the reason for the oxygen debt (and the need to breathe heavily) after exercise.

Lactate fermentation cannot occur indefinitely for two main reasons:

- the reduced quantity of ATP produced would not be enough to maintain vital processes for a long period of time
- the accumulation of lactic acid causes a fall in pH leading to proteins denaturing. Respiratory enzymes and muscle filaments are made from proteins and will cease to function at low pH.

Lactic acid is removed from muscles and taken to the liver in the bloodstream. One of the main aims when improving physical fitness is to increase the blood supply and flow, through muscles. This increases the rate of lactic acid removal allowing the intensity and duration of exercise to be increased.

Alcoholic fermentation in yeast (and many plants)

Alcoholic fermentation is not a reversible process like lactate fermentation. Pyruvate is first converted to ethanal, catalysed by the enzyme pyruvate decarboxylase. Ethanal can then accept a hydrogen atom from reduced NAD, becoming ethanol. The regenerated NAD can then continue to act as a coenzyme and glycolysis can continue.

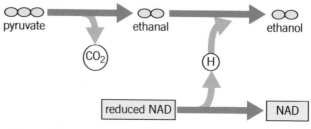
▲ Figure 2 *Summary of alcoholic fermentation in yeast and plant cells*

This is not a short-term process and can continue indefinitely in the absence of oxygen. Ethanol is a toxic waste product to yeast cells and they are unable survive if the ethanol accumulates above approximately 15%. This is allowed to happen during the production of alcohol in brewing or wine making.

Investigation into respiration rates in yeast

The apparatus shown could be used to measure the rate of carbon dioxide production of a yeast suspension. This will be equivalent to the rate of anaerobic respiration or alcoholic fermentation of the yeast cells.

The glucose in solution provides a respiratory substrate. The flask is sealed during the experiment to ensure anaerobic conditions.

As the yeast respires carbon dioxide is released increasing the volume of gas in the flask. As the volume of gas in the tube increases the pressure will increase causing the coloured liquid to move along the capillary tube. The distance moved by the liquid together with diameter of the tube can be used to calculate the increase in volume of gas (carbon dioxide) in the flask over a certain time. This is a measure of the rate of respiration.

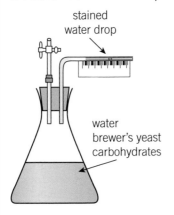

◀ **Figure 3** *Diagram showing the apparatus used to measure the anaerobic respiration in yeast*

Data logging

Respiration is not 100% efficient and energy is lost as heat when organisms respire.

When yeast respires it produces heat which will increase the temperature of a solution containing yeast. Sensors can be used to measure changes in temperature.

A student carried out an investigation into respiration in yeast using a data logger to measure the changes in carbon dioxide concentration as a measure of the rate of respiration.

The student set up the apparatus to be used as shown in Figure 4.

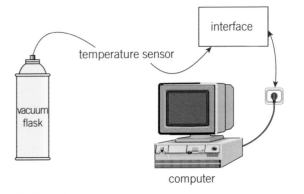

▲ **Figure 4**

The student placed a solution containing yeast and glucose in the flask and inserted a carbon dioxide sensor.

The solution was covered with a layer of liquid paraffin.

The software was set up to record readings every 50 seconds for 1 600 seconds.

The readings were displayed in graphical form shown in Figure 5.

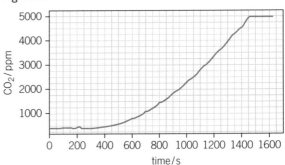

▲ **Figure 5**

1 Explain why the student did the following:
   Carried out the investigation in a vacuum flask
   Covered the solution containing the yeast with liquid paraffin
2 Calculate, using Figure 5, the fastest rate of respiration observed during the investigation.
3 Suggest why the graph eventually reached a plateau.
4 Describe how the apparatus could be adapted to investigate aerobic respiration.

Small-scale and large-scale adaptations to low oxygen environments

Many animals live in or around water and spend time underwater to hunt for food. These animals are adapted in a variety of ways to survive periods of anaerobic respiration while they cannot breathe air. Many bacteria also live in low oxygen environments. There are many adaptations that have evolved in different organisms to overcome the problems of a temporary or permanent lack of oxygen:

Bacterial adaptations

Different groups of bacteria have evolved to use nitrate ions, sulphate ions, and carbon dioxide as final electron acceptors in anaerobic respiration. This enables them to live in very low, or zero, oxygen environments.

Anaerobic bacteria present in the digestive systems of animals play an essential role in the breakdown of food and absorption of minerals. Methanogens are a type of bacteria found in the digestive system of ruminants, such as cows. They digest cellulose from grass cell walls into products that can be further digested, absorbed and used by the ruminants. The final electron acceptor in the respiratory pathway of these bacteria is carbon dioxide, and methane and water are produced. The methane builds up and eventually has to be released – it has been estimated that a cow produces around 500 L of methane per day.

Mammalian adaptations

Marine mammals that dive for long periods, such as seals and whales, have a range of different types of adaptations for surviving when they cannot take in more oxygen:

1 *Biochemical adaptations* include greater concentrations of haemoglobin and myoglobin than land mammals, particularly in the muscles used in swimming. This maximises their oxygen stores, delaying the onset of anaerobic metabolism. Whales have a higher tolerance to lactic acid than human beings, so they can respire anaerobically much longer without suffering tissue damage. They also have a greater tolerance of high carbon dioxide levels – they have very effective blood buffering systems that prevent a catastrophic rise in pH.

▲ Figure 6 *Sperm whale (*Physeter macrocephalus*) tail*

2 *Physiological adaptations* in many diving mammals include a modified circulatory system. When they dive they show peripheral vasoconstriction, so blood is shunted to the brain, heart, and muscles. The heart slows by up to 85% – this is known as bradycardia and reduces the energy demand of the heart muscle. Whales also exchange 80–95% of the air in the lungs when they breathe – in humans, that figure is around 15%. In some species dives can last up to two hours, so the adaptations are very effective.

3 *Physical adaptations* include streamlining to reduce drag due to friction from water while swimming, therefore reducing the energy demand during a dive. The limbs of marine mammals are 'fin-shaped' to maximise the efficient use of energy in propulsion.

1 The lungs of whales are proportionally no larger than humans but some whales can stay under water for two hours. Suggest how the lungs might be adapted to enable these long dives and why larger lungs would be a disadvantage to a whale.

2 Summarise the adaptations of whales for making long underwater dives.

Synoptic link

It may be useful to look back at Topic 3.11, ATP.

Study tip

Lactate fermentation is important because it regenerates NAD which allows glycolysis to continue. Lactate, or lactic acid, is not a waste product, it is recycled as glucose.

Summary questions

1   Explain why yeast cells are described as facultative anaerobes.   (*2 marks*)

2   Describe why alcoholic fermentation can be described as having more in common with aerobic respiration than with fermentation.   (*3 marks*)

3   Explain why the build-up of lactic acid eventually stops muscle contraction which we experience as fatigue.   (*4 marks*)

4   Glycolysis, the anaerobic stage of respiration, is the only source of ATP in red blood cells. Cardiac muscle is adapted to reduce the chances of anaerobic respiration ever being needed.
    Outline the benefits to red blood cells and cardiac muscle of the different types of respiration they undertake.   (*6 marks*)

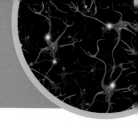

Different respiratory substrates

Glucose is not the only organic molecule that is broken down to release energy for the synthesis of ATP. There are many other **respiratory substrates**. Triglycerides are hydrolysed to fatty acids, which enter the Krebs cycle via acetyl CoA and glycerol.

Glycerol is first converted to pyruvate before undergoing oxidative decarboxylation, producing an acetyl group which is picked up by coenzyme A, forming acetyl CoA. The fatty acids in a triglyceride molecule can lead to the formation of as many as 50 acetyl CoA molecules, resulting in the synthesis of up to 500 ATP molecules.

Gram for gram, lipids store and release about twice as much energy as carbohydrates. Alcohol contains more energy than carbohydrates but less than lipids. Proteins are roughly equivalent to carbohydrates.

Proteins first have to be hydrolysed to amino acids and then the amino acids have to be deaminated (removal of amine groups) before they enter the respiratory pathway, usually via pyruvate. These steps require ATP, reducing the net production of ATP.

The **respiratory quotient (RQ)** of a substrate is calculated by dividing the volume of carbon dioxide released by the volume of oxygen taken in during respiration of that particular substrate. This is measured using a simple piece of apparatus called a respirometer (Figure 2).

$$RQ = \frac{CO_2 \text{ produced}}{O_2 \text{ consumed}}$$

It takes six oxygen molecules to completely respire one molecule of glucose and this results in the production of six molecules of carbon dioxide (and six molecules of water). This results in an RQ of 1.0.

Lipids contain a greater proportion of carbon–hydrogen bonds than carbohydrates which is why they produce so much more ATP in respiration. Due to the greater number of carbon–hydrogen bonds, lipids require relatively more oxygen to break them down and release relatively less carbon dioxide. This results in RQs of less than one for lipids. The structure of amino acids leads to RQs somewhere between carbohydrates and lipids.

- carbohydrates = 1.0
- protein = 0.9
- lipids = 0.7

So, by measuring the volume of oxygen taken in and carbon dioxide released, and calculating RQ, the type of substrate being used for respiration at that point can be roughly determined.

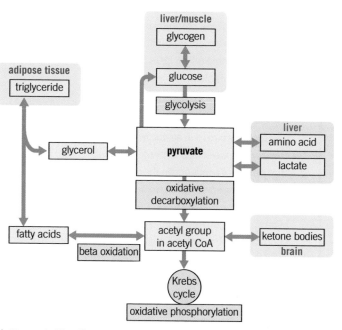

▲ Figure 1 *The diagram shows how different biological molecules, and fragments of these molecules, can be 'fed' into the respiratory pathway at different points. Some tissues, or organs, are better adapted to use certain substrates. For example, the brain is well adapted to use ketone bodies as a respiratory substrate*

During normal activity, the RQ is in the range of 0.8 to 0.9, showing that carbohydrates and lipids (and probably some proteins) are being use as respiratory substrates.

During anaerobic respiration, the RQ increases above 1.0, although this not easy to measure as the point at which anaerobic respiration begins is not easy to pinpoint.

Calculating the respiratory quotient

The oxygen taken in and carbon dioxide produced by a desert locust was measured experimentally under different conditions:

Conditions	at rest	in flight
Oxygen absorbed / au	10.5	160.0
Carbon dioxide produced / au	10.0	113.6

1 Calculate the RQ of the insect at rest and in flight.
2 What does this suggest about the substrate being respired at rest and in flight?

Low carbohydrate diets

Many people choose low carbohydrate diets when they want to lose weight – and in particular to lose some body fat. The diets can work – but the science suggests that you need to think carefully before cutting out the molecules that are most commonly used as fuel in your body. Here are some of the facts about low carbohydrate diets for you to consider:

1 Triglycerides are hydrolysed into fatty acids and glycerol. The fatty acids are broken down in the mitochondria to give many two-carbon acetyl groups that combine with coenzyme A molecules and enter the Krebs cycle.

2 Triglycerides cannot act as the only respiratory substrate. Carbohydrates are needed to keep the Krebs cycle going so that acetyl groups from the breakdown of fatty acids can be 'fed in'. If carbohydrates are in short supply the body will make them using a process called gluconeogenesis. This process often uses glycerol, but it may also use pyruvate from glycolysis.

3 Oxaloacetate from the Krebs cycle can be used to make glucose when carbohydrate levels are low. Reducing the number of oxaloacetate molecules in the Krebs cycle reduces the rate at which the acetyl groups produced during the breakdown of lipids can be fed into the cycle and produce ATP.

4 Oxaloacetate can be replaced by the conversion of pyruvate from carbohydrate breakdown in the mitochondria. Pyruvate is also synthesised using glycerol from the breakdown of lipids. However, the breakdown of a lipid molecule provides a relatively small quantity of glycerol and so a relatively small amount of pyruvate. This means carbohydrates are still needed to ensure the continued respiration of fat.

5 Proteins can be hydrolysed into amino acids which are then deaminated in the liver. The remaining keto acids can be converted into glucose molecules. Lean muscle is the protein of choice in this process, so a low carbohydrate diet can lead to the breakdown of muscle tissue. The liver and kidneys also have to remove the nitrogenous waste.

6 If the level of acetyl CoA increases because it is not being taken into the Krebs cycle, the liver starts converting it into ketone bodies. Brain cells normally require glucose as an energy source. They cannot use fatty acids as a respiratory substrate but they can use ketone bodies.

7 When the body is producing more ketone bodies than usual, it is said to be in ketosis. This can lead to a dangerous condition known as ketoacidosis. Ketoacidosis is the result of an accumulation of ketone bodies which cause the pH level of the blood to drop to dangerous, or even fatal, levels. This condition is often seen in alcoholics, untreated diabetes, and during starvation. It is often diagnosed by the fruity smell of propanone (acetone), a breakdown product of ketone bodies, on the breath of an affected person.

1 The term 'fats burn in the flame of carbohydrates' was coined more than a century ago. Discuss the accuracy of this statement.
2 Evaluate the benefits and drawbacks of a low carbohydrate diet.

Practical investigations into the factors affecting rate of respiration using respirometers

A student carried out an experiment to investigate the effect of temperature on the rate of respiration in soaked (germinating) pea seeds and dry (dormant) pea seeds.

A respirometer was used, shown in Figure 2. The potassium hydroxide solution in this apparatus absorbs carbon dioxide. If the apparatus is kept at a constant temperature, any changes in the volume of air in the respirometer will be due to oxygen uptake. The student set up three respirometers, A, B and C in water baths at two different temperatures. The respirometers were left for 10 minutes to equilibrate.

The contents of each respirometer are shown in Table 1.

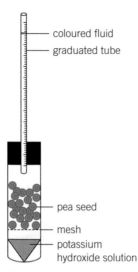

coloured fluid
graduated tube

pea seed
mesh
potassium hydroxide solution

▲ Figure 2

▼ Table 1

Temperature (°C)	Respirometer	Contents
15	A	30 soaked pea seeds
	B	glass beads + 30 dry pea seeds
	C	glass beads
25	A	30 soaked pea seeds
	B	glass beads + 30 dry pea seeds
	C	glass beads

At each temperature, respirometer C, which contained only glass beads, was a control. After the student had left each respirometer to equilibrate, a small volume of coloured fluid was introduced into each graduated tube.

The respirometers were then left in the appropriate water baths for 20 minutes and maintained at the correct temperature. During this time, the coloured fluid in the graduated tube moved. The level of the coloured fluid in each respirometer was recorded at the start of the experiment and after 20 minutes. The results are summarised in Table 2.

▼ Table 2

Temperature °C	Respirometer	Reading at start (cm^3)	Reading after 20 minutes (cm^3)	Difference (cm^3)	Corrected difference (cm^3)	Rate of oxygen uptake ($cm^3\,min^{-1}$)
15	A	0.93	0.74	0.19	0.16	0.008
	B	0.93	0.86	0.07	0.04	0.002
	C	0.91	0.88	0.03		
25	A	0.94	0.63	0.31	0.27	
	B	0.93	0.84	0.09	0.05	0.003
	C	0.95	0.91	0.04		

1 Copy and complete the table.
2 Suggest why, at each temperature, respirometer B contained some glass beads.
3 Suggest how the student determined the quantity of glass beads to place in respirometer B at each temperature.
4 Explain why there is an increased rate of respiration in soaked seeds at 25°C compared with soaked seeds at 15°C.
5 Suggest a reason for the difference in the rate of respiration between soaked and dry pea seeds.

Study tip

A respirometer is not the same as a spirometer. A spirometer measures volume changes during breathing but a respirometer measures the change in volume of oxygen or carbon dioxide.

Summary questions

1 Outline the respiration pathway of a triglyceride. (4 marks)

2 Describe the difference between a respirometer and a spirometer. (3 marks)

3 Consider the three respiratory substrates shown here.
Calculate the percentage of hydrogen in each molecule and use these to compare the relative energy values of the substrates.

A

B

C CH₃(CH₂)₁₆COOH

(6 marks)

Practice questions

1 a (i) State the meaning of the term phosphorylation. *(1 mark)*

(ii) Outline the similarities and differences between oxidative phosphorylation and substrate level phosphorylation. *(3 marks)*

b NADH and FADH transfer electrons to the electron transport chain.

10 protons are pumped due to the electrons from NADH

6 protons are pumped due to the electrons from FADH

4 protons are needed by ATP synthase to make one ATP molecule

(i) Calculate the number of ATP molecules produced due to each reduced coenzyme. Show your working. *(3 marks)*

(ii) Calculate the total number of ATP molecules produced due to oxidative phosphorylation. Show your working in your answer. *(3 marks)*

2 a The diagram represents the first stage of respiration.

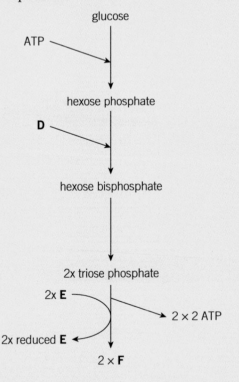

(i) Name the stage represented by the figure. *(1 mark)*

(ii) State precisely where in the cell this stage takes place. *(1 mark)*

(iii) Identify the compounds **D**, **E**, and **F**. *(3 marks)*

b In anaerobic conditions, compound **F** does not proceed to oxidative decarboxylation.

Describe the fate of compound **F** during anaerobic respiration in an animal cell **and** explain the importance of this reaction. *(5 marks)*

c The diagram shows a common seal, *Phoca vitulina*, an aquatic mammal.

The seal comes to the surface of the water to obtain air and it can then stay underwater for over 20 minutes.

The figure shows a seal at the surface of the water and the following diagram shows the same animal when submerging again.

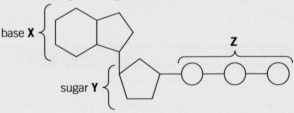

Suggest how the seal is adapted to respire for such a long time underwater. *(3 marks)*

OCR F214 2010

3 The diagram represents a molecule of ATP.

a (i) Name the parts of the ATP molecule labelled **X**, **Y**, and **Z**. *(3 marks)*

(ii) With reference to the figure, describe and explain the role of ATP in the cell. *(3 marks)*

b The electron micrograph shows a mitochondrion from an animal cell (×31 400 magnification).

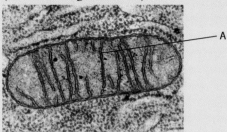

    A

(i) Name the structure labelled **A**.

(1 mark)

(ii) Name the specific process that is carried out by structure **A** in the mitochondrion. *(1 mark)*

c Some animals conserve energy by entering a state of torpor (a short period of dormancy), in which they allow their body temperature to fall below normal for a number of hours.

In an investigation into torpor in the Siberian hamster, *Phodopus sungorus*, the animal's respiratory quotient (RQ) was measured before and during the period of torpor.

The respiratory quotient is determined by the following equation:

$$RQ = \frac{\text{volume of carbon dioxide produced}}{\text{volume of oxygen consumed in the same time}}$$

RQ values for different respiratory substrates have been determined and are shown in the table.

substrate	RQ
carbohydrate	1.0
lipid	0.7
protein	0.9

Initially, the RQ value determined for the hamster was 0.95, but as the period of torpor progressed, its RQ value decreased to 0.75.

What do these values suggest about the substrates being respired by the hamster during the period of investigation? *(3 marks)*

OCR F214 2010

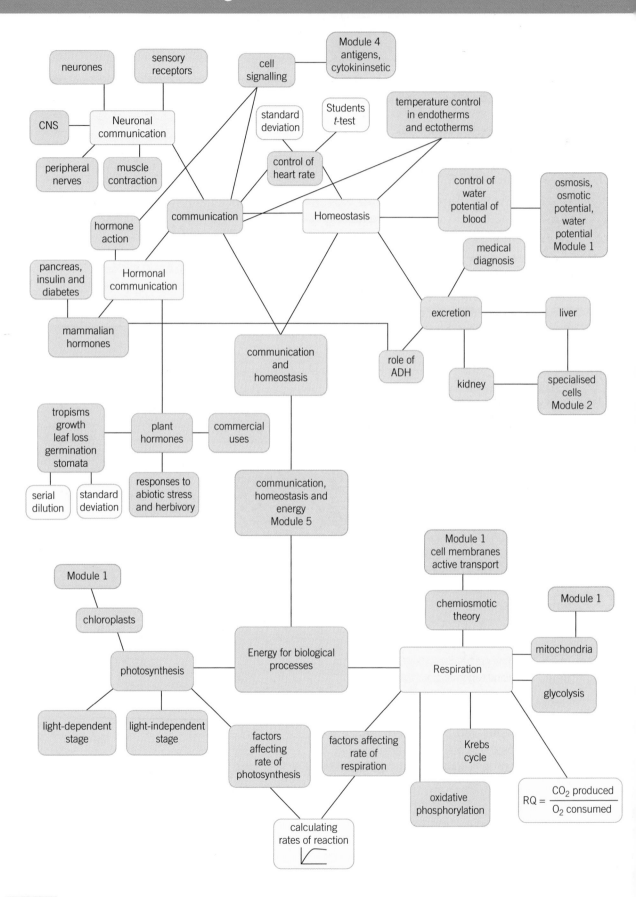

Application

A major drawback of many brain imaging techniques is that individuals need to keep at least their heads completely still. In quantitative electroencephalograms (QEEGs) sensors are attached to the outside of the skull to measure the activity of the brain as people carry out different actions. It allows scientists to build up brain maps indicating which areas are used in different activities and skills. QEEGs are not as spatially accurate as fMRI can be, however, combined with other types of brain imaging they are increasing our knowledge of how the brain works.

Recent work using QEEGs to look at the changes in the brain in children affected by autism has produced some interesting findings, and a new form of therapy. They have used QEEG to show the patterns of different brain waves in autistic brains.

While QEEGs show that the activity in some regions is particularly high, the overall rate of brain activity in people affected by autism has been shown to be lower than that in unaffected people. In fact, levels of brain activity are highest in anxious people, and lowest in people with traumatic brain injuries, but autistic patients are not much above them.

A whole range of therapies is used to help people with autism cope with everyday life. A new tool is the use of neurofeedback training, which uses information from

QEEGs to help children and adults 'retrain' their brains and control the levels of activity in different regions. There is growing evidence that for some people affected by autism this can enable them to function far more effectively and interact successfully with the people and the world around them.

1 a Far less is understod about the brain than, for example, the heart or kidney. Suggest why our understanding of the brain lags behind some other organs.
  b How have we found out what happens in the brain?
2 a What is a feedback system? Explain how they work and give examples from the various control and communication systems in the body.
  b Investigate what is meant by a neurofeedback system and discuss how this might be used to help people retrain their brains to work in different ways.
3 When QEEGs and other recordings of brain activity are taken it is often noted whether the eyes are open or closed.
  a Summarise how information from open eyes reaches the brain.
  b Suggest why it is important to record whether the eyes are open or closed.

Extension

Either investigate the main methods used for investigating the brain and make a table or poster to summarise the technology and the information it gives about the structure and function of the brain

OR investigate our current understanding of autism, including the areas and activity of the brain most

affected, the impact on functioning and examples of therapies used to help affected individuals, including at least one both drug-based intervention and neuro-feedback based on QEEGs.

Module 5 practice questions

1 Respiratory inhibitors acting on different structures within a mitochondrion were used to determine the sequence of events in respiration.

Basal respiration or basal metabolic rate (BMR) is the minimum rate of metabolism needed to maintain basic body functions.

a Name two essential functions that require energy. (*2 marks*)

Proton leak respiration is a measure of the energy (released by electron transfer) lost as protons leak through the membrane at points other than ATP synthase.

b Explain why isolated mitochondria still undergo a slow rate respiration in the absence of ADP. (*2 marks*)

ATP linked respiration is a measure of the proportion of the energy released in respiration that is used in the production of ATP in mitochondria.

The reserve capacity is the quantity of additional ATP that can be produced by oxidative phosphorylation when there is an increase in energy demand.

The human mitochondrial genome exclusively encodes 13 of the essential subunits of the ETC and ATP synthase.

c (i) State the name of the molecules coded for by the mitochondrial genome. (*1 mark*)

(ii) Explain the term genome. (*2 marks*)

(iii) Suggest why the mitochondrial reserve capacity reduces with age. (*2 marks*)

The diagram shows the changes in mitochondrial respiration after the addition of different inhibitors.

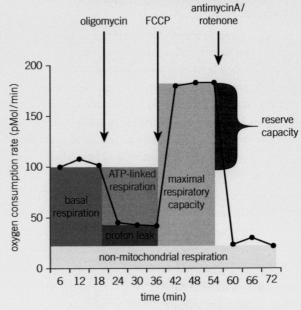

Oligomycin inhibits the final protein carrier in the electron transport chain.

d Describe the effect of Oligomycin. (*2 marks*)

Carbonyl cyanide-p-trifluoromethoxyphenyl-hydrazon (FCCP) is a protonophore, a lipid soluble molecule that transports ions across phospholipid bilayers.

e (i) Explain, with reference to the structure of the phospholipid bilayer, why protons can only pass through cell membranes by facilitated diffusion or active transport. (*4 marks*)

(ii) Describe the effect of FCCP. (*3 marks*)

f Name the process(es) involved and describe how ATP is produced in non-mitochondrial respiration. (*5 marks*)

g Describe how the following are calculated using the diagram. (*5 marks*)

Basal respiration

ATP-linked respiration

Proton leak respiration

Maximal respiratory capacity

Mitochondrial reserve capacity

Antimycin A and rotenone inhibit two more protein carriers in the electron transport chain which stops the flow of electrons.

h State what can be calculated after Antimycin A and rotenone have been added. (*2 marks*)

2 A short time after the death of a human, or animal, the joints within the organism become locked in place. This process is called rigor mortis and begins with the muscles partially contracting.

The graph shows how the amount of ATP changes in the time after death.

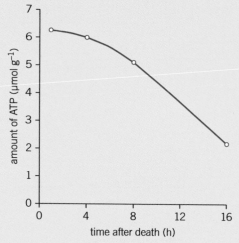

a (i) Describe the changes in ATP shown in the graph. (*4 marks*)

(ii) Explain why the amount of ATP falls after death. (*2 marks*)

(iii) Suggest the reason for the onset of rigor mortis. (*3 marks*)

(iv) Suggest why a small amount of muscle contraction can occur after death. (*3 marks*)

The graph shows how the pH changes in muscle tissue after death.

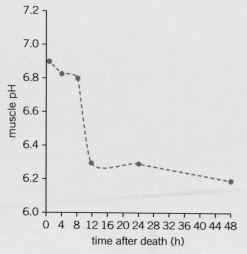

b Suggest a reason for the change in pH in muscle tissue after death. (*2 marks*)

Depending on various factors such as the temperature of the environment, rigor mortis can last up to about three days.

c (i) Suggest an explanation for why rigor mortis is only temporary. (*4 marks*)

(ii) Suggest, with explanations, which factors affect how long rigor mortis lasts. (*4 marks*)

The diagram shows a part of the sliding filament theory.

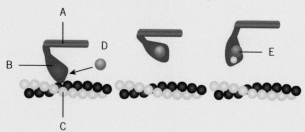

d (i) State the names of the structures A, B, and C (*3 marks*)

(ii) State the name of the molecule labelled D. (*1 mark*)

(iii) Roughly draw the diagram, and indicate, using an arrow, in which direction structure A moves relative to structure C. (*1 mark*)

(iv) Suggest, using your knowledge of the structure of proteins, the role of the ion labelled E in muscle contraction.
(*3 marks*)

(v) Outline how myosin acts as a molecular ratchet in the sliding filament theory. (*3 marks*)

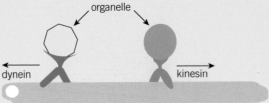

Kinesin is a motor protein that has a structure very similar to myosin. Dynein is another motor protein with a very different structure to myosin.

e Describe what determines the 3D structure of a protein and how this differs in kinesin and dynein. (*3 marks*)

Both dynein and kinesin have regions that binds to microtubules, a region that binds and hydrolyses ATP and a region that binds to an organelle.

The energy released during the hydrolysis of ATP changes their three-dimensional structures resulting in the motor proteins moving along microtubules and transporting organelles within cells. Dynein and kinesin move in opposite directions.

f (i) Compare and contrast the ways in which dyneins, kinesins, and myosin lead to movement. (*4 marks*)

(ii) State two other processes, other than the movement of organelles, in which microtubules also have an important role. (*2 marks*)

3 The chemical structure of the hormone thyroxine, which is lipid soluble, is shown in the diagram.

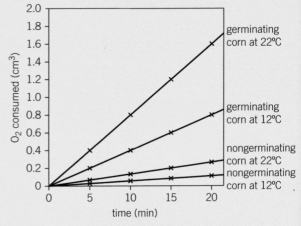

a Explain why secondary messengers are not required for thyroxine to produce a response within target cells. (*4 marks*)

b Suggest why hormones, like thyroxine, travel in the blood stream bound to proteins. (*2 marks*)

4 The oxygen consumption of germinating and non-germinating corn seeds was measured over a period of 20 minutes.

The results are shown in the graph.

a Describe the relationship between temperature and consumption of oxygen. (*3 marks*)

b Calculate the rate of oxygen consumption for germinating and non-germinating corn at 22°C. (*2 marks*)

c Discuss whether or not non-germinating seeds respire. (*4 marks*)

5 a The rate of photosynthesis is determined mainly by environmental limiting factors. These are light intensity, the availability of carbon dioxide, and temperature. Water supply has indirect effects by influencing the availability of carbon dioxide.

(i) Define the term limiting factor. *(1 mark)*

(ii) Explain how water shortage could have an indirect effect on photosynthesis by influencing the availability of carbon dioxide. *(2 marks)*

b Some plants, such as wood sorrel, and Oxalis acetosella, nearly always grow in shade where light intensity is commonly a limiting factor for photosynthesis. They are known as shade plants.

Plants that live in open habitats, for example the daisy, Bellis perennis, are called sun plants.

The graph shows the net rate of photosynthesis of sun and shade plants in response to increasing light intensity. The net rate of photosynthesis is defined as:

mass of CO_2 fixed in photosynthesis minus mass of CO_2 produced in respiration, per unit time

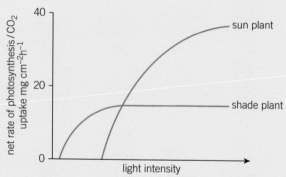

Using the graph, describe the responses of sun and shade plants to increasing light intensity. *(2 marks)*

c Leaves of wood sorrel have been shown to have very low respiration rates per unit leaf area.

The diagram shows sections through the leaves of typical sun and shade plants.

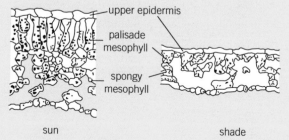

(i) With reference to the diagram, suggest why the leaves of wood sorrel have low rates of respiration. *(1 mark)*

(ii) Explain why a low rate of respiration in leaves is an adaptation to low light intensities. *(3 marks)*

d A large number of seedlings of common orache, Atriplex patula, were grown for several weeks in different environmental conditions as follows:

● group 1 – high light intensities

● group 2 – intermediate light intensities

● group 3 – low light intensities.

When fully grown, the net rate of photosynthesis at different light intensities was measured for each group. The results obtained are shown in the graph.

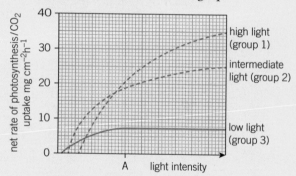

(i) State the net rate of photosynthesis for plants in groups 1 and 3 at light intensity A on the graph. *(2 marks)*

(ii) Suggest how adaptation to light intensity is controlled in A. patula. *(3 marks)*

OCR 2804 2008

165

MODULE 6
Genetics, evolution, and ecosystems

Introduction

In this module you will learn how genes regulate and control cell function. This will lead on to the study of heredity, evolution, and DNA manipulation. Finally, you will consider the impacts of human activities on the natural environment and biodiversity.

Genetics of living systems introduces how the genetic control of metabolic reactions determines an organism's growth, development, and function. This also includes looking at the effects of gene mutations on protein function.

Patterns of inheritance and variation allows you to study how genetic and environmental factors contribute to variation within a population. Over a prolonged period, organisms can change so significantly that new species have formed, whereas others have become extinct. Humans can use artificial selection to produce similar changes in plants and animals.

Manipulating genomes has many potential benefits such as the treatment of disease but the implications of genetic techniques are of public debate. You will find out how genomes are sequenced as well as how DNA profiling is used in forensics and to determine the risk of certain diseases.

Cloning and biotechnology explores how farmers and growers exploit natural vegetative propagation in the production of uniform crops, as well as the role of scientists in the production of artificial plant and animal clones. The use of microorganisms in biotechnology to produce food, drugs, and other products is also studied.

Ecosystems are dynamic and influenced by both biotic and abiotic factors. You will study the complex interactions which occur between organisms and their environment as well as finding out how materials are passed on and recycled.

Populations and sustainability investigates the factors that determine population size and the economic, social, and ethical reasons why ecosystems may need to be managed. This includes looking at how biological resources can be used sustainably to support an increasing human population.

Knowledge and understanding checklist

From your Key Stage 4 study you should be able to answer the following questions. Work through each point, using your Key Stage 4 notes and the support available on Kerboodle.

☐ Describe the process of mitosis in growth, including the cell cycle.

☐ Explain how some abiotic and biotic factors affect communities.

☐ Describe the importance of interdependence and competition in a community.

☐ Explain the role of microorganisms in the cycling of materials through an ecosystem.

☐ Describe human interactions within ecosystems and explain their impact on biodiversity.

☐ Explain some of the benefits and challenges of maintaining local and global biodiversity.

☐ Explain the following terms – genome, gamete, chromosome, gene, allele/variant, dominant, recessive, homozygous, heterozygous, genotype, and phenotype.

☐ Explain single gene inheritance and predict the results of single gene crosses.

☐ State that there is usually extensive genetic variation within a population of a species.

☐ Explain the impact of the selective breeding of food plants and domesticated animals.

☐ Describe the main steps in the process of genetic engineering.

Maths skills checklist

In this module, you will need to use the following maths skills. You can find support for these skills on Kerboodle and through MyMaths.

☐ **Ratios.** You will need to use phenotypic ratios to identify linkage and epistasis.

☐ **Chi-squared test.** You will need to use the chi-squared test to determine the significance of the difference between observed and expected results.

☐ **Hardy–Weinberg principle.** You will need to use the Hardy–Weinberg principle to calculate allele frequencies in populations.

☐ **Correlation coefficient.** You will need to use this test to consider the relationship between two sets of data. This will determine if and how the data are correlated.

MyMaths.co.uk
Bringing Maths Alive

Gene mutations (when DNA goes wrong)

A **mutation** is a change in the sequence of bases in DNA. Protein synthesis can be disrupted if the mutation occurs within a gene. The change in sequence is caused by the **substitution**, **deletion**, or **insertion** of one or more nucleotides (or base pairs) within a gene. If only one nucleotide is affected it is called a point mutation.

The substitution of a single nucleotide changes the codon in which it occurs. If the new codon codes for a different amino acid this will lead to a change in the primary structure of the protein. The degenerate nature of the genetic code may mean however that the new codon still codes for the same amino acid leading to no change in the protein synthesised.

The position and involvement of the amino acid in R group interactions within the protein will determine the impact of the new amino acid on the function of the protein. For example, if the protein is an enzyme and the amino acid plays an important role within the active site, then the protein may no longer act as a biological catalyst.

The insertion or deletion of a nucleotide, or nucleotides, leads to a frameshift mutation (Figure 1). The triplet code means that sequences of bases are transcribed (or read) consecutively in non-overlapping groups of three. This is the reading frame of a sequence of bases. Each group of three bases corresponds to one amino acid. The addition or deletion of a nucleotide moves, or shifts, the reading frame of the sequence of bases. This will change every successive codon from the point of mutation.

The same effect is seen however many nucleotides are added or deleted, unless the number of nucleotides changed is a multiple of three. Multiples of three correspond to full codons and therefore the reading frame will not be changed – but the protein formed will still be affected as a new amino acid is added.

normal DNA sequence: | AGT | CGA | TAG |
codon 1 codon 2 codon 3

point mutations:

base substitution (A substituted instead of C in codon 2): | AGT | AGA | TAG |
codon 1 codon 2 codon 3

frameshift mutations:
insertion (T inserted before G in codon 1): | ATG | TCG | ATA |
codon 1 codon 2 codon 3

deletion (G deleted from codon 1): | ATC | GAT | AGC |
codon 1 codon 2 codon 3

▲ Figure 1 *The diagram shows the effect of different point mutations on codons and sequences of codons*

Effects of different mutations

- No effect – there is no effect on the phenotype of an organism because normally functioning proteins are still synthesised.

- Damaging – the phenotype of an organism is affected in a negative way because proteins are no longer synthesised or proteins synthesised are non-functional. This can interfere with one, or more, essential processes.

- Beneficial – very rarely a protein is synthesised that results in a new and useful characteristic in the phenotype. For example, a mutation in a protein present in the cell surface membranes of human cells means that the human immunodeficiency virus (HIV) cannot bind and enter these cells. People with this mutation are immune to infection from HIV.

Causes of mutations

Mutations can occur spontaneously, often during DNA replication, but the rate of mutation is increased by **mutagens**. A mutagen is a chemical, physical, or biological agent which causes mutations (Table 1).

The loss of a purine base (depurination) or a pyrimidine base (depyrimidination) often occurs spontaneously. The absence of a base can lead to the insertion of an incorrect base through complementary base pairing during DNA replication.

Free radicals, which are oxidising agents, can affect the structures of nucleotides and also disrupt base pairing during DNA replication. Antioxidants, such as vitamins A, C, and E (found in fruit and vegetables), are known anticarcinogens because of their ability to negate the effects of free radicals.

▼ Table 1 *Summary of some of the main mutagens and what they do*

Physical mutagens	ionizing radiations such as X-rays	break one or both DNA strands – some breaks can be repaired but mutations can occur in the process
Chemical mutagens	deaminating agents	chemically alter bases in DNA such as converting cytosine to uracil in DNA, changing the base sequence
	alkylating agents	methyl or ethyl groups are attached to bases resulting in the incorrect pairing of bases during replication
Biological agents	base analogs	incorporated into DNA in place of the usual base during replication, changing the base sequence
	viruses	viral DNA may insert itself into a genome, changing the base sequence

Effects of mutation

Silent mutations

The vast majority of mutations are silent (or neutral) which means they do not change any proteins, or the activity of any proteins synthesised. Therefore, they have no effect on the phenotype of an organism. They can occur in the non-coding regions of DNA (introns) or code for the same amino acid due to the degenerate nature of the genetic code. They may also result in changes to the primary structure but do not change the overall structure or function of the proteins synthesised.

Nonsense mutations

Nonsense mutations result in a codon becoming a stop codon instead of coding for an amino acid. The result is a shortened protein being synthesised which is normally non-functional. These mutations normally have negative or harmful effects on phenotypes.

Missense mutations

Missense mutations result in the incorporation of an incorrect amino acid (or amino acids) into the primary structure when the protein is synthesised. The result depends on the role the amino acid plays in the structure and therefore function of the protein synthesised. The mutation could be silent, beneficial, or harmful. A conservative mutation occurs when the amino acid change leads to an amino acid being coded for which has similar properties to the original, this means the effect of the

mutation is less severe. In contrast, a non-conservative mutation is when the new amino acid coded for has different properties to the original, this is more likely to have an effect on protein structure, and may cause disease.

> 1   Different mutations have a range of effects on proteins synthesised. These are categorised as follows:
> Amorph – mutation that results in the loss of function of a protein
> Hypomorph – mutation that results in a reduction of function of a protein
> Hypermorph – mutation that results in a gain in function of a protein.
> Suggest whether amorphic mutations result in dominant or recessive alleles. Give the reason(s) for your answer.

▼ **Table 2** *Point mutations can be silent, nonsense, or missense depending on their effect on the primary structure of the protein coded for by the gene*

Level where effect occurs	No mutations	Point mutations			
		Silent	Nonsense	Missense	
				conservative	non-conservative
DNA level	TTC	TTT	ATC	TCC	TGC
mRNA level	AAG	AAA	UAG	AGG	ACG
protein level	**Lys**	Lys	STOP	Arg	Thr

Beneficial mutations

The ability to digest lactose, the sugar present in milk, is thought to be the result of a relatively recent mutation. The majority of mammals in the world become lactose intolerant after they cease to suckle. The ability to digest lactose is found primarily in European populations who are more likely to farm cattle. The ability to drink milk and process lactose as an adult helps prevent diseases such as osteoporosis – this could also have prevented individuals with the mutation from starving during famines. This mutation appears to have arisen spontaneously more than once – it has also been found in people in East Africa.

Sickle-cell anaemia

Sickle-cell anaemia is a blood disorder where erythrocytes develop abnormally. The disorder is cause by a mutation in the gene coding for haemoglobin. There is a substitution of just one base. Thymine replaces adenine, making the sixth amino acid valine instead of glutamic acid on the beta haemoglobin chain.

Glutamic acid is a hydrophilic amino acid but valine is a hydrophobic amino acid. When the partial pressure of oxygen is low, and haemoglobin is dissociated from oxygen, the hydrophobic valine amino acids bind to hydrophobic regions on adjacent haemoglobin molecules.

The aggregation of haemoglobin molecules deforms the shape of erythrocytes causing them to become sickle shaped. The erythrocytes are less flexible and have difficulty moving through capillaries resulting in reduced oxygen delivery to tissues, which causes anaemia.

In homozygous individuals, there are two copies of the mutant alleles present. Heterozygous individuals only get mild symptoms of the condition, but are resistant to malaria. Hundreds of millions of people get malaria each year with over a million fatal cases.

1 Evaluate the benefits of being heterozygous for sickle cell anaemia.

2 Explain how the change of one amino acid in haemoglobin could reduce the oxygen-carrying ability of the blood.

Chromosome mutations

Gene mutations occur in single genes or sections of DNA whereas chromosome mutations affect the whole chromosome or number of chromosomes within a cell. They can also be caused by mutagens and normally occur during meiosis. As with gene mutations, the mutations can be silent but often lead to developmental difficulties.

Changes in chromosome structure include:

- **Deletion** – a section of chromosome breaks off and is lost within the cell.
- Duplication – sections get duplicated on a chromosome.
- **Translocation** – a section of one chromosome breaks off and joins another non-homologous chromosome.
- Inversion – a section of chromosome breaks off, is reversed, and then joins back onto the chromosome.

Synoptic link

To remind yourself of the process of meiosis, look back at Topic 6.3, Meiosis.

Summary questions

1 The development of lactose tolerance is thought to have spread over approximately 20 000 years, which in evolutionary terms is very quick.

Explain why the percentage of adults with the ability to digest lactose increased at such a rate.

2 Outline why the majority of mutations do not have an influence phenotype. (*2 marks*)

3 Discuss why beneficial mutations are rare and suggest a process that beneficial mutations underpin. (*4 marks*)

Gene regulation

Enzymes which are necessary for reactions present in metabolic pathways like respiration are constantly required, and the genes that code for these are called housekeeping genes. Protein-based hormones (required for the growth and development of an organism or enzymes) are only required by certain cells at certain times to carry out a short-lived response. They are coded for by tissue-specific genes.

The entire genome of an organism is present in every prokaryotic cell, or eukaryotic cell that contains a nucleus. This includes genes not required by that cell so the expression of genes and the rate of synthesis of protein products like enzymes and hormones has to be regulated. Genes can be turned on or off, and the rate of product synthesis increased or decreased depending on demand.

Bacteria are able to respond to changes in the environment because of gene regulation. Expressing genes only when the products are needed also prevents vital resources being wasted.

Gene regulation is fundamentally the same in both prokaryotes and eukaryotes. However, the stimuli that cause changes in gene expression and the responses produced are more complex in eukaryotes. Multicellular organisms not only have to respond to changes in the external environment but also the internal environment. Gene regulation is required for cells to specialise and work in a coordinated way.

There are a number of different ways in which genes are regulated, categorised by the level at which they operate:

- Transcriptional – genes can be turned on or off
- Post-transcriptional – mRNA can be modified which regulates translation and the types of proteins produced
- Translational – translation can be stopped or started
- Post-translational – proteins can be modified after synthesis which changes their functions.

Transcriptional control

There are a number of mechanisms that can affect the transcription of genes.

Chromatin remodelling

As you have already learnt, DNA is a very long molecule and has to be wound around proteins called histones in eukaryotic cells, in order to be packed into the nucleus of a cell. The resulting DNA/protein complex is called a chromatin.

Heterochromatin is tightly wound DNA causing chromosomes to be visible during cell division whereas **euchromatin** is loosely wound DNA present during interphase. The transcription of genes is not possible when DNA is tightly wound because RNA polymerase cannot access the genes. The genes in euchromatin, however, can be freely transcribed. Protein synthesis does not occur during cell division but during interphase between cell divisions. This is a simple form of regulation that ensures the proteins necessary for cell division are synthesised in time. It also prevents the complex and energy-consuming process of protein synthesis from occurring when cells are actually dividing.

Histone modification

DNA coils around histones because they are positively charged and DNA is negatively charged. Histones can be modified to increase or decrease the degree of packing (or condensation).

The addition of acetyl groups (**acetylation**) or phosphate groups (phosphorylation) reduces the positive charge on the histones (making them more negative) and this causes DNA to coil less tightly, allowing certain genes to be transcribed. The addition of methyl groups (**methylation**) makes the histones more hydrophobic so they bind more tightly to each other causing DNA to coil more tightly and preventing transcription of genes.

Epigenetics is a term that is increasingly used to describe this control of gene expression by the modification of DNA. It is sometimes used to include all of the different ways in which gene expression is regulated.

Lac operon

An **operon** is a group of genes that are under the control of the same regulatory mechanism and are expressed at the same time. Operons are far more common in prokaryotes than eukaryotes owing to the smaller and simpler structure of their genomes. They are also a very efficient way of saving resources because if certain gene products are not needed, then all of the genes involved in their production can be switched off.

Glucose is easier to metabolise and is the preferred respiratory substrate of *Escherichia coli* and many other bacteria. If glucose is in short supply, lactose can be used as a respiratory substrate. Different enzymes are needed to metabolise lactose.

The **lac operon** is a group of three genes, lacZ, lacY, and lacA, involved in the metabolism of lactose. They are **structural genes** as they code for three enzymes (β-galactosidase, lactose permease, and transacetylase) and they are transcribed onto a single long molecule of mRNA. A **regulatory gene**, lacI, is located near to the operon and codes for a **repressor protein** that prevents the transcription of the structural genes in the absence of lactose (Figure 1).

Synoptic link

You learnt about the structure of DNA in Topic 3.9, DNA replication and the genetic code and in Topic 6.1, Cell cycle.

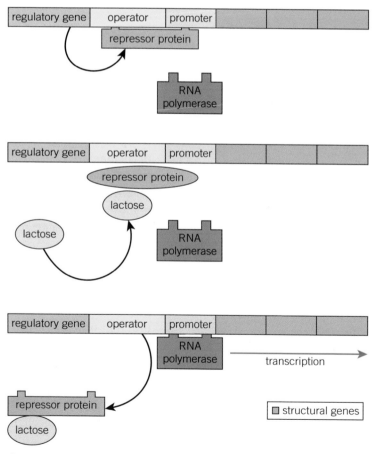

▲ **Figure 1** *A repressor protein coded for by a regulatory gene binds to the operator and prevents the binding of RNA polymerase, preventing transcription of the structural genes*

The repressor protein is constantly produced and binds to an area called the operator, which is also close to the structural genes. The binding of this protein prevents RNA polymerase binding to DNA and beginning transcription. This is called down regulation. The section of DNA that is the binding site for RNA polymerase is called the promoter.

When lactose is present, it binds to the repressor protein causing it to change shape so it can no longer bind to the operator. As a result RNA polymerase can bind to the promoter, the three structural genes are transcribed, and the enzymes are synthesised.

Role of cyclic AMP

The binding of RNA polymerase still only results in a relatively slow rate of transcription that needs to be increased or up-regulated to produce the required quantity of enzymes to metabolise lactose efficiently. This is achieved by the binding of another protein, cAMP receptor protein (CRP), that is only possible when CRP is bound to cAMP (a secondary messenger that you are already familiar with).

The transport of glucose into an *E. coli* cell decreases the levels of cAMP, reducing the transcription of the genes responsible for the metabolism of lactose. If both glucose and lactose are present then it will still be glucose, the preferred respiratory substrate, that is metabolised.

Post-transcriptional/pre-translational control

RNA processing

The product of transcription is a precursor molecule, **pre-mRNA**. This is modified forming **mature mRNA** before it can bind to a ribosome and code for the synthesis of the required protein.

A cap (a modified nucleotide) is added to the 5′ end and a tail (a long chain of adenine nucleotides) is added to the 3′ end. These both help to stabilise mRNA and delay degradation in the cytoplasm. The cap also aids binding of mRNA to ribosomes. Splicing also occurs where the RNA is cut at specific points – the introns (non-coding DNA) are removed and the exons (coding DNA) are joined together. Both processes occur within the nucleus.

RNA editing

The nucleotide sequence of some mRNA molecules can also be changed through base addition, deletion, or substitution. These have the same effect as point mutations and result in the synthesis of different proteins which may have different functions. This increases the range of proteins that can be produced from a single mRNA molecule or gene.

Translational control

These following mechanisms regulate the process of protein synthesis:

- degradation of mRNA – the more resistant the molecule the longer it will last in the cytoplasm, that is, a greater quantity of protein synthesised.

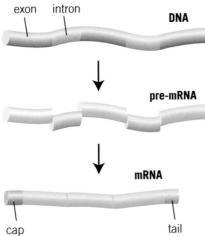

▲ **Figure 2** *This diagram summarises the modification of a pre-mRNA molecule*

- binding of inhibitory proteins to mRNA prevents it binding to ribosomes and the synthesis of proteins.

- activation of initiation factors which aid the binding of mRNA to ribosomes (the eggs of many organisms produce large quantities of mRNA which is not required until after fertilisation, at which point initiation factors are activated).

Protein kinases

Protein kinases are enzymes that catalyse the addition of phosphate groups to proteins. The addition of a phosphate group changes the tertiary structure and so the function of a protein. Many enzymes are activated by phosphorylation. Protein kinases are therefore important regulators of cell activity. Protein kinases are themselves often activated by the secondary messenger cAMP.

Post-translational control

Post-translational control involves modifications to the proteins that have been synthesised. This includes the following:

- addition of non-protein groups such as carbohydrate chains, lipids, or phosphates

- modifying amino acids and the formation of bonds such as disulfide bridges

- folding or shortening of proteins.

- modification by cAMP – for example, in the lac operon cAMP binds to the cAMP receptor protein increasing the rate of transcription of the structural genes.

Summary questions

1 The lac operon if often referred to as being 'leaky' meaning that it is still transcribed to a limited extent even in the absence of lactose.
   a Using your knowledge of how the lac operon works, explain why this is necessary. (3 marks)
   b Suggest the functions of β-galactosidase and lactose permease synthesised by the lac operon. (3 marks)

2 Another example of gene regulation in prokaryotes is the trp operon. This operon codes for the production of tryptophan, an essential amino acid for the bacterium *E. coli*. When tryptophan is available in the environment the structural genes in the trp operon are not expressed.

   Suggest a mechanism for the genetic regulation of this operon. (5 marks)

3 Using your knowledge of enzymes, explain how enzyme cofactors could play a role in gene regulation. (4 marks)

19.3 Body plans

Specification reference: 6.1.1

▲ **Figure 2** *False-colour scanning electron micrograph of a mutant fruit fly,* Drosophila melanogaster, *with four wings. The normal fly has one pair of wings. Fruit flies have been used for many years in genetic research because it is easy to raise in large numbers in the laboratory, reproduces rapidly, and many of its mutations are easy to see under a low-powered microscope, approx ×9.8 magnification*

Synoptic link

You learnt about evolution in Chapter 10, Classification and evolution.

Body plans

Living organisms come in all shapes and sizes from tulips to mosquitoes to humans. It is the same small group of genes, however, that control the growth and development of these vastly different living forms.

▲ **Figure 1** *Coloured scanning electron micrograph of the head of a mutant fruit fly* Drosophila melanogaster *with leg antennae, approx ×58 magnification*

The regulation of the pattern of anatomical development is called *morphogenesis*.

These genes were discovered by scientists investigating strange mutations observed in fruit flies such as legs on the head in place of antennae or extra pairs of wings. Fruit flies are small flies belonging to the genus *Drosophila* that feed and reproduce on rotting fruit. They are small, easy to keep, and have a short life cycle so have always been a popular choice for use in genetic studies.

Homeobox genes

Homeobox genes are a group of genes which all contain a homeobox. The homeobox is a section of DNA 180 base pairs long coding for a part of the protein 60 amino acids long that is highly conserved (very similar) in plants, animals, and fungi. This part of the protein, a *homeodomain*, binds to DNA and switches other genes on or off. Therefore, homeobox genes are regulatory genes.

The common ancestor of the mouse and human is thought to have lived about 60 million years ago. Mutations have been accumulating ever since and evolution has led to two very different organisms. Many of the homeobox genes present in the mouse and human, however, still have identical nucleotide sequences.

▲ **Figure 3** *This two-headed calf is an example of what can go wrong when there is a mutation in a regulatory Hox gene. Although this calf has two heads, it only has one brain and so the heads react simultaneously*

Pax6 is one of the homeobox genes. When mutated it causes a form of blindness (due to underdevelopment of the retina) in humans. Mice and fruit flies also have this gene and disruption of the gene causes blindness in these organisms as well. These findings suggest that Pax6 is a gene involved in the development of eyes in all three species.

Hox genes

Hox genes (often used interchangeably with homeobox genes) are one group of homeobox genes that are only present in animals. They are responsible for the correct positioning of body parts. In animals the Hox genes are found in gene clusters – mammals have four clusters on different chromosomes.

The order in which the genes appear along the chromosome is the order in which their effects are expressed in the organism. Human beings have 39 Hox genes in total that are all believed to have arisen from one ancient homeobox gene by duplication and accumulated mutations over time.

The layout of living organisms

Body plans are usually represented as cross-sections through the organism showing the fundamental arrangement of tissue layers. *Diploblastic* animals have two primary tissue layers and *triploblastic* animals have three primary tissue layers.

A common feature of animals is that they are segmented, that is, the rings of a worm or the less obvious back bone of vertebrates. These segments have multiplied over time and are specialised to perform different functions. Hox genes in the head control the development of mouthparts and Hox genes in the thorax control the development of wings, limbs, or ribs.

The individual vertebrae and associated structures have all developed from segments in the embryo called somites. The somites are directed by Hox genes to develop in a particular way depending on their position in the sequence.

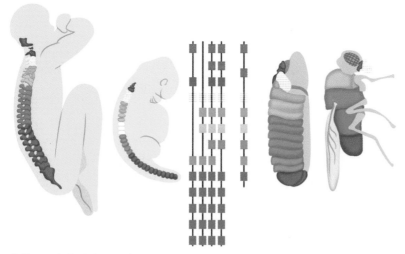

▲ **Figure 4** *Body layout from head to tail in both flies and human beings is controlled by Hox genes. The fly has a set of eight Hox genes but human beings have four comparable sets*

Symmetry

The body shape of most animals shows symmetry.

- Radial symmetry is seen in diploblastic animals like jellyfish. They have no left or right sides, only a top and a bottom.
- Bilateral symmetry which is seen in most animals means the organisms have both left and right sides and a head and tail rather than just a top and bottom.
- Asymmetry is seen in sponges which have no lines of symmetry.

Mitosis and apoptosis

Mitosis, which results in cell division and proliferation, and apoptosis (Figure 6), which is programmed cell death, are both essential in shaping organisms.

The role of mitosis is to increase the number of cells leading to growth. The role of apoptosis is not so immediately obvious. Consider how a sculptor works to give shape to a block of wood or stone. The shape is revealed as material is removed bit by bit. This is one of the ways in which apoptosis shapes different body parts – by removing unwanted cells and tissues. Cells undergoing apoptosis can also release chemical signals which stimulate mitosis and cell proliferation leading to the remodelling of tissues. A sculptor working with clay will often add and remove clay during the reshaping process. Hox genes regulate both mitosis and apoptosis.

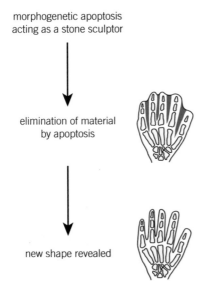

morphogenetic apoptosis acting as a stone sculptor

↓

elimination of material by apoptosis

↓

new shape revealed

▲ **Figure 5** *The development of the hand is an example of morphogenetic apoptosis. The red area shows how apoptosis is used to sculpt the individual fingers*

Synoptic link

You learnt about mitosis in Topic 6.2, Mitosis.

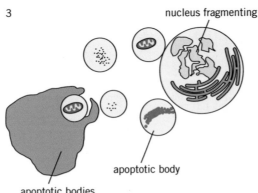

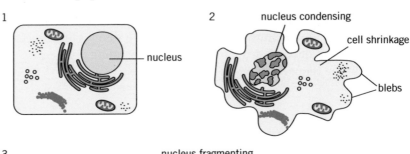

▲ **Figure 6** *The process of apoptosis*

Factors affecting the expression of regulatory genes

The expression of regulatory genes can be influenced by the environment, both internal and external. Stress can be defined as the condition produced when the homeostatic balance within an organism is upset. This can be due to external factors such as a change in temperature or intensity of light. Internal factors can change due to the release of hormones or psychological stress.

These factors will have a greater impact during the growth and development of an organism.

Drugs can also affect the activity of regulatory genes. An example of this was the drug thalidomide. Thalidomide was given to pregnant women in the 1950s and 1960s to treat morning sickness. It was later discovered that it prevented the normal expression of a particular Hox gene. This resulted in the birth of babies with shortened limbs.

Thalidomide is currently used in the treatment of some forms of cancer. The property of this drug that has previously caused problems during pregnancy is being exploited to stop the development of some tumours. It is believed that thalidomide prevents the formation of networks of capillaries which are necessary for some tumours to grow and develop.

▲ **Figure 7** *A girl with birth defects caused by the drug thalidomide. This drug was given to pregnant women in the 1950s and 1960s to treat morning sickness, until it was discovered to cause severe malformations in babies*

 ## Ontology doesn't mimic phylogeny

The theory of recapitulation states that as organisms develop from fertilised egg to embryo they repeat the evolutionary process that they have been through. In biology, this can be summarised by the phrase ontology (the development of an organism) mimics phylogeny (the evolutionary history of an organism).

This theory is not accepted by modern biologists but a modern theory states that 'oncology (the study of cancer) recapitulates ontology'. This refers to the discovery that genes originally expressed in the development of the embryo are expressed again by cancerous cells.

Use your knowledge of Hox genes to suggest their possible roles in the development of certain tumours.

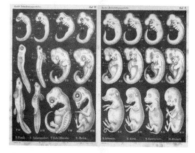

▲ **Figure 8** *The drawing above was made in the 19th century as evidence supporting this theory.*

Summary questions

1 Explain, with reference to their body shape, why human beings are referred to as bilaterally symmetrical but jellyfish are radially symmetrical. *(3 marks)*

2 The Hox gene Pax6 is necessary for the normal development of the retina in humans. A mutation in this gene can lead to blindness. Pax6 mutations also cause blindness in mice and fruit flies. Describe how scientists could have tested the idea that Pax6 plays a role in eye development in all three species. *(5 marks)*

3 Consider the statement.

   All Hox genes are homeobox genes but not all homeobox genes are Hox genes.

   Discuss the validity of this statement. *(5 marks)*

Practice questions

1 The common groundsel, *Senecio vulgaris*, is a weed that is often found in large numbers on cultivated land. It was the first plant species to develop resistance to triazine herbicides. This resistance is the result of a gene mutation in the chloroplast DNA.

Since its first appearance, triazine resistance has spread very rapidly in groundsel populations.

a Explain the rapid spread of herbicide resistance in a weed such as groundsel.
(*5 marks*)

b DNA was extracted from the chloroplasts of triazine-susceptible and triazine-resistant groundsel plants. Equivalent lengths of DNA, including the site of the mutation, were isolated from each extract and then treated with the restriction enzyme MaeI prior to electrophoresis.

The resulting electrophoresis gel, after staining the DNA, is shown in the diagram.

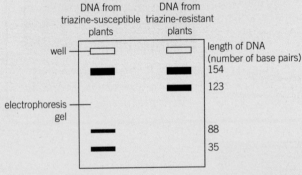

(i) State, giving a reason, whether the mutation giving resistance to triazine is a deletion, a substitution or an addition.
(*2 marks*)

(ii) Explain the difference in banding pattern between DNA from triazine-susceptible and triazine-resistant plants.
(*2 marks*)

(iii) Suggest one way in which the mutation could give resistance to triazine.
(*2 marks*)

OCR 2805/02 2010

2 a The diagram shows a family's history of Huntington's disease.

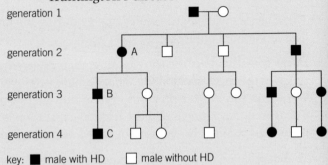

Explain how the diagram provides evidence that Huntington's disease results from the inheritance of an autosomal dominant allele.
(*3 marks*)

b Genetic screening for Huntington's disease can be carried out, using a process similar to genetic fingerprinting, to find the length of a repeated triplet, or 'stutter', in an allele. After treatment with a restriction enzyme, fragments of DNA of different lengths are separated by gel electrophoresis.

(i) Describe the role of a restriction enzyme in this technique.
(*4 marks*)

(ii) Explain why fragments of DNA of different lengths can be separated by gel electrophoresis
(*5 marks*)

c DNA from individuals A, B and C from the family shown in the diagram was analysed.

The resulting banding patterns are shown in the diagram.

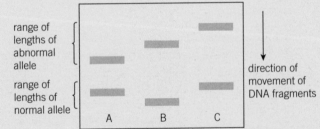

Explain why the DNA of the following bands in the diagram are not the same length for:

(i) the three normal alleles
(*1 mark*)

(ii) the three abnormal alleles.
(*1 mark*)

OCR 2805/02 2008

3 Homeobox genes show astonishing similarity across widely different species of animal, from fruit flies, which are insects, to mice and humans, which are mammals. The sequences of these genes have remained relatively unchanged throughout evolutionary history and the same genes control embryonic development in flies and mammals.

a State what is meant by a homeobox gene.
(*2 marks*)

b Homeobox genes show 'astonishing similarity across widely different species of animal'.

Explain why there has been very little change by mutation in these genes.
(*2 marks*)

c Frogs reproduce by laying eggs in water, each egg develops into a tadpole, which has external gills to extract oxygen from the water, and a tail to help it swim. The tadpole gradually changes into an adult frog as it grows. During this time its gills and tail disappear.

List two cellular processes that must occur during the development of a tadpole into a frog. (*2 marks*)

d Name another kingdom of organisms, other than animals, that have similar homeotic genes. (*1 mark*)

OCR F215 2012

4 An enhancer of a regulatory gene responsible for limb development in mammals, called the sonic hedgehog gene, is located in the intron of a neighbouring gene. This regulatory gene was first investigated using Drosophila – genetically modified flies. Without this gene the fruit flies grew small projections (known as denticles) all over their bodies.

Point mutations in the enhancer sequence for this gene result in polydactyly, extra fingers or toes.

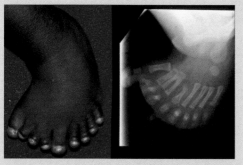

The photograph and X-ray show a child's foot with polydactyl – a deformity in which more than the usual number of digits are present. The condition is genetic in origin, and in most cases causes no harm. The extra digits are often underdeveloped (those at lower left in this case), and if removed surgically soon after birth cause no long term complications.

a Explain how an enhancer works.
(*2 marks*)

b State the meaning of the term intron.
(*1 mark*)

c Describe what is meant by a point mutation. (*2 marks*)

d Explain why the sonic hedgehog gene is an example of a Hox gene. (*2 marks*)

e The sonic hedgehog gene is usually expressed in cells close to tissue that eventually develops into the small fingers or toes.

Suggest how different rates of transcription of this gene leads to the formation of fingers with different sizes and shapes. (*3 marks*)

20 PATTERNS OF INHERITANCE AND VARIATION

20.1 Variation and inheritance

Specification reference: 6.1.2

Learning outcomes

Demonstrate knowledge, understanding, and application of:

→ the contribution of both environmental and genetic factors to phenotypic variation

→ how sexual reproduction can lead to genetic variation within a species

→ the meaning of the terms continuous and discontinuous variation and their genetic basis.

Members of different species are, usually, clearly different from each other and even members of the same species are rarely identical so variation is an important feature of living organisms. Variation arises as a result of mutations – changes to the genetic code which are random and constantly taking place. Variation is essential for the process of natural selection – and therefore evolution.

You have already learnt about how organisms within a species vary. It is important to remember that variation can occur both as a result of environmental variation and genetic variation. In the majority of cases both play a role in determining an organism's characteristics – examples of this include chlorosis in plants and the body mass of an animal.

Chlorosis

Most plants are genetically coded to produce large quantities of chlorophyll, the green pigment that is vital for photosynthesis and gives leaves their green colour. Some plants, however, suffer from a condition known as chlorosis, when the leaves look pale or yellow. This occurs because the cells are not producing the normal amount of chlorophyll. This lack of chlorophyll reduces the ability of the plant to make food by photosynthesis.

Most plants which show chlorosis have normal genes coding for chlorophyll production. The change in their phenotype is the result of environmental factors. There are many different environmental factors which cause chlorosis, each having a different effect on the physiology of the plant but causing the same change in phenotype. Examples include:

- Lack of light – for example, when a toy or gardening tool is left on a lawn. In the absence of light, plants will turn off their chlorophyll production to conserve resources. In this case, chlorosis only occurs where the plant gets no light.

- Mineral deficiencies – for example, a lack of iron or magnesium. Iron is needed as a cofactor by some of the enzymes that make chlorophyll, and magnesium is found at the heart of the chlorophyll molecule. If either of these elements are lacking in the soil, a plant simply cannot make chlorophyll and gradually all the leaves will become yellow.

- Virus infections – when viruses infect plants, they interfere with the metabolism of cells. A common symptom is yellowing in the infected tissues as they can no longer support the synthesis of chlorophyll.

▲ **Figure 1** *Normal leaf (left), leaf with chlorosis (right)*

In summary, even though genetic factors in a plant are likely to code for green leaves, the environment plays a key role in the final leaf appearance.

Animal body mass

Within a species, the body mass of individual animals varies. An organism's body mass is determined by a combination of both genetic and environmental factors. In the majority of cases dramatic variations in size such as obesity and being severely underweight are a result of environmental factors. For example, the amount (and quality) of foods eaten, the quantity of exercise which the organism gets, or the presence of disease can affect the body mass. Being extremely overweight or underweight can result in significant health problems for an animal.

▲ **Figure 2** *This severely obese cat has an extremely large body mass as a result of overfeeding and lack of exercise*

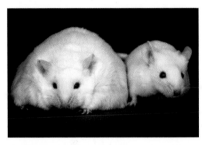

Occasionally obesity can be a result of the genetic make-up of an organism. The obese mouse in Figure 3 has a mutation on chromosome 7. This mutation causes the pattern of fat deposition in its body to be altered. Scientific studies have shown that this gene acts in conjunction with other genes that regulate the energy balance of the body, and as a result mice possessing the mutation grow 35–50% fatter by middle age than a normal mouse would.

▲ **Figure 3** *Left – obese mouse with a mutated chromosome 7. Right – normal mouse with normal chromosome 7*

Creating genetic variation

Genetic variation is created by the versions of genes you inherit from your parents. For most genes there are a number of different possible alleles or variants. The individual mixture of alleles an organism inherits influences the characteristics they will display. This combination is determined by sexual reproduction involving meiosis (the formation of gametes), and the random fusion of gametes at fertilisation. This results in the vast genetic variation seen between individuals of the same species.

Synoptic link

You learnt about genetic and environmental variation and examples of these in Topic 10.5, Types of variation. In this topic you also learnt about the detail of how sexual reproduction and meiosis lead to genetic variation.

For most genes in your body two alleles are inherited (one from each parent). These alleles may be the same or different versions of the gene. The combination of alleles an organism inherits for a characteristic is known as their **genotype** – this is the genetic make-up of an organism in respect of that gene. The observable characteristics of an organism are known as its **phenotype** (you will learn more about two special cases of inheritance known as codominance and sex linkage in Topic 20.2, Monogenetic inheritance). The actual characteristics that an organism displays are also often influenced by the environment.

▲ **Figure 4** *Doctors monitor the height of children at routine developmental check-ups. The genotype of this child may determine that they have the potential to grow to 1.9 m, but the child's final height is influenced greatly by their diet. For example, a shortage of calcium when the child is young can restrict bone development, therefore preventing them reaching their maximum potential height*

Study tip

Remember all individuals of the same species have the same genes but not necessarily the same alleles of these genes. Many people refer to alleles and genes interchangeably – make sure you understand the difference between these words, and use the terms correctly.

▲ **Figure 5** *Dimples in your cheeks are caused by a dominant allele. By simply looking at this man you cannot tell if he is heterozygous or homozygous for this dominant allele*

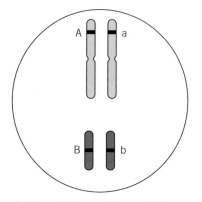

▲ **Figure 6** *In sexually reproducing organisms, most chromosomes occur in pairs known as homologous chromosomes (where one of each pair comes from each parent). The allele of a particular gene occurs in the same position on each of the homologous chromosomes. This is known as its locus. The diagram shows a pair of homologous chromosomes from a pea plant*

Synoptic link

To remind yourself how meiosis creates variation, for example through crossing over and independent assortment, look back at Topic 6.3, Meiosis.

Any changes the environment makes to a person's phenotype are not inherited – these are referred to as modifications. Only mutations (changes to the DNA) in the gametes can be passed on to the offspring.

It is not always possible to determine an organism's genotype from its phenotype owing to the dominance of particular alleles. A **dominant allele** is the version of the gene that will always be expressed if present in an organism. This means an individual showing the dominant characteristic in their phenotype could have one or two copies of the dominant gene – you can't tell from their appearance. A **recessive allele,** however, will only be expressed if two copies of this allele are present in an organism. This means if an individual has a recessive phenotype, you also know their genotype – they must have two alleles coding for the recessive phenotype.

A number of key terms are used to describe an organism's genotype for a particular characteristic:

- **Homozygous** – they have two identical alleles for a characteristic. The organism could be *homozygous dominant* (contain two alleles for the dominant phenotype) or *homozygous recessive* (contain two alleles for the recessive phenotype).

- **Heterozygous** – they have two different alleles for a characteristic. In this case the allele for the dominant phenotype will be expressed.

Continuous and discontinuous variation

The variation of a characteristic displayed within a species can be divided into two groups – those which show continuous variation, and those which show discontinuous variation. Remember – in discontinuous variation individuals fall into distinct groups (for example, blood groups) and normally only one gene is involved and the environment has little, if any, effect (Figure 7). In continuous variation there are two extremes, with every degree of variation possible in between (Figure 8). Examples would include your height or weight.

Many genes will be involved, and the environment has a large effect.

Table 1 summarises some of the important differences between continuous and discontinuous variation.

▼ **Table 1**

	Continuous variation	Discontinuous variation
definition	a characteristic that can take any value within a range	a characteristic that can only appear in specific (discrete) values
cause of variation	genetic and environmental	mostly genetic
genetic control	polygenes – controlled by a number of genes	one or two genes
examples	leaf surface area animal mass skin colour	blood group albinism round and wrinkled pea shape

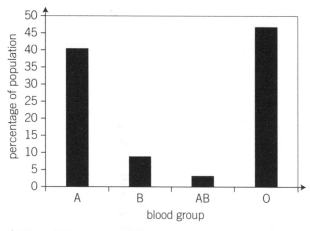

▲ **Figure 7** *An example of discontinuous variation showing individuals falling into distinct categories*

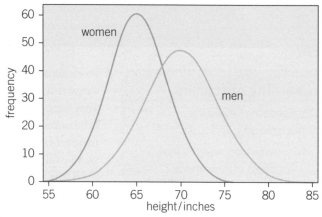

▲ **Figure 8** *Graph showing the range of heights of a large sample of men and women. It should be noted that sex is a discontinuous characteristic but within each category (male or female) there is continuous variation (height)*

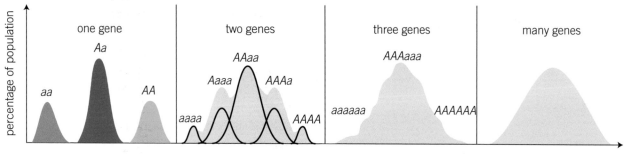

▲ **Figure 9** *Series of graphs showing how the type of variation changes as the number of genes responsible for the production of a characteristic increases*

Summary questions

1. State the difference between a homozygous and heterozygous genotype. (*1 mark*)

2. Explain the difference between the phenotype and genotype of an oak tree. (*2 marks*)

3. Using named examples, state and explain the difference between continuous and discontinuous variation. (*6 marks*)

Synoptic link

Look back at Topic 10.6, Representing variation graphically for more examples of continuous and discontinuous variation.

20.2 Monogenic inheritance

Specification reference: 6.1.2

To explain how characteristics are inherited you need to be able to show how genes are passed on from one generation to the next. This is normally shown using a genetic cross. Most commonly the inheritance of a single gene is shown, this is known as **monogenic inheritance**. The basic laws by which characteristics are inherited were established by Gregor Mendel, a scientist and monk working in the 19th century.

Performing a genetic cross

There are a number of key steps you should follow when analysing a genetic cross. This ensures that your diagram explains fully what is happening to the genes of an organism during fertilisation (and helps you to avoid making errors):

Step 1 – State the phenotype of both the parents.

Step 2 – State the genotype of both parents. To do this, assign a letter code to represent the alleles of the gene being studied. A capital letter should be used to represent the dominant allele and its lower case form to represent the recessive allele. For example, if studying the inheritance of an animal's fur colour, you may choose B to represent brown fur (dominant) and b to represent white fur (recessive).

Step 3 – State the gametes of each parent. It is common practice to circle the letters, for example, Ⓖ.

Step 4 – Use a Punnett Square to show the results of the random fusion of gametes during fertilisation. Remember to label the gametes on the edges of the square.

Step 5 – State the proportion of each genotype which are produced among the offspring. This can be in the form of a percentage, ratio, or 'x out of y offspring ...'.

Step 6 – State the corresponding phenotype for each of the possible genotypes. It must be clear that you know which phenotype results from each genotype.

Homozygous genetic cross

Mendel carried out many of his famous experiments on pea plants. Pea pods come in two colours – green and yellow. The cross in Figure 1 shows what happens when a homozygous green pea pod plant is crossed with a homozygous yellow pea pod plant. The allele for green pea pods is dominant. Organisms that contain homozygous alleles for a particular gene are known as true breeding or pure breeding individuals. Therefore in this experiment Mendel was studying what happened when two true breeding individuals are crossed.

All of the offspring are heterozygous. This means that all plants will have green pods as this is the dominant allele. These offspring are known as the F_1 generation.

Heterozygous genetic cross

The cross in Figure 2 shows what happens if you take two of the heterozygous offspring from the first generation and cross them together.

G = allele for green pods
g = allele for yellow pods

parental phenotypes	green pods GG (arbitrarily assumed as male♂)
parental genotypes	yellow pods gg (arbitrarily assumed as female♀)

meiosis · meiosis

gametes   G   G   g   g

offspring (F_1) genotypes

♀gametes	♂ gametes	
	G	G
g	Gg	Gg
g	Gg	Gg

offspring (F_1) phenotypes   all plants have green pods (Gg)

▲ **Figure 1**

F_1 offspring phenotypes   green pods   green pods
F_1 offspring genotypes   Gg   Gg

meiosis   meiosis

gametes   G   g   G   g

offspring (F_2) genotypes

♀gametes	♂ gametes	
	G	g
G	GG	Gg
g	Gg	gg

offspring (F_2) phenotypes   ratio of 3 plants with green (GG and Gg) pods to 1 plant with yellow (gg) pods

▲ **Figure 2**

The offspring produced from this cross are known as the F_2 generation. Offspring will be produced in a ratio of three pea plants with green pods to one pea plant with yellow pods.

Codominance

Codominance occurs when two different alleles occur for a gene – both of which are equally dominant. As a result both alleles of the gene are expressed in the phenotype of the organism if present.

One example of this condition is the colour of snapdragon flowers. Two equally dominant alleles exist, each of which codes for the colour of the flower:

An allele that codes for red flowers – the allele codes for the production of an enzyme which catalyses the production of red pigment from a colourless precursor.

An allele that codes for white flowers – the allele codes for an altered version of the enzyme which does not catalyse the production of the pigment, therefore the flowers are white.

Study tip

Potential outcomes from genetic crosses can be expressed in different forms which are mathematically equivalent:

Probability	Ratio	Percentage
4 in 4	4:0	100%
3 in 4	3:1	75%
2 in 4 (or 1 in 2)	2:2 (or 1:1)	50%
1 in 4	1:3	25%
0 in 4	0:4	0%

▲ **Figure 3** *Red snapdragon flowers are produced by a plant which is homozygous for the allele coding for the enzyme that catalyses the production of red pigment*

In this example of codominance, three colours of flower can be produced:

1 Red flowers – the plant is homozygous for the allele coding for the production of red pigment.

2 White flowers – the plant is homozygous for the allele coding for no pigment production.

3 Pink flowers – the plant is heterozygous. The single allele present which codes for red pigmentation produces enough pigment to produce pink flowers.

The genetic cross in Figure 6 shows how pink flowers are produced. Two of the pink flowers produced in the F_1 generation are then crossed in the second cross.

When studying codominance, upper and lower case letters are not used to represent the alleles, as this would imply one allele is dominant and the other recessive. Instead a letter is chosen to represent the gene, in this example C for colour of flowers. The different alleles are then represented using a second letter which is shown as a superscript. In this example C^R is used to represent the allele coding for red flowers and C^W for the allele coding for white flowers.

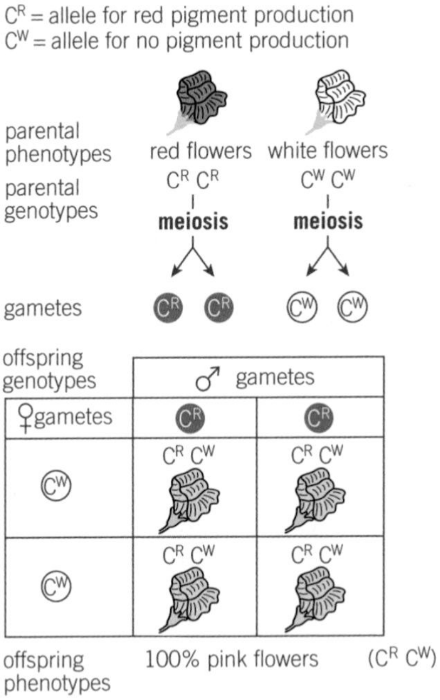

▲ **Figure 4** *White snapdragon flowers are produced by a plant which is homozygous for the allele coding for the enzyme which does not catalyse the production of pigment*

▲ **Figure 5** *Pink snapdragon flowers are produced by a plant which is heterozygous for both alleles*

C^R = allele for red pigment production
C^W = allele for no pigment production

Figure 6 cross:

parental phenotypes — red flowers / white flowers
parental genotypes — $C^R C^R$ / $C^W C^W$
meiosis / meiosis
gametes — C^R C^R / C^W C^W

offspring genotypes — ♂ gametes

♀gametes	C^R	C^R
C^W	$C^R C^W$	$C^R C^W$
C^W	$C^R C^W$	$C^R C^W$

offspring phenotypes — 100% pink flowers ($C^R C^W$)

▲ **Figure 6** *Production of pink snapdragon flowers*

Figure 7 cross:

parental phenotypes — pink flowers / pink flowers
— $C^R C^W$ / $C^R C^W$
parental genotypes — meiosis / meiosis
gametes — C^R C^W / C^R C^W

offspring genotypes — ♂ gametes

♀ gametes	C^R	C^W
C^R	$C^R C^R$	$C^R C^W$
C^W	$C^R C^W$	$C^W C^W$

offspring phenotypes — 50% pink flowers ($C^R C^W$)
25% red flowers ($C^R C^R$)
25% white flowers ($C^W C^W$)

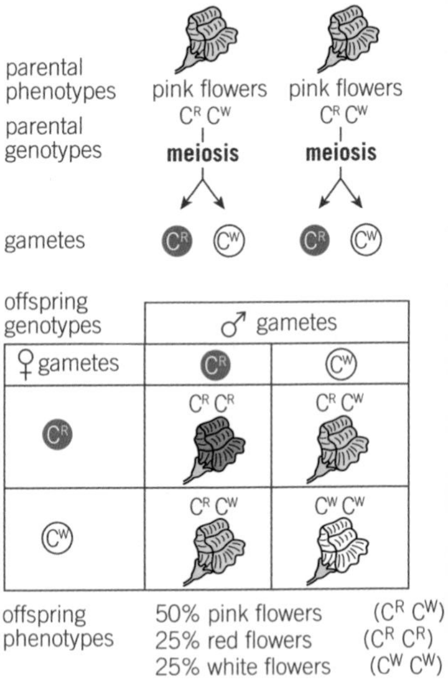

▲ **Figure 7** *Genetic cross between two snapdragons displaying pink flowers*

Multiple alleles

In the previous examples, you have learnt about characteristics coded by a gene with two alleles. Some genes have more than two versions, they have **multiple alleles**. However, as an organism carries only two versions of the gene (one on each of the homologous chromosomes) only two alleles can be present in an individual.

Your blood group is determined by a gene with multiple alleles. The immunoglobulin gene (Gene I) codes for the production of different antigens present on the surface of red blood cells. There are three alleles of this gene:

- I^A – results in the production of antigen A
- I^B – results in the production of antigen B
- I^O – results in the production of neither antigen

I^A and I^B are codominant whereas I^O is recessive to both of the other alleles. Different combinations of these alleles result in the four blood groups:

- Blood group A – $I^A I^A$ or $I^A I^O$
- Blood group B – $I^B I^B$ or $I^B I^O$
- Blood group AB – $I^A I^B$
- Blood group O – $I^O I^O$

There are many different possible crosses. Figure 8 shows how parents of blood group A and B can reproduce to produce children who may display any of the four blood groups.

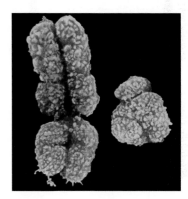

parental phenotypes	Group A	Group B
parental genotypes	$I^A I^O$	$I^B I^O$
	|	|
	meiosis	meiosis
gametes	I^A I^O	I^B I^O

offspring genotypes

♀ gametes	♂ gametes	
	I^A	I^O
I^B	$I^A I^B$	$I^B I^O$
I^O	$I^A I^O$	$I^O I^O$

offspring phenotypes	25% blood group A	$(I^A I^O)$
	25% blood group B	$(I^B I^O)$
	25% blood group AB	$(I^A I^B)$
	25% blood group O	$(I^O I^O)$

▲ **Figure 8**

Determining sex

In humans and other mammals, as well as many other species, sex is genetically determined. Humans have 23 pairs of chromosomes of varying sizes and shapes. In 22 of the pairs, both members of the pair are the same but the 23rd pair, known as the sex chromosomes, are different. Human females have two X chromosomes, whereas a male has an X and a Y chromosome. The X chromosome is large and contains many genes not involved in sexual development. The Y chromosome is very small, containing almost no genetic information, but it does carry a gene that causes the embryo to develop as a male.

Therefore the sex of the offspring will be determined by whether the sperm fertilising the egg contains a Y chromosome or an X.

Sex linkage

Some characteristics are determined by genes carried on the sex chromosomes – these genes are called **sex linked.** As the Y chromosome is much smaller than the X chromosome, there are a number of genes in the X chromosome that males have only one copy of. This means that any characteristic caused by a recessive allele on the section of the X chromosome, which is missing in the Y chromosome, occurs more frequently in males. This is because many females will also have a dominant allele present in their cells.

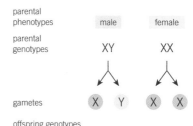

▲ **Figure 9** *An X and Y human chromosome viewed through a scanning electron microscope*

parental phenotypes	male	female
parental genotypes	XY	XX
gametes	X Y	X X

offspring genotypes

♀ gametes	♂ gametes	
	X	Y
X	XX	XY
X	XX	XY

offspring phenotypes
50% male (XY)
50% female (XX)

▲ **Figure 11** *Sex inheritance in humans*

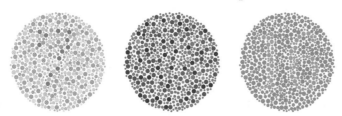

▲ **Figure 10** *This is a test used to determine whether a person is colour blind. Red green colour blindness is a sex-linked disorder. It is caused by a recessive allele carried on the X chromosome. It is therefore much more common in males*

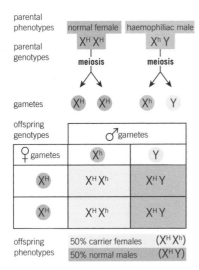

offspring phenotypes — 50% carrier females (X^H X^h)
50% normal males (X^H Y)

▲ **Figure 12** *Inheritance of the haemophiliac allele from a haemophiliac male*

H = allele for production of clotting protein (rapid blood clotting)
h = allele for non-production of clotting protein (slow blood clotting)

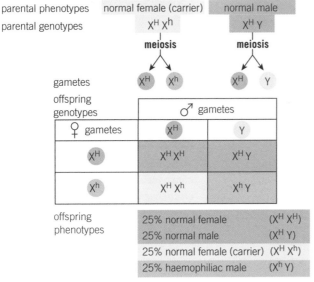

offspring phenotypes

25% normal female (X^H X^H)
25% normal male (X^H Y)
25% normal female (carrier) (X^H X^h)
25% haemophiliac male (X^h Y)

▲ **Figure 13** *Inheritance of haemophilia from a normal female (carrier)*

Haemophilia

Haemophilia is an example of a sex-linked genetic disorder. Patients with haemophilia have blood which clots extremely slowly due to the absence of a protein blood-clotting factor (in the majority of cases this is factor VIII). As a result injury can result in prolonged bleeding which, if left untreated, is potentially fatal.

If a male inherits the recessive allele that codes for haemophilia (on their X chromosome) they cannot have a corresponding dominant allele on their Y chromosome, and so develop the condition. As a result the vast majority of haemophilia sufferers are males. Females who are heterozygous for the haemophilia coding gene are known as **carriers**. They do not suffer from the disorder, however they may pass on the allele to their children. This can result in the birth of a son who suffers from haemophilia.

When showing the inheritance of a sex-linked condition the alleles are shown linked to the sex chromosome they are found on. In this example, haemophilia is linked to the X chromosome, therefore:

X^H is used to represent the dominant 'healthy' allele.

X^h is used to represent the recessive allele coding for haemophilia (through the non-production of a blood-clotting protein). This is often known as the faulty allele.

Y is used to represent the Y chromosome – it has no allele attached to it as it does not carry the gene which produces the specific blood-clotting protein (Figure 12).

If a carrier female and a normal male have children, then in theory half of the male offspring produced will have the disorder (Figure 13). Half of the female offspring will be carriers. As male offspring only inherit an X chromosome from their mother, sons can only inherit the condition from their mother.

However, an affected male can pass on the faulty allele to his daughters, resulting in them becoming carriers of the disorder (Figure 12).

Summary questions

1 What is special about the phenotypes formed as a result of a gene which has codominant alleles? *(2 marks)*

2 Using a genetic cross show the blood groups of the potential offspring of a father with blood group AB and a mother with blood group O. *(6 marks)*

3 Some individuals cannot distinguish between the colours red and green. This form of colour blindness is a sex-linked condition, linked to the X chromosome. The allele for red-green colour blindness is recessive to the normal allele. Show how colour blindness can be present in the son of two parents who do not display colour blindness. *(6 marks)*

20.3 Dihybrid inheritance

Specification reference: 6.1.2

In the previous topic you learnt how genes were inherited through monogenic inheritance. In reality, thousands of genes are inherited during fertilisation. Dihybrid crosses are used to show the inheritance of two genes and this is known as **dihybrid inheritance**.

Dihybrid cross

A dihybrid cross is used to show the inheritance of two different characteristics, caused by two genes, which may be located on different pairs of homologous chromosomes. Each of these genes can have two or more alleles.

A dihybrid cross is set out in a very similar format to the one used when studying a monohybrid cross – however, four alleles (two for each characteristic) are shown at each stage instead of two.

A classic example is the inheritance of seed phenotype in pea plants. The seeds a pea plant produces can be produced in two different colours – yellow or green. They are also produced in two different shapes – round or wrinkled.

The following codes can be used to represent the alleles:

Y – allele coding for yellow seeds (this is the dominant allele)

y – allele coding for green seeds (this is the recessive allele)

R – allele coding for round seeds (this is the dominant allele)

r – allele coding for wrinkled seeds (this is the recessive allele)

In the dihybrid cross in Figure 2, a true breeding homozygous pea plant with yellow round seeds is crossed with a true breeding homozygous pea plant with green wrinkled seeds.

All of the offspring produced in the F_1 generation will have a heterozygous genotype – YyRr. Therefore, they will all produce yellow round seeds as they will all have inherited one of each of the dominant alleles from the yellow round seeded parent.

In the second example in Figure 2, two of the pea plants grown from the seeds of the F_1 generation are now crossed.

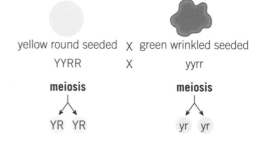

Learning outcomes

Demonstrate knowledge, understanding, and application of:

→ genetic diagrams to show patterns of inheritance including dihybrid inheritance.

▲ **Figure 1** *Pea seeds can be round or wrinkled. Round seeds are caused by the dominant allele*

parental phenotype	yellow round seeded	X	green wrinkled seeded
parental genotype	YYRR	X	yyrr
	meiosis		meiosis
gametes	YR YR		yr yr

offspring F_1 genotypes	♂ gametes	♀ gametes	
		YR	YR
	Yr	YyRr	YyRr
	Yr	YyRr	YyRr

offspring F_1 phenotypes   all yellow round seeded

▲ **Figure 2**

F₁ offspring phenotype	yellow round seeded	X	yellow round seeded

F₁ offspring phenotype: yellow round seeded X yellow round seeded

F₁ offspring genotype: YyRr X YyRr

Meiosis   **Meiosis**

gametes: YR Yr yR yr   YR Yr yR yr

offspring (F₂) genotypes

	♀ gametes			
♂ gametes	YR	Yr	yR	yr
YR	YYRR	YYRr	YyRR	YyRr
Yr	YYRr	YYrr	YyRr	Yyrr
yR	YyRR	YyRr	yyRR	yyRr
yr	YyRr	Yyrr	yyRr	yyrr

offspring F₂ phenotypes

9 yellow round seeded : 3 yellow wrinkled seeded: 3 green round seeded: 1 green wrinkled

(YYRR, YYRr, YyRR, YyRr) (YYrr, Yyrr) (yyRR, yyRr) (yyrr)

▲ **Figure 3**

Synoptic Link

Look back at Topic 6.3, Meiosis to remind yourself of the different stages of meiosis.

When the phenotypes of the 16 possible combinations of alleles are identified (as shown in Figure 3), it is found that the expected ratio in the F₂ generation is nine pea plants producing yellow round seeds, to three pea plants producing yellow wrinkled seeds to three pea plants producing green round seeds, to one pea plant producing green wrinkled seeds.

It is worth remembering that this is the expected ratio of the four different phenotypes. As with all genetic crosses the actual ratio of offspring produced can differ from the expected. This may because:

- The fertilisation of gametes is a random process so in a small sample a few chance events can lead to a skewed ratio.
- The genes being studied are both on the same chromosome. These are known as linked genes. If no crossing over occurs the alleles for the two characteristics will always be inherited together.

Summary questions

1 State the difference between monogenic inheritance and dihybrid inheritance. *(1 mark)*

2 State and explain two reasons why the offspring produced from a particular genetic cross may differ from the expected ratio. *(2 marks)*

3 Work through a dihybrid genetic cross between a pea plant which produces green wrinkled seeds and a heterozygous pea plant which produces yellow round seeds to determine the expected ratio of the F₁ offspring. *(6 marks)*

20.4 Phenotypic ratios

Specification reference: 6.1.2

The ratios of phenotypes that you would expect to see in the offspring produced from a dihybrid cross can be easily calculated as long as you know which alleles are dominant and which are recessive. The actual numbers may vary from those expected to some extent because the process is random but the differences should not be large. The larger the sample the closer the numbers will be to the expected ratio.

Linkage

The ratios observed in many dihybrid crosses differ significantly from those expected. This is often due to linkage meaning that the genes are located on the same chromosome. In Topic 20.2 you met the idea of sex linkage, when a particular gene is located on the sex chromosomes. The effects of sex linkage can be seen in conditions such as haemophilia and red-green colour blindness.

When the genes that are linked are found on one of the other pairs of chromosomes it is called **autosomal linkage**. **Linked genes** are inherited as one unit – there is no independent assortment during meiosis unless the alleles are separated by chiasmata. They tend to be inherited together.

Linked genes cannot undergo the normal random 'shuffling' of alleles during meiosis and the expected ratios will not be produced in the offspring. The linked genes are inherited effectively as a single unit. Body colour and wing length, for example, are linked characteristics in fruit flies. The allele B is responsible for a brown body and is dominant to b which results in a black body. The allele V is responsible for long wings and is dominant to the allele v which results in short wings.

A heterozygous brown bodied, long winged fly (BbVv) was crossed with a homozygous black bodied, short winged fly (bbvv). The expected and observed phenotypic ratios from this cross are shown in Figure 1.

The homozygous parent can only produce (bv) gametes and the heterozygous parent produces mainly (BV) and (bv) gametes as the alleles are linked. The heterozygous parent will produce a few (bV) gametes and (Bv) gametes due to crossing over, which results in the separation of some of the linked genes. This is the reason for the small number of brown flies with short wings and black flies with long wings. These are called **recombinant** offspring (they have different combinations of alleles than either parent). The closer the genes are on a chromosome the less likely they are to be separated during crossing over and the fewer recombinant offspring produced.

Learning outcomes

Demonstrate knowledge, understanding, and application of:

→ the use of phenotypic ratios to identify linkage (autosomal and sex linkage) and epistasis

→ using the chi-squared (χ^2) test to determine the significance of the difference between observed and expected results.

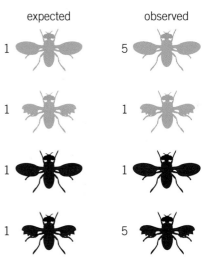

▲ **Figure 1** A test cross involving linked genes does not produce the expected ratio of phenotypes in the offspring. There is a larger proportion of brown flies with long wings and black flies with short wings

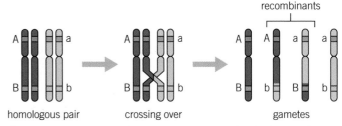

▲ **Figure 2** During crossing over in prophase I of meiosis, non-sister chromatids of a homologous pair of chromosomes may break and reform. As these chromatids break at the same point, any gene loci below the point of the break will be exchanged as a result of recombination creating new gene combinations

The **recombination frequency** is a measure of the amount of crossing over that has happened in meiosis.

$$\text{Recombination frequency} = \frac{\text{number of recombinant offspring}}{\text{total number of offspring}}$$

A recombination frequency of 50% indicates that there is no linkage and the genes are on separate chromosomes. Less than 50% indicates that there is gene linkage and the random process of independent assortment has been hindered. As the degree of crossing over reduces, the recombination frequency also gets smaller. The degree of crossing over is determined by how close the genes are on a chromosome. The closer they are the less likely they will be separated during crossing over and vice versa.

The recombination frequencies for a number of characteristics coded for by genes on the same chromosome can be used to map the genes on the chromosome. A recombination frequency of 1% relates to a distance of one map unit on a chromosome.

Chi-squared test

The observed results from a genetic cross will almost always differ to some extent from the expected results and this will be due to chance. If you toss a coin 10 times you would be unlikely to get five heads and five tails. The observed ratio of heads to tails will probably be quite different from the expected ratio. This does not mean there is anything wrong with the coin. If the same coin were tossed a thousand times you would see less relative difference between the expected and observed ratios. The number of observations made, therefore, determines how chance affects the results.

It is important when making comparisons between observed and expected results that it is known whether any differences are due to chance or if there is a reason for the differences (they are significant).

The **chi-squared (χ^2) test** is a statistical test that measures the size of the difference between the results you actually get (observe) and those you expected to get. It helps you determine whether differences in the expected and observed results are significant or not, by comparing the sizes of the differences and the numbers of observations.

The chi-squared test is conventionally used to test the null hypothesis. The null hypothesis is that there is no significant difference between what we expect and what we observe – in other words any differences we do see are due to chance. Calculated chi squared values are used to find the probability of the difference being due to chance alone.

The chi-squared formula is:

$$\chi^2 = \Sigma \frac{(O - E)^2}{E}$$

Where:

χ^2 = the test statistic

Σ = the sum of

O = observed frequencies

E = expected frequencies

Large chi-squared values mean there is a statistically significant difference between the observed and expected results and the probability that these differences are due to chance is low. There must be a reason, other than chance, for the unexpected results.

The number of categories being compared in an investigation affects the size of the chi-squared value calculated. The degrees of freedom is the number of comparisons being made and is calculated as $n-1$, where n is the number of categories or possible outcomes (phenotypes in the case of phenotypic ratios) present in the analysis. For example, if you were looking at yellow and green peas there would be two categories and therefore one degree of freedom.

> **Study tip**
>
> A full table of chi-squared values can be found in the appendix.

▼ **Table 1** *Table of chi-squared values and the probability of each occurring at different degrees of freedom (df)*

df	p values								df
	0.99	0.95	0.90	0.50	0.10	0.05	0.01	0.001	
1	0.0001	0.0039	0.016	0.46	2.71	3.84	6.63	10.83	1
2	0.02	0.10	0.21	1.39	4.60	5.99	9.21	13.82	2
3	0.12	0.35	0.58	2.37	6.25	7.81	11.35	16.27	3
4	0.30	0.71	1.06	3.36	7.78	9.49	13.28	18.46	4
5	0.55	1.41	1.61	4.35	9.24	11.07	15.09	20.52	5

■ = null hypothesis accepted

■ = null hypothesis rejected – another factor involved

If the calculated χ^2 value is greater than the critical value found in a table at 5% significance ($p = 0.05$) we do not have sufficiently strong evidence to reject our null hypothesis. Therefore we accept the null hypothesis – there is no significant difference between what we observed and what we expected. However if the calculated χ^2 value is less than the critical value we reject the null hypothesis – some other factor, outside our original expectation, is likely to be causing a significant difference between expectation and observation.

The minimum χ^2 value that gives a 5% probability is called the critical value. The critical value increases as the degrees of freedom increase.

If χ^2 is less than the critical value there is no significant difference

If χ^2 is greater than or equal to the critical value there is a significant difference

▦ Worked example: Calculating the chi-squared value

A monohybrid cross involving two plants heterozygous for purple flowers was carried out.

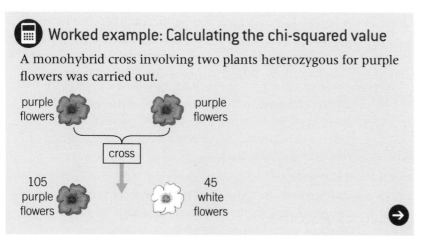

purple flowers   purple flowers

cross

105 purple flowers   45 white flowers

→

In this case, the null hypothesis is that any differences in the observed numbers of purple and white phenotypes from the expected numbers are due to chance.

Step 1: The probability of getting each expected phenotype is calculated.

Purple is dominant to white, so there should be a 3:1 ratio in the offspring.

The chi-squared value is calculated based on this 3:1 hypothesis.

Step 2: The expected number of each phenotype based on the probabilities and total number of offspring is calculated by multiplying the expected proportion by the total:

There are 150 offspring so $150 \times \dfrac{3}{4} = 112.5$

$$150 \times \dfrac{1}{4} = 37.5$$

Step 3: Next, put your calculated and observed values into a table:

Phenotype	Observed	Expected
purple	105	$\dfrac{3}{4} \times 150 = 112.5$
white	45	$\dfrac{1}{4} \times 150 = 37.5$
total	150	

Step 4: Then the chi-squared value can be calculated:

$$\chi^2 = \Sigma \dfrac{(O - E)^2}{E}$$

$$\chi^2 = \dfrac{(105 - 112.5)^2}{112.5} + \dfrac{(45 - 37.5)^2}{37.5}$$

$$\chi^2 = \dfrac{56.25}{112.5} + \dfrac{56.25}{37.5}$$

$$\chi^2 = 0.5 + 1.5 = 2.0$$

Step 5: The chi-square value is determined from the chi-squared distribution table at the 0.05 probability level (p) for the correct degrees of freedom. Remember this is $n - 1$. In this case there are only two phenotypes so $n = 2$ and the degrees of freedom is $2 - 1 = 1$.

The table value is 3.84.

Step 6: Because 2.0 < 3.84, the null hypothesis is accepted. The difference between the predicted and actual cross results is not significant (at the 5% probability level). Any difference between the expected results and the actual results is due to chance.

If the χ^2 value had been greater than 3.84, it would indicate that the differences in the observed results were due to another factor such as gene linkage or a new or different way of inheriting alleles.

Study tip

You do not need to learn the chi-squared formula but you need to know how apply it.

Corn and the chi-squared (χ^2) test

There will almost always be differences between expected and observed results because of the random nature of the processes involved. Statistical tests like the chi-squared test are performed to determine whether these differences are due to chance alone or caused by some other factor that may not have been considered.

Maize plants have been used for many years to study genetic crosses. An ear of corn contains around 500 kernels, or seeds. The seeds are produced as the result of the cross-fertilisation of two maize plants. The colour of the seeds is controlled by one gene with a dominant (P, purple) allele and a recessive (p, yellow) allele. Another gene determines the texture of the seeds and there is again a dominant (R, round) allele and recessive (r, wrinkled) allele.

◀ Figure 3

A genetic study was carried out to determine if these two genes are linked.

Two maize plants that were heterozygous for both colour and texture were cross-fertilised and an ear of corn produced as a result of this cross was analysed.

The results are shown in Table 2.

1 State the ratio of phenotypes that you expect from this cross if the genes are not linked.
2 Copy and complete the table.
3 Show how you would use the chi-squared distribution table, shown in Table 1, to determine the critical value for these results.
4 State, giving your reasons, the conclusions you would make from your statistical analysis of these results.

▼ **Table 2** *Shows the number of each of the different types of kernel produced from a cross between two maize plants which are both heterozygous for colour and texture*

Grain phenotype	Observed number	Observed ratio	Expected ratio	Expected number	(observed – expected)^2 / expected
purple and smooth	216			$381 \times \dfrac{9}{16} = 214$	
purple and shrunken	79			$381 \times \dfrac{3}{16} = 71$	
yellow and smooth	65				
yellow and shrunken	21				
				chi-squared value:	

Epistasis

Epistasis is the interaction of genes at different loci. Gene regulation is a form of epistasis with regulatory genes controlling the activity of structural genes, for example, the lac operon. Gene interaction also occurs in biochemical pathways involving only structural genes.

It was originally thought that all genes were expressed independently, and therefore their effects on the phenotype seen. Now it is known that many genes interact *epistatically*. It is the results of these interactions that we see in the phenotypes of living organisms. The characteristics of plants and animals that show continuous variation involve multiple genes and epistasis occurs frequently.

Synoptic Link

You learnt about the lac operon in Topic 19.2, Control of gene expression.

Consider the biochemical pathway shown in Figure 3, where four enzymes a, b, c, and d are required to produce a pigment responsible for the colour of a flower petal. Four genes a, b, c, and d need to be expressed to produce these enzymes. If one of these genes is not expressed then one step will be missing, and the petal will not have the expected colour.

The lack of the enzyme normally produced when gene a, b, or c is expressed means the intermediate molecule necessary for the next reaction in the sequence is not produced. This results in a lack of substrate for the next enzyme in the pathway and so the expression of this gene will not be observed in the phenotype. The gene is effectively 'masked' by the lack of expression of the previous gene in the pathway.

If enzyme d is not produced then precursor **D** will not be converted to a pigment. This means it is often hard to observe the expression of genes **a**, **b**, and **c** in the phenotype. The disabled gene **d** is likely to mask their expression.

A gene that is affected by another gene is said to be hypostatic and a gene that affects the expression of another gene is said to be epistatic.

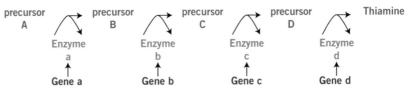

▲ **Figure 3** *The genes a, b, and c code for enzymes that catalyse the production of the substrate needed for the next reaction in the sequence. Gene d produces an enzyme that catalyses the production of the pigment from precursor D*

Dominant and recessive epistasis
An epistatic gene may influence the activity of other genes as result of the presence of dominant or recessive alleles.

In the example previously, if the presence of two recessive alleles at a gene locus led to the lack of an enzyme then it would be called recessive epistasis.

Dominant epistasis occurs if a dominant allele results in a gene having an effect on another gene. This would happen if an epistatic gene (not present in this pathway) coded for an enzyme that modified one of the precursor molecules in the pathway. The next enzyme in the pathway would then lack a suitable substrate molecule and so the pigment would again not be produced. All of the genes in the sequence would be effectively 'masked'.

Labrador colours
The colour of Labrador dogs is produced as a result of the pigment melanin being deposited in the skin and fur. One gene codes for the production of the pigment and has the alleles B (dominant, black pigment produced) and b (recessive, brown pigment produced). A second gene codes for where the pigment is deposited and, again, has two alleles E (dominant, pigment deposited in the skin and fur) and e (recessive, pigment deposited in the skin only).

The colour of a Labrador varies depending on which alleles are present at each locus. The genes are not expressed independently and so this is an example of epistasis. The gene at the E locus is epistatic to the hypostatic gene at the B locus.

The different phenotypes and genotypes are shown in Table 3.

▼ Table 3

B locus genotype	E locus genotype	Fur colour	Skin colour	Overall colour	Phenotype
BB/Be	EE/Ee	black	black	black	
BB/Be	ee	yellow	black	brown	
bb	EE/Ee	brown	brown	yellow	
bb	ee	yellow	brown	yellow	

There are only three registered colours, black, brown (chocolate), and yellow (golden). The yellow coat is an example of recessive epistasis and it ranges from deep gold to pale blond.

Summary questions

1 Explain what the chi-squared (χ^2) test is used to measure. (3 marks)

2 Horse coat colour is an example of epistasis. Two genes are involved. The different colours and the genotypes responsible are summarised here.

G___ produces a grey horse

gg E_ produces a black horse

gg ee produces a chestnut horse.

Explain, giving your reasons, which form of epistasis this represents. (5 marks)

3 A biologist test crosses a plant that is heterozygous for the alleles Xx and Yy in order to see how far apart the gene loci are on a chromosome. The offspring of this cross contains 5.2% recombinant individuals.
   a State the distance between these alleles. (1 mark)
   b Describe how the biologist could determine where a third gene with the alleles Zz is on the chromosome relative to Xx and Yy. (4 marks)
   c Suggest why this sort of investigation is best carried out with genes that are relatively close to each other. (2 marks)

20.5 Evolution

Demonstrate and apply knowledge and understanding of:

→ factors that can affect the evolution of a species

→ the use of the Hardy–Weinberg principle to calculate allele frequencies in populations.

Evolution, the change in inherited characteristics of a group of organisms over time, occurs due to changes in the frequency of different alleles within a population.

Population genetics

Population genetics investigates how allele frequencies within populations change over time. The sum total of all the genes in a population at any given time is known as the **gene pool**. The gene pool of a population includes millions of genes, but you will look at the variation in the different alleles of a single gene within the gene pool. The relative frequency of a particular allele in a population is the **allele frequency**.

The frequency with which an allele occurs in a population is not linked to whether it codes for a dominant or a recessive characteristic, and it is not fixed. It can change over time in response to changing conditions. Evolution involves a long-term change in the allele frequencies of a population, for example, alleles for antibiotic resistance have increased in many bacteria populations over time. Biologists have developed ways of determining allele frequencies and use them in models to determine whether evolution is taking place.

Calculating allele frequency

Imagine a population of 100 diploid organisms that can all breed successfully. You are going to look at a gene that has two possible alleles, A and a. The frequency of allele A in the population is represented by the letter p. The frequency of allele a in the population is represented by q. If every individual in your population of 100 is a heterozygote (Aa), then the frequency of each allele is 100/200 or 0.5 (50%) so $p + q = 1$

In a diploid breeding population with two potential alleles, the frequency of the dominant allele plus the frequency of the recessive allele will *always* equal 1. This simple formula is very important when using the Hardy–Weinberg principle.

The Hardy–Weinberg principle

The Hardy–Weinberg principle models the mathematical relationship between the frequencies of alleles and genotypes in a theoretical population that is stable and not evolving. The Hardy–Weinberg principle states: **in a stable population with no disturbing factors, the allele frequencies will remain constant from one generation to the next and there will be no evolution**. A completely stable population is not common in the real world, but this is still a useful tool. The Hardy–Weinberg principle provides a simple model of a theoretical stable population that allows us to measure and study evolutionary changes when they occur.

The Hardy–Weinberg principle is expressed as:

$$p^2 + 2pq + q^2 = 1$$

where p^2 = frequency of homozygous dominant genotype in the population

$2pq$ = frequency of heterozygous genotype in the population

q^2 = frequency of homozygous recessive genotype in the population

How do you use this information? Recessive phenotypes are often easy to observe. As a result you can find the frequency of the recessive genotype and use it to measure the equivalent allele frequency.

Frequency of the recessive genotype = q^2

So, the frequency of the recessive allele is $\sqrt{q^2} = q$

You can then use this to find p because you know $p + q = 1$

Finally, you can substitute these values back into the equation of the Hardy–Weinberg principle to find the frequencies of the three different genotypes.

 ## Hardy–Weinberg worked example

The peppered moth, *Biston betularia*, comes in two forms, light coloured and dark coloured. The light colour is inherited through a recessive allele. Students investigated a population in an area of woodland and found that 48 of the 50 peppered moths they captured were light in colour.

This gives the frequency of the homozygous recessive genotype (q^2) that results in a light colouration as 48/50, or 0.96 (96%). Now you can calculate the value of q, the frequency of the allele in the population.

$q^2 = 0.96$

so $q = \sqrt{0.96} =$ **0.98 (98%)** (2 s.f.)

You know that $p + q = 1$, so $p + 0.98 = 1$, so $p = 1 - 0.98 =$ **0.02 (2%)**

Now substitute these values into the equation for the Hardy–Weinberg principle to work out the frequency of the homozygous dominant genotype and the heterozygous genotype in this population of *Biston betularia*.

Frequency of homozygous dominant genotype (p^2) = 0.02^2 = **0.0004 (0.04%)**

Frequency of the heterozygous genotype (**2pq**) = 2 × 0.02 × 0.98 = **0.039 (3.9%)** (2 s.f.)

This gives you the frequencies for the three main genotypes of the *Biston betularia* population in the woodland studied. Around 96% of the moths are homozygous recessive and therefore light coloured, 3.9% are heterozygous and so dark coloured, and 0.04% are homozygous dominant and dark in colour.

Remember allele frequencies must add up to 1 and population percentages to 100% (allowing for rounding numbers up or down).

Disturbing the equilibrium

The Hardy–Weinberg principle assumes a theoretical breeding population of diploid organisms that is large and isolated, with random mating, no mutations, and no selection pressure of any type. In a natural environment these conditions virtually never occur. Species are continuously changing. In the peppered moths of the worked example, the light alleles were dominant historically but the allele frequencies changed dramatically after the Industrial Revolution, when the dark alleles gave individuals an advantage. Now the allele frequencies have changed again as cities and woodlands have become cleaner again. These changes in allele frequencies can be illustrated using the Hardy–Weinberg principle and upsetting the equilibrium may eventually result in evolution.

Factors affecting evolution

There are a number of factors that lead to changes in the frequency of alleles within a population and so they affect the rate of evolution:

● Mutation is necessary for the existence of different alleles in the first place, and the formation of new alleles leads to genetic variation.

● Sexual selection leads to an increase in frequency of alleles which code for characteristics that improve mating success.

● Gene flow is the movement of alleles between populations. Immigration and emigration result in changes of allele frequency within a population.

● Genetic drift occurs in small populations. This is a change in allele frequency due to the random nature of mutation. The appearance of a new allele will have a greater impact (is more likely to increase in number) in a smaller population than in a much larger population where there is a greater number of alleles present in the gene pool.

● Natural selection leads to an increase in the number of individuals that have characteristics that improve their chances of survival. Reproduction rates of these individuals will increase as will the frequency of the alleles coding for the characteristics. This is how changes in the environment can lead to evolution.

The impact of small populations

The gene pool of a large population ensures lots of genetic diversity owing to the presence of many different genes and alleles. Genetic diversity leads to variation within a population which is essential in the process of natural selection. Selection pressures such as changes in the environment, the presence of new diseases, prey, competitors, or even human influences lead to evolution. The population can adapt to change over time.

Small populations with limited genetic diversity cannot adapt to change as easily and are more likely to become extinct. A new strain of pathogen could wipe out a whole population.

The size of a population can be affected by many factors. Factors which limit or decrease the size of a population are called limiting factors. There are two types of limiting factors:

1   Density-dependent factors are dependent on population size and include competition, predation, parasitism, and communicable disease.

2   Density-independent factors affect populations of all sizes in the same way including – climate change, natural disasters, seasonal change, and human activities (for example, deforestation).

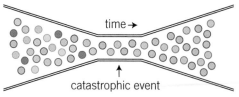

Large reductions in population size which last for at least one generation are called population bottlenecks (Figure 1). The gene pool, along with genetic diversity, is greatly reduced and the effects will be seen in future generations. It takes thousands of years for genetic diversity to develop in a population through the slow accumulation of mutations.

▲ **Figure 1** *A natural disaster or epidemic can drastically reduce a population. The gene pool will be greatly reduced and the remaining individuals may not be representative of the original population as some rarer alleles may not have been present in any of the survivors. The 'founder effect' and genetic drift will influence genetic variation as the population grows again*

Northern elephant seals were almost hunted to extinction in the 19th century. There were probably only about 20 seals left by the time hunting stopped. They now have a population of about 30 000 but show much less genetic diversity than southern elephant seals that did not experience a genetic bottleneck.

Cheetahs are thought to have experienced an initial population bottleneck about 10 000 years ago with other bottlenecks happening more recently. The species now shows low genetic diversity. Cheetahs face the same threats as many other African animals such as habitat loss and poaching, but while the population sizes of other animals are increasing thanks to the efforts of conservationists, cheetahs are not recovering as quickly. They are, in fact, close to extinction.

▲ **Figure 2** *Cheetah mother and cubs, Acinonyx jubatus, Masai Mara Reserve, Kenya*

The reduced genetic diversity of cheetahs means that they share around 99% of their alleles with other members of the species, more than we share with members of our own family. Mammals usually share about 80% of their alleles with other members of a species. As a result they are showing problems of inbreeding including reduced fertility.

Humans and chimpanzees split from a common ancestor about six million years ago. A small group of chimpanzees are likely to show more genetic diversity than all the humans alive today. It is believed that humans have experienced at least one genetic bottleneck, reducing our genetic diversity, as we have evolved into our present form.

A positive aspect of a genetic bottleneck is that a beneficial mutation will have a much greater impact and lead to the quicker development of a new species. This is thought to have a played a role in the evolution of early humans.

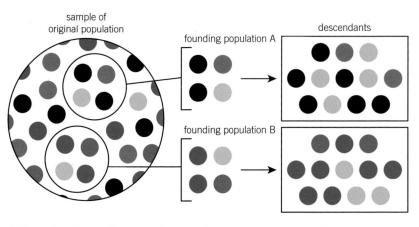

sample of
original population

founding population A

descendants

founding population B

▲ **Figure 3** *Diagram illustrating how small samples from a population can lead to populations with very different, and reduced, gene pools*

Founder effect

Small populations can arise due to the establishment of new colonies by a few isolated individuals, leading to the **founder effect**. The founder effect is an extreme example of genetic drift.

These small populations have much smaller gene pools than the original population and display less genetic variation. If carried to the new population, the frequency of any alleles that were rare in the original population will be much higher in the new, smaller population and so they will have a much bigger impact during natural selection.

The Afrikaner population in South Africa is descended mainly from a few Dutch settlers. The population today has an unusually high frequency of the allele that causes Huntington's disease. It is thought that just one of the original settlers carried the disease-causing allele.

The Amish people of America have descended from 200 Germans who settled in Pennsylvania in the 18th century. They rarely marry and have children outside their own religion and are therefore a closed community. The Amish have unusually high frequencies of alleles that cause the normally rare genetic disorder Ellis–van Creveld syndrome. People with the syndrome are short, they often have polydactyly (extra fingers or toes), abnormalities of nails and teeth, and a hole between the two upper chambers of the heart. Ellis–van Creveld syndrome is an example of founder effect caused by one couple, Samuel King and his wife, who settled in the area in 1744.

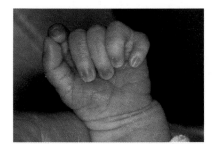

▲ **Figure 4** *Close-up of a baby's hand showing an extra finger. This condition is called polydactylism*

Evolutionary forces

The traits or characteristics of all living organisms show variation within populations. The distribution of the different variants will take the form of a bell-shaped curve if plotted on a graph. This is known in statistics as a **normal distribution**.

Stabilising selection

Taking the birth weight of babies as an example, babies with an average birth weight will be the most common and therefore form the peak of the graph. Babies with very low birth weight are more prone to infections and very large babies result in difficult births. Both of these extremes in weight reduce the survival chances of babies so the numbers of survivals of very small or very large babies remains low forming the tails on Figure 5.

This is natural selection, or survival of the fittest, at work. Babies with average birth weights are more likely to survive and reproduce than

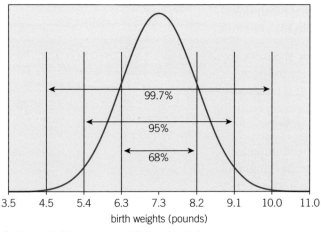

99.7%

95%

68%

3.5   4.5   5.4   6.3   7.3   8.2   9.1   10.0   11.0

birth weights (pounds)

▲ **Figure 5** *Birth weights of baby girls in Europe*

underweight or overweight babies. It is an example of **stabilising selection** because the norm or average is selected for (positive selection) and the extremes are selected against (negative selection). Stabilising selection therefore results in a reduction in the frequency of alleles at the extremes, and an increase in the frequency of 'average' alleles.

Directional selection

Directional selection occurs when there is change in the environment and the normal (most common) phenotype is no longer the most advantageous. Organisms which are less common and have more extreme phenotypes are positively selected. The allele frequency then shifts towards the extreme phenotypes and evolution occurs.

The changes seen in peppered moths during the industrial revolution are a good example of directional selection. During this period of time a lot of smoke was released from factories, which killed lichens growing on barks of trees, and the soot made the bark black. Peppered moths were originally light coloured meaning they were camouflaged by the lichen from predation by birds. There were always a few darker moths present, due to variation, but these were quickly eaten and the allele frequency maintained.

When the lichens died and the trees became black the situation was reversed. The light-coloured moths were very visible and were eaten and the darker moths were camouflaged. Over time the allele frequency shifted due to natural selection and the majority of the peppered moths had the darker colour. The allele frequency had been shifted towards an extreme (less common) phenotype.

As pollution has decreased again the allele frequency of the lighter coloured moths has increased.

Disruptive selection

In **disruptive selection** the extremes are selected for and the norm selected against. The finches observed by Darwin in the Galapagos Islands had been subjected to disruptive selection. This is opposite to stabilising selection when the norm is positively selected.

▲ **Figure 6** *Light and dark-coloured peppered moths*

▲ **Figure 8** *Top: lazuli bird with bright, blue plumage, middle: lazuli bird with intermediate plumage, and bottom: lazuli bird with dull, brown plumage*

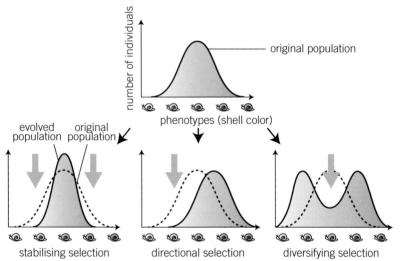

▲ **Figure 7** *Graphs showing the different forms of selection. The arrows indicate a selection pressure*

Although examples of disruptive selection are relatively rare, a well-documented example involves feather colour in male lazuli buntings (*Passerina amoena*), birds which are native to North America. The feather colour of young males can range from bright blue to dull brown.

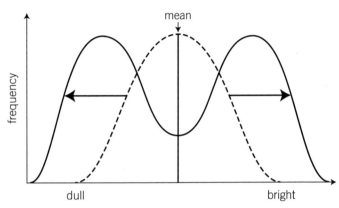

▲ **Figure 9** *The distribution of phenotypes as a result of disruptive selection pressures on lazuli buntings*

There are limited nesting sites in their habitat and so there is a lot of competition between male birds to establish territories and attract female birds. Dull, brown males are seen as non-threatening and bright, blue males too threatening by adult males. Both the brown and blue birds are therefore left alone but birds of intermediate colour are attacked by adult birds and so fail to mate or establish territories.

The extremes are selected for and the distribution of phenotypes shows two peaks as in Figure 9.

Synoptic link

You learnt about natural selection in Topic 10.4, Evidence for evolution.

Summary questions

1 Explain why evolution does not occur within single organisms but groups of organisms. (*3 marks*)

2 Around the world, humans choose their partners for a wide variety of reasons. Explain why this might affect any conclusions about human evolution drawn using the Hardy–Weinberg principle. (*3 marks*)

3 In cats, the short-haired allele L is dominant to the long-haired allele l. In a population of feral cats, 10 out of 90 animals had long hair. Give the expected frequency for the homozygous recessive, homozygous dominant, and heterozygote genotypes in this population of cats. (*6 marks*)

4 Explain why the allele frequency is changing so quickly in Figure 10. (*4 marks*)

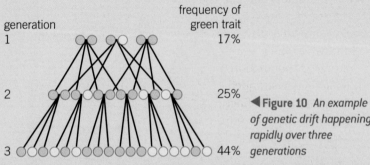

◀ **Figure 10** *An example of genetic drift happening rapidly over three generations*

5 Eukaryotic organisms have large quantities of non-coding DNA whereas most prokaryotic organisms have very little.

Suggest, with reference to the different forms of reproduction in eukaryotes and prokaryotes, why eukaryotes may have evolved to have more non-coding DNA. (*4 marks*)

20.6 Speciation and artificial selection

Specification reference: 6.1.2

Speciation is the formation of new species through the process of evolution. The organisms belonging to the new species will no longer be able to interbreed to produce fertile offspring with organisms belonging to the original species. A number of events happen leading to speciation:

- Members of a population become isolated and no longer interbreed with the rest of the population resulting in no gene flow between the two groups.

- Alleles within the groups continue to undergo random mutations. The environment of each group may be different or change (resulting in different selection pressures) so different characteristics will be selected for and against.

- The accumulation of mutations and changes in allele frequencies over many generations eventually lead to large changes in phenotype. The members of the different populations become so different that they are no longer able to interbreed (to produce fertile offspring). They are now reproductively isolated and are different species.

Learning outcomes

Demonstrate knowledge, understanding, and application of:

→ the role of isolating mechanisms in the evolution of new species

→ the principles of artificial selection and its uses

→ the ethical considerations surrounding the use of artificial selection.

Allopatric speciation

Allopatric speciation is the more common form of speciation and happens when some members of a population are separated from the rest of the group by a physical barrier such as a river or the sea – they are geographically isolated. The environments of the different groups will often be different and so will the selection pressures resulting in different physical adaptations. Separation of a small group will often result in the founder effect leading to genetic drift further enhancing the differences between the populations.

A famous example of allopatric speciation is the finches inhabiting the Galapagos Islands located in the Pacific Ocean off the coast of South America. For about two million years, small groups of finches, from an original population on the mainland, have flown to, and been stranded on, different islands. The finches, separated from finches on other islands and the mainland by the sea, have formed new colonies on the different islands.

The finches have evolved and adapted to the different environments, particularly food sources, present on the islands and are an example of *adaptive radiation* – where rapid organism diversification takes place. As the finches are unable to breed with each other, new species have evolved with unique beaks adapted to the type of food available. Some species

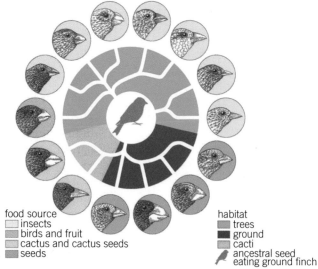

food source
☐ insects
■ birds and fruit
☐ cactus and cactus seeds
■ seeds

habitat
■ trees
■ ground
☐ cacti
✦ ancestral seed eating ground finch

▲ **Figure 1** *The range of beaks seen in species of finches found in the Galapagos Islands*

▲ **Figure 2** *Snapping shrimp*

have large, blunt beaks that can crack nuts, some have long, thin beaks for getting to the nectar in flowers, and some have medium-sized beaks which are ideal for catching insects.

The honeycreepers (family Drepanidinae) of the islands of Hawaii are birds that are an even larger example of adaptive radiation. A single ancestor species has led to the evolution of at least 54 species that have filled every available niche in the different islands.

Panama is a narrow strip of land (isthmus) that joins North and South America and separates the Atlantic and Pacific oceans, and was formed about three million years ago. This was due to the movement of tectonic plates and resulted in the separation of the organisms that had originally occupied the same habitat when the two oceans were joined.

There were originally about 15 species of snapping shrimp present, now there are 15 species present on one side of the isthmus and 15 different species present on the other side. Although the shrimp from either side appear to be identical if males and females are mixed they will snap at each other rather than mate.

In 1995 15 iguanas, *Iguana iguana*, survived a hurricane in the Caribbean on a raft of uprooted trees. They eventually reached the Caribbean island of Anguilla. These iguanas were the first of their species to reach the island. If these iguanas are successful in colonising the island it could be the start of an allopatric speciation. Of course, it could take thousands, if not millions, of years before this is known.

Sympatric speciation

Sympatric speciation occurs within populations that share the same habitat. It happens less frequently than allopatric speciation and is more common in plants than animals. It can occur when members of two different species interbreed and form fertile offspring – this often happens in plants. The hybrid formed, which is a new species, will have a different number of chromosomes to either parent and may no longer be able to interbreed with members of either parent population. This stops gene flow and reproductively isolates the hybrid organisms. Examples of sympatric speciation include fungus-farming ants and blind mole rats.

Fungus-farming ants cultivate the growth of fungi, which is their source of nutrition, by supplying organic material to keep the fungi growing. Parasitic ants have been found in one colony of these industrious ants. Instead of helping in the growth of this fungus these parasitic ants spend their time eating the fungi and reproducing. The parasitic ants are sometimes ignored and at other times attacked and killed. Genetic analysis has shown that although genetically different from the fungus-farming ants the parasitic ants are, in fact, their descendants. They are not a species of ant that has evolved in geographic isolation but within the same habitat as a result of a change in behaviour. It is believed that the genetic division of the original species of ant only happened 37 000 years ago, not long in evolutionary terms.

Blind mole rats live in a small area of northern Israel that is part igneous basalt rock and part chalk bedrock. The different types of soils formed above the bedrock support a different range of plants. Blind mole rats found in both types of soil are sometimes only separated by a few metres of the loose soil. Mole rats will only interbreed with mole rats living in the same type of soil. DNA analysis has shown that the lack of gene flow between the two species is already resulting in genetic differences even though members of the different groups often come into contact with each other as there is no physical barrier. Over time the genetic differences could accumulate to the point that the mole rats from different soil types will no longer be able to interbreed and they will be separate species.

Plants cross with plants of different species forming hybrids much more frequently than animals. The indiscriminate release of large numbers of pollen grains by plants is one reason for this. The hybrids are reproductively isolated from each parent species but could still be present in the same habitat. The evolution of modern wheat has involved at least two hybridisation events and the formation of new species along the way.

Disruptive selection, mating preferences, and other behavioural differences can all result in individuals or small groups becoming reproductively isolated. They will, however, still be living in the same habitat so gene flow, even if reduced, often interferes with the process of speciation.

Reproductive barriers

Barriers to successful interbreeding can form within populations before or after fertilisation has occurred. Prezygotic reproductive barriers prevent fertilisation and the formation of a zygote. Postzygotic reproductive barriers, often produced as a result of hybridisation, reduce the viability or reproductive potential of offspring.

Artificial selection

Populations are usually **polymorphic** (display more than one distinct phenotype) for most characteristics. The allele coding for the most common, or normal, characteristic is called the *wild type* allele. Other forms of that allele, resulting from mutations, are called mutants.

Artificial selection (or **selective breeding**) is fundamentally the same as natural selection except for the nature of the selection pressure applied. Instead of changes in the environment leading to survival of the fittest, it is the selection for breeding of plants or animals with desirable characteristics by farmers or breeders.

Farmers have been selectively breeding plants and animals since before genes were discovered or the theory of evolution was proposed. Individuals with the desired characteristics are selected and interbred.

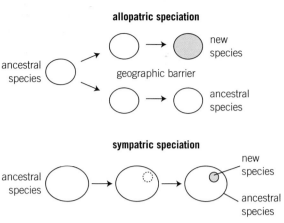

▲ **Figure 3** *Diagram showing the formation of a new species both with and without a geographical barrier*

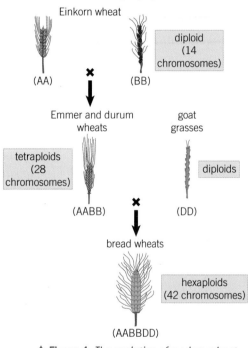

▲ **Figure 4** *The evolution of modern wheat including two hybridisation events which would have both led to sympatric speciation*

Offspring from this cross showing the best examples of the desired traits are then selected to breed. This breeding of closely related individuals is called **inbreeding**. The process is repeated over many generations resulting in changes to the frequency of alleles within the population and eventually speciation.

Brassica oleracea is a wild mustard which has been selectively bred for many centuries producing a number of common vegetables.

Problems caused by inbreeding

Limiting the gene pool and so decreasing genetic diversity reduces the chances of a population of inbred organisms evolving and adapting to changes in their environment.

Many genetic disorders are caused by recessive alleles, for example, cystic fibrosis, a condition where the digestive system and lungs become clogged with mucus. Recessive alleles are not uncommon in most populations but two recessive alleles are needed before they are expressed and most individuals will be heterozygous.

Organisms that are closely related are genetically similar and are likely to have the same recessive alleles. The breeding of closely related organisms therefore results in offspring which have a greater chance of being homozygous for these recessive traits and being affected by genetic disorders. Over time this reduces the ability of these organisms to survive and reproduce. This results in the organisms being less biologically fit – in other words, less likely to survive and produce two surviving offspring to replace themselves.

 ### Pedigree dogs and the ethics of artificial selection

Domesticated dogs are all members of the same species, *Canis familiaris*. They are another, sometimes controversial, example of how selective breeding has created a variety of different-looking individuals from one wild species. The wild species was the grey wolf and the process began between 18 000 and 32 000 years ago when humans were still leading a hunter-gatherer lifestyle.

It is thought that wolves starting 'hanging around' human hunting parties because of the availability of scraps of leftover food. Over time, the more social wolves would have become integrated into the human groups and so the process of selective breeding began. The wolves that became integrated were eventually used to help catch animals during the hunts or served a protective role like modern-day guard dogs.

Hunter-gatherers eventually started forming settled communities and began the practice of farming. The evolving wolves would have had new roles such as herding animals. Different traits were selected for depending on whether they were used for hunting, fighting, herding, or even as status symbols, and so a range of dogs with different characteristics evolved.

Many of those characteristics have been exaggerated by continued selective breeding and are most obvious in pedigree dogs seen today. Rather than being selected for the role they performed which was not dissimilar to natural selection, they began to be selected for their looks which took no account of any impact on their health. Interbreeding with wild wolves would have been common which can make tracing the evolution of dogs difficult.

Dachshunds were selected for small size and short legs so that they could follow prey such as foxes and badgers into burrows. Great Danes were selected for large size and strength for hunting and fighting.

The breeding of pedigree dogs is restricted to the descendants of dogs that were registered by the Kennel club in 1873 after the different types or breeds of dog had been developed by breeders and the standard characteristics of each breed identified. With the limited gene pool and lack of outbreeding it is inevitable that unwanted traits are selected for also. Big dogs often have hip or heart problems. The skull of the King Charles spaniel is too small to accommodate the brain comfortably leading to pain and discomfort. Bulldogs usually have breathing difficulties due to the shape of their noses. Dachshunds often have back problems and suffer from epilepsy. Diseased dogs have effectively been deliberately interbred. There are now moves to change some of the breed descriptions to prevent some of the worst examples of this practice.

Selective breeding has been used for centuries in farming to improve the quality and yield of crops and animal produce.

> Discuss the ethical considerations of using selective breeding to produce aesthetically pleasing show dogs and high-yielding plants and animals for agricultural use.

Gene banks

Seed banks keep samples of seeds from both wild type and domesticated varieties. They are an important genetic resource. **Gene banks** store biological samples, other than seeds, such as sperm or eggs. They are usually frozen.

Owing to the problems caused by inbreeding, alleles from gene banks are used to increase genetic diversity in a process called outbreeding. Breeding unrelated or distantly related varieties is also a form of outbreeding. This reduces the occurrence of homozygous recessives and increases the potential to adapt to environmental change.

> **Synoptic link**
>
> You learnt about seed banks in Topic 11.8, Methods of maintaining biodiversity.

Summary questions

1 Describe, with examples, the difference between pre-zygotic and post-zygotic reproductive barriers. *(4 marks)*

2 A small population of adders in Sweden underwent inbreeding depression when farming activities isolated them from other adder populations. The numbers of stillborn and deformed offspring increased as compared to the original population. Researchers introduced adders from other population and the isolated population recovered and produced a higher proportion of viable offspring.
   a Name the type of breeding carried out by the researchers. *(1 mark)*
   b Explain how this type of breeding reduces the problems caused by inbreeding. *(3 marks)*

3 Discuss why variation within a species has to be present for speciation to occur. *(5 marks)*

4 Allopatric speciation is considered by most biologists to be the most common way in which new species evolve.

   Outline the differences between sympatric speciation and allopatric speciation and suggest why some biologists think sympatric speciation is a rare event. *(5 marks)*

Practice questions

1 **a** The presence or absence of red pigmentation in the outer scales of onion bulbs is controlled by two genes, A/a and B/b.

- The dominant allele, A, codes for the production of a red anthocyanin pigment.
- Onion bulbs homozygous for the recessive allele, a, produce no pigment and are white.
- The dominant allele, B, inhibits the expression of allele A.
- The recessive allele, b, allows anthocyanin production.

  (i) Describe the effect of allele B on allele A. *(3 marks)*

  (ii) Two onion plants with the genotypes AABB and aabb were cross-pollinated and the resulting F1 generation interbred to give an F2 generation.

  Draw a genetic diagram of this cross to show:

  - the phenotypes of the parent plants
  - the gametes
  - the genotypes and phenotypes of the F1 and F2 generations
  - the ratio of phenotypes expected in the F2 generation. *(8 marks)*

b Most red-scaled onion bulbs produce a colourless substance which makes them resistant to a fungal infection called 'smudge'. Most white onion bulbs are susceptible to 'smudge'. Suggest why:

  (i) Resistance to 'smudge' is almost always inherited together with red pigmentation *(2 marks)*

  (ii) Some white onion bulbs are resistant to 'smudge'. *(2 marks)*

  OCR 2805/02 2008

2 A pure-breeding variety of tomato plant, variety A, produced red fruit which remained green at their bases even when ripe.

Plants of variety A were crossed with another pure-breeding variety, B, with orange fruit which have no green bases when ripe.

The F1 generation plants all had red fruit with green bases.

a Describe the interaction of the alleles,

  (i) At the locus G/g, controlling green-based or not green-based fruit *(1 mark)*

  (ii) At the locus R/r, controlling red or orange fruit. *(1 mark)*

b Using the symbols given in (a), state the genotype of variety B. *(1 mark)*

c Plants from the F1 generation were test crossed (backcrossed) to variety B. The ratio of phenotypes expected in a dihybrid test cross such as this is 1 : 1 : 1 : 1.

Using the symbols given in (a), draw a genetic diagram of the test cross to show that the expected ratio of offspring phenotypes is 1 : 1 : 1 : 1. *(4 marks)*

d Two hundred randomly chosen offspring from the test cross described in c had the following phenotypes:

- red fruit with green bases 55
- red fruit with no green bases 45
- orange fruit with green bases 43
- orange fruit with no green bases 57

The χ^2 (chi-squared) test was performed on these data, giving a calculated value for χ^2 of 3.2.

  (i) State the number of degrees of freedom applicable to these data. *(1 mark)*

Distribution of χ^2 values

degrees of freedom	probability p				
	0.10	0.05	0.02	0.01	0.001
1	2.71	3.84	5.41	6.63	10.83
2	4.60	5.99	7.82	9.21	13.82
3	6.25	7.81	9.84	11.34	16.27
4	7.78	9.49	11.67	13.28	18.46

  (ii) Use the calculated value of χ^2 and the table of probabilities provided in the table to find the probability of the results of the test cross departing significantly by chance from the expected ratio. *(1 mark)*

(iii) State what conclusions may be drawn from the probability found in (d)(ii).
(3 marks)

e Experiments have shown that loci G/g and R/r are on the same chromosome of the tomato plant genome. The two loci are 44 map units apart.

Explain how the results of the test cross shown in (d) could occur when the two loci are on the same chromosome.
(3 marks)

OCR 2805/02 2008

3 Bread wheat is a hexaploid (6n) plant, with three sets of paired chromosomes.

The likely origin of hexaploid bread wheat from diploid wild grasses is shown in the diagram.

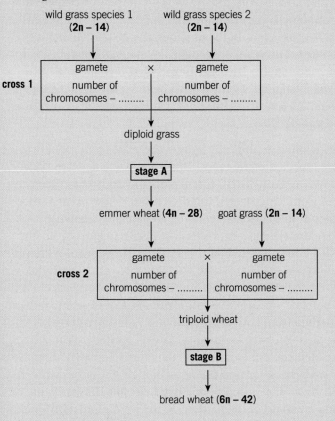

a (i) Copy the diagram and write into the spaces the numbers of chromosomes in the gametes involved in cross 1 and cross 2.
(2 marks)

(ii) With reference to the diagram, explain what happened at stages A and B to give emmer wheat and bread wheat respectively.
(2 marks)

b For many years, plant breeders have selectively bred wheat varieties with progressively higher yields. However, bread wheat cannot be interbred with diploid species of grass to establish a variety of wheat with new traits.

Explain why bread wheat cannot be interbred with diploid species of grass to establish new varieties of wheat. *(2 marks)*

c The International Maize and Wheat Improvement Centre (CIMMYT) in Mexico has re-created cross 2 and stage B of the diagram by interbreeding different varieties of wild emmer wheat and goat grass.

In this way more than 1000 varieties of hexaploid 'synthetic wheat' have been produced which can then be interbred with bread wheat.

Explain the need for CIMMYT to maintain seed banks of emmer wheat and goat grass.
(8 marks)

OCR 2805/02 2008

▲ **Figure 1** *All three of these girls have identical DNA because they were formed from a single fertilised egg – and this included both coding DNA and introns. However, the DNA of the great majority of human beings is unique to each individual*

Every person has a unique combination of DNA in the chromosomes of their cells, unless they are an identical twin or triplet. Your DNA is more similar to your family members' than to other people's, and more similar to other people's than to a gorilla's, a butterfly's or a banana's. It is this combination of individual uniqueness yet similarity to other family or species members that makes DNA so useful in solving crimes, predicting disease, and classifying organisms.

The human genome

The **genome** of an organism is all of the genetic material it contains – for eukaryotes including ourselves, that is the DNA in the nucleus and the mitochondria combined. The chromosomes are made up of hundreds of millions of DNA base pairs, but your genes, the 20–25 000 regions of the DNA that code for proteins, only make up about 2% of your total DNA. They are called exons. The large non-coding regions of DNA that are removed from messenger (m)RNA before it is translated into a polypeptide chain are known as **introns**. Scientists are still investigating the role of non-coding DNA.

Within introns, telomeres, and centromeres there are short sequences of DNA that are repeated many times. This is known as satellite DNA. In a region known as a minisatellite, a sequence of 20–50 base pairs will be repeated from 50 to several hundred times. These occur at more than 1000 locations in the human genome and are also known as variable number tandem repeats (VNTRs). A microsatellite is a smaller region of just 2–4 bases repeated only 5–15 times. They are also known as short tandem repeats (STRs). These satellites always appear in the same positions on the chromosomes, but the number of repeats of each mini- or microsatellite varies between individuals, as different lengths of repeats are inherited from both parents. So just as in the coding DNA, only identical twins will have an identical satellite pattern, although the more closely related you are to someone, the more likely you are to have similar patterns. These patterns in the non-coding DNA were discovered by Professor Sir Alec Jeffreys and his team at Leicester University in 1984. Producing an image of the patterns in the DNA of an individual is known as **DNA profiling** and is a technique employed by scientists to assist in the identification of individuals or familial relationships.

Producing a DNA profile

The process of producing a DNA profile has five main stages:

1   Extracting the DNA

The DNA must be extracted from a tissue sample. When DNA profiling was first discovered, relatively large samples were needed – about 1 μg of

DNA, equivalent to the DNA from the nuclei of about 10 000 human cells. Now, using a technique called the **polymerase chain reaction** (PCR), the tiniest fragment of tissue can give scientists enough DNA to develop a profile.

2 Digesting the sample

The strands of DNA are cut into small fragments using special enzymes called **restriction endonucleases**. Different restriction endonucleases cut DNA at a specific nucleotide sequence, known as a restriction site or recognition site. All restriction endonucleases make two cuts, once through each strand of the DNA double helix. There are many different restriction endonucleases – the recognition sequences and cut sites of three examples are given in Table 1.

Restriction endonucleases give scientists the ability to cut the DNA strands at defined points in the introns. They use a mixture of restriction enzymes that leave the repeating units or satellites intact, so the fragments at the end of the process include a mixture of intact mini- and microsatellite regions.

▼ **Table 1** *Three restriction enzymes and the nucleotide sequences they recognise. The black triangles show where they cut*

Enzyme	Recognition sequence
Sau3A1	5′…▼GATC…3′ 3′…CTAG▲…5′
Not1	5′…GC▼GGCCGC…3′ 3′…CGCCGG▲CG…5′
Alu1	5′…AG▼CT…3′ 3′…TC▲GA…5′

3 Separating the DNA fragments

To produce a DNA profile, the cut fragments of DNA need to be separated to form a clear and recognisable pattern. This is done using **electrophoresis**, a technique that utilises the way charged particles move through a gel medium under the influence of an electric current. The gel is then immersed in alkali in order to separate the DNA double strands into single strands. The single-stranded DNA fragments are then transferred onto a membrane by Southern blotting. (See the Application box for more detail).

4 Hybridisation

Radioactive or fluorescent DNA probes are now added in excess to the DNA fragments on the membrane. DNA probes are short DNA or RNA sequences complementary to a known DNA sequence. They bind to the complementary strands of DNA under particular conditions of pH and temperature. This is called hybridisation. DNA probes identify the microsatellite

①

extraction
DNA is extracted from the sample

②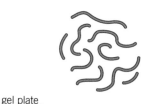

digestion
restriction endo-nucleases cut the DNA into fragments

③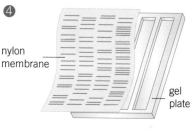

gel plate
large fragments
small fragments
direction of movement

separation
fragments are separated using gel electrophoresis

④

nylon membrane
gel plate

separation (cont.)
DNA fragments are transferred from the gel to nylon membrane in a process known as Southern blotting

⑤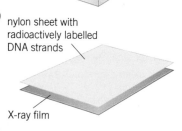

DNA probes

hybridisation
DNA probes are added to label the fragments. These radioactive probes attach to specific fragments

⑥

nylon sheet with radioactively labelled DNA strands

X-ray film

development
membrane with radioactively labelled DNA fragments is placed onto an X-ray film

⑦

development (cont.)
development of the X-ray film reveals dark bands where the radioactive or fluorescent DNA probes have attached

▲ **Figure 2** *Summary of DNA profiling*

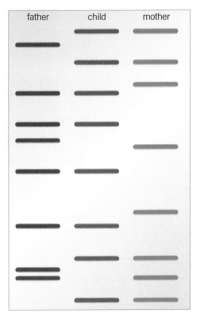

father   child   mother

▲ **Figure 3** *DNA profile of a child and its parents. How many fragments were inherited from each parent?*

regions that are more varied than the larger minisatellite regions. The excess probes are washed off.

5   Seeing the evidence

If radioactive labels were added to the DNA probes, X-ray images are taken of the paper/membrane. If fluorescent labels were added to the DNA probes, the paper/membrane is placed under UV light so the fluorescent tags glow. This is the method most commonly used today. The fragments give a pattern of bars – the DNA profile – which is unique to every individual except identical siblings.

Separation of nucleic acid fragments by electrophoresis

DNA fragments are put into wells in agarose gel strips (Figure 4 and 5), which also contain a buffering solution to maintain a constant pH. In one or more wells (usually the first and last), DNA fragments of known length are used to provide a reference for fragment sizing.

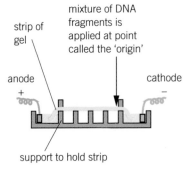

strip of gel

mixture of DNA fragments is applied at point called the 'origin'

anode +

cathode −

support to hold strip

▲ **Figure 4** *Gel electrophoresis apparatus*

▲ **Figure 5** *Preparing a gel plate for electrophoresis*

When an electric current is passed through the electrophoresis plate, the DNA fragments in the wells at the cathode end move through the gel towards the positive anode at the other end. This is due to the negatively charged phosphate groups in the DNA fragments. The rate of movement depends on the mass or length of the DNA fragments – the gel has a mesh-like structure that resists the movement of molecules. Smaller fragments can move through the gel mesh more easily than larger fragments. Therefore, over a period of time, the smaller fragments move further than the larger fragments. When the faster smallest fragments reach the anode end of the gel, the electric current is switched off.

The gel is then placed in an alkaline buffer solution to denature the DNA fragments. The two DNA strands of each fragment separate, exposing the bases.

In a technique called Southern blotting (named after its inventor, Edwin Southern), these strands are transferred to a nitrocellulose paper or a nylon membrane, which is placed over the gel. The membrane is covered with several sheets of dry absorbent paper, drawing the alkaline solution containing the DNA through the membrane by capillary action (Figure 6).

The single-stranded fragments of DNA are transferred to the membrane, as they are unable to pass through it. They are transferred in precisely the same relative positions as they had on the gel. They are then fixed in place using UV light or heated at 80°C.

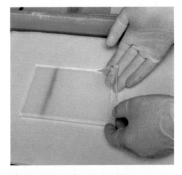

▲ **Figure 6** *Southern blot apparatus*

Gel electrophoresis can also be used to separate proteins. Explain why.

Polymerase chain reaction (PCR)

DNA profiling is often used in solving crimes and only very tiny amounts of DNA may be available. The PCR is a version of the natural process by which DNA is replicated, and allows scientists to produce a lot of DNA from the tiniest original sample.

The DNA sample to be amplified, an excess of the four nucleotide bases A, T, C, and G (in the form of deoxynucleoside triphosphates), small primer DNA sequences, and the enzyme DNA polymerase are mixed in a vial that is placed in a PCR machine (also called a thermal cycler). The temperature within the PCR machine is carefully controlled and changes rapidly at programmed intervals, triggering different stages of the process (Figure 7). The reaction can be repeated many times by the PCR machine, which cycles through the programmed temperature settings. About 30 repeats gives around one billion copies of the original DNA sample – more than enough to carry out DNA profiling.

Step 1 Separating the strands:
The temperature in the PCR machine is increased to 90–95 °C for 30 seconds, this denatures the DNA by breaking the hydrogen bonds holding the DNA strands together so they separate.

Step 2 Annealing of the primers:
The temperature is decreased to 55–60 °C and the primers bind (anneal) to the ends of the DNA strands. They are needed for the replication of the strands to occur.

Step 3 Synthesis of DNA:
The temperature is increased again to 72–75 °C for at least 1 minute, this is the optimum temperature for DNA polymerase to work best. DNA polymerase adds bases to the primer, building up complementary strands of DNA and so producing double-stranded DNA identical to the original sequence. The enzyme Taq polymerase is used, which is obtained from thermophilic bacteria found in hot springs.

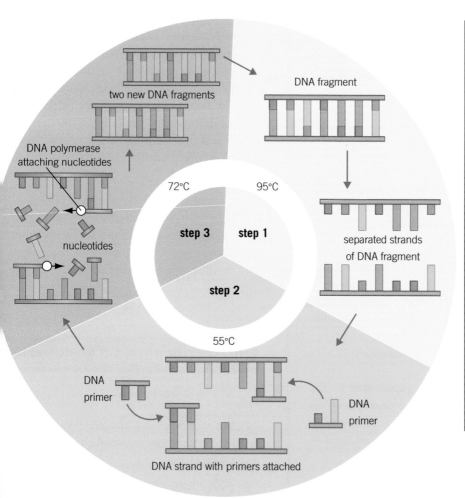

▲ **Figure 7** *The main stages of the polymerase chain reaction*

The uses of DNA profiling

DNA profiling has many uses. Its best known use is in the field of forensic science, especially criminal investigations. PCR and DNA profiling is performed on traces of DNA left at the crime scene. These DNA traces can be obtained from blood, semen, saliva, hair roots, and skin cells. The DNA profile is compared to that of a sample taken from a suspect, or can be identified from a criminal DNA database. DNA profiling is an extremely useful tool in providing evidence for either the guilt or innocence of a suspect.

DNA profiling is also used to prove paternity of a child when it is in doubt. It is used in immigration cases to prove or disprove family relationships. Identifying the species to which an organism belongs can also now be done by DNA profiling, which is much more accurate than any of the older methods. It is also increasingly used to demonstrate the evolutionary relationships between different species.

Another valuable use of DNA profiling is in identifying individuals who are at risk of developing particular diseases. Certain non-coding microsatellites, or the repeating patterns they make, have been found to be associated with an increased risk/incidence of particular diseases, including various cancers and heart disease. These specific gene markers can be identified and observed in DNA profiles.

The information that scientists can obtain from DNA profiling is often used together with the more detailed information obtained from DNA sequencing (Topic 21.2) to make more confident risk assessments for different diseases.

Pitfalls of profiling

One of the earliest recorded cases of mistaken DNA identity occurred in the UK in 2000. Raymond Easton was in the advanced stages of Parkinson's disease – he could hardly dress himself yet he was arrested and charged with a burglary that happened over 200 miles from his home. The arrest was based solely on DNA evidence.

Four years earlier, Raymond had been involved in a family dispute that had got out of hand. He received a police caution and his DNA was taken and kept on file. DNA from the 2000 burglary scene appeared to match Raymond's profile.

Raymond protested his innocence and had a strong alibi for the time of the burglary, so eventually a more rigorous DNA test, looking at satellites in 10 loci rather than the original six, was carried out. None of the additional satellites matched Raymond's DNA and so the charges were dropped. This mis-identification had been caused by an extremely improbable, but not impossible, coincidental DNA profile match.

1   In a UK court of law, until recently DNA was taken as evidence if 11 loci and a sex marker matched. Recently this was put up to 17 loci. Suggest reasons why 11 loci were acceptable and why the number has now risen to 17.

2   DNA evidence at this level is not absolute. The probability that full siblings will have the same DNA profile for these 10 loci is quoted by the Forensic Science Service as 1 in 10 000, and of first cousins, 1 in 100 million. Suggest explanations for these observations.

Summary questions

1   What is an intron?   (3 marks)

2   State the purpose of the polymerase chain reaction and explain how it has advanced DNA profiling.   (4 marks)

3   Discuss the benefits and limitations of DNA profiling.   (6 marks)

4   Produce a flow diagram to show the main stages of the modern DNA profiling process.   (6 marks)

21.2 DNA sequencing and analysis
Specification reference: 6.1.3

DNA sequencing is the process of determining the precise order of nucleotides within a DNA molecule. This knowledge is invaluable in various scientific applications, from diagnostics to biotechnology, and its uses are explored further in the next topic. In recent years, scientists have not only discovered how to read the information in the genome – they have developed ways of reading it fast.

The beginning of DNA sequencing

DNA sequencing was just an aspiration for scientists until Frederick Sanger and his team developed some techniques for sequencing nucleic acids from viruses and then bacteria. The technique involved radioactive labelling of bases and gel electrophoresis on a single gel. The processes were carried out manually, so it took a long time, but eventually, in the 1970s, the technique now known as Sanger sequencing enabled Sanger and his team to read sequences of 500–800 bases at a time. The first entire genome that they sequenced was just over 5000 bases long and belonged to phiX174, a virus that attacks bacteria. They went on to sequence many other genomes, including the 16 000 base pairs of human mitochondrial DNA. In 1980, Frederick Sanger was awarded the Nobel Prize for his work on sequencing DNA – this was his second Nobel Prize, his first was in 1958 for determining the sequence of the amino acids in insulin. These DNA sequencing techniques are continually being refined. One such development was the swapping of radioactive labels for coloured fluorescent tags, which led to scaling up and automation of the process. This in turn led to the capillary sequencing version of the Sanger sequencing method that was used during the Human Genome Project (HGP), and similar techniques that are used today.

The human genome – pushing the boundaries

In 1990, the HGP was established. It was a massive international project in which scientists from a number of countries worked to map the entire human genome, making the data freely available to scientists all over the world. The early work involved sequencing the DNA of smaller, simpler organisms to refine and develop the techniques. In 1995, after 18 months of work, scientists completed the 1.8 million base pair genome of the bacterium *Haemophilus influenza*. By 1998, the UK team at the Sanger Centre and a US team at Washington University had sequenced the genome of *Caenorhabditis elegans* (*C. elegans*), a nematode worm widely used in scientific experiments, before applying the technique to the three billion base pairs of the human genome itself.

The aim was to complete the HGP in 15 years but the automation of sequencing techniques and the development of more powerful, faster computers meant that the first draft of the human genome was

Learning outcomes

Demonstrate knowledge, understanding, and application of:

→ the principles of DNA sequencing

→ the development of new DNA sequencing techniques.

Synoptic link

You learnt about the structure of proteins and nucleic acids in Chapter 3, Biological molecules and about the ultrastructure of mitochondria in Topic 2.4, Eukaryotic cell structure.

▲ Figure 1 *From* C. elegans *to a human being may seem a giant leap – but our genomes are surprisingly similar. Scanning electron micrograph, approximately ×650 magnification*

▲ **Figure 2** *The contribution of Fred Sanger to our knowledge and understanding of genomes earned him his second Nobel Prize*

Synoptic link

You learnt about the structure of DNA in Topic 3.8, Nucleic acids.

ready in 2000, and the first complete human genome sequence was published in 2003, two years ahead of schedule and under budget.

Principles of DNA sequencing

Sequencing the genome of an organism involves a number of different processes. The DNA is chopped into fragments and each fragment is sequenced. The process involves terminator bases, modified versions of the four nucleotide bases, adenine (A), thymine (T), cytosine (C), and guanine (G), which stop DNA synthesis when they are included. An A terminator will stop DNA synthesis at the location that an A base would be added, a C terminator where a C base would go, and so on. The terminator bases are also given coloured fluorescent tags – A is green, G is yellow, T is red and C is blue. The description of the sequencing process (capillary method) explained here is a simplified version of a technique, has largely been overtaken by much more complex methods – but the basic principles remain the same:

1   The DNA for sequencing is mixed with a primer, DNA polymerase, an excess of normal nucleotides (containing bases A, T, C, and G) and terminator bases.

2   The mixture is placed in a thermal cycler – a piece of equipment as used for PCR (Topic 21.1, DNA profiling) that rapidly changes temperature at programmed intervals in repeated cycles – at 96°C the double-stranded DNA separates into single strands, at 50°C the primers anneal to the DNA strand.

3   At 60°C DNA polymerase starts to build up new DNA strands by adding nucleotides with the complementary base to the single-strand DNA template.

4   Each time a terminator base is incorporated instead of a normal nucleotide, the synthesis of DNA is terminated as no more bases can be added. As the chain-terminating bases are present in lower amounts and are added at random, this results in many DNA fragments of different lengths depending on where the chain terminating bases have been added during the process. After many cycles, all of the possible DNA chains will be produced with the reaction stopped at every base. The DNA fragments are separated according to their length by capillary sequencing, which works like gel electrophoresis in minute capillary tubes. The fluorescent markers on the terminator bases are used to identify the final base on each fragment. Lasers detect the different colours and thus the order of the sequence.

5   The order of bases in the capillary tubes shows the sequence of the new, complementary strand of DNA which has been made. This is used to build up the sequence of the original DNA strand.

The data from the sequencing process is fed into a computer that reassembles the genomes by comparing all the fragments and finding the areas of overlap between them. Once a genome is assembled, scientists want to identify the genes or parts of the genome that code for specific characteristics. Medical researchers want to identify regions that are linked with particular diseases. Many genomes are freely available online – anyone who chooses to can have a look at them.

Stages 1–2: DNA strand chopped up, mixed with primer, bases, DNA polymerase + terminator bases

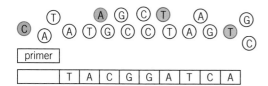

Stage 3: Each time a terminator base is added a strand terminates until all possible chains produced

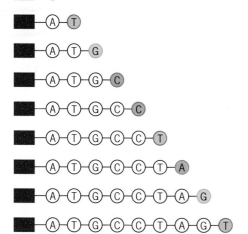

Stage 4: Readout from capillary tubes: DNA fragments separated by electrophasis in capillary tubes by mass and lasers detect the colours and the sequence

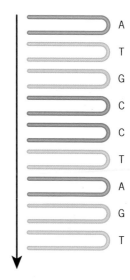

Stage 5: Computer analysis of all data to give original DNA sequence

▲ **Figure 3** *The main steps in the DNA sequencing process*

Next-generation sequencing

Early on, working out the base sequence of even short strands of DNA was difficult and time-consuming for scientists using the original Sanger sequencing method. DNA sequencing technologies have become faster and more automated as they have been developed. Recently, technological developments have led to new, automated, high-throughput sequencing processes. Instead of using a gel or capillaries, the sequencing reaction takes place on a plastic slide known as a flow cell. Millions of fragments of DNA are attached to the slide and replicated in situ using PCR to form clusters of identical DNA fragments. The sequencing process still uses the principle of adding a coloured terminator base to stop the reaction so an image can be taken. As all of the clusters are being sequenced and imaged at the same time, the technique is known as 'massively parallel sequencing' and sometimes referred to as 'next-generation sequencing'.

The process of massively parallel sequencing is integrated with state-of-the-art computer technology and is constantly being refined and developed. These high-throughput methods are extremely efficient and very fast – the 3 billion base pairs of the human genome can be sequenced in days and those of a bacterium in less than 24 hours. High-throughput sequencing also means that the cost has fallen, so more genomes can be sequenced. These techniques are being used by projects such as the 100 000 Genomes Project. They open up the range of questions that scientists can ask and enable us to use the information from the genome in many new and different ways (21.3, Using DNA sequencing).

▲ **Figure 4** *The Wellcome Trust Sanger Institute is the biggest DNA sequencing facility in Europe – on this site, the genomes of organisms from bacteria and viruses to humans and even human cancers are sequenced continuously in a highly automated process*

Summary questions

1 Produce a flow chart to summarise the main stages of DNA sequencing. *(5 marks)*

2 a ⚙ What is the difference in the time it takes to sequence the genetic material of a bacterium today compared to the first complete bacterial genome in 1995? *(1 mark)*
   b Explain the reasons for the difference in these times. *(2 marks)*

3 a What are terminator bases? *(1 mark)*
   b Explain why terminator bases are so important in both the Sanger method of sequencing and in the more modern high-throughput sequencing methods. *(6 marks)*

21.3 Using DNA sequencing

Specification reference: 6.1.3

The development of DNA profiling and DNA sequencing has led to the development of new areas of bioscience that help us analyse, understand, and make use of all the data generated.

Computational biology and bioinformatics

People often use the terms **computational biology** and **bioinformatics** interchangeably. In fact they describe different aspects of the application of computer technology to biology.

Bioinformatics is the development of the software and computing tools needed to organise and analyse raw biological data, including the development of algorithms, mathematical models, and statistical tests that help us to make sense of the enormous quantities of data being generated.

Computational biology then uses this data to build theoretical models of biological systems, which can be used to predict what will happen in different circumstances. Computational biology is the study of biology using computational techniques, especially in the analysis of huge amounts of biodata. For example, it is important in the analysis of the data from sequencing the billions of base pairs in DNA, for working out the 3D structures of proteins, and for understanding molecular pathways such as gene regulation. It helps us to use the information from DNA sequencing – for example in identifying genes linked to specific diseases in populations and in determining the evolutionary relationships between organisms.

Genome-wide comparisons

As whole genome sequencing has become increasingly automated, it has become cheaper and faster, leading to some amazing advances in biology. The field of genetics that applies DNA sequencing methods and computational biology to analyse the structure and function of genomes is called genomics.

Analysing the human genome

Since the first complete draft of the human genome was published in 2003 (Topic 21.2, DNA sequencing and analysis), tens of thousands of human genomes have been sequenced as part of research projects such as the 10 000 Genomes Project UK10K, and most recently the 100 000 Genomes Project.

Computers can analyse and compare the genomes of many individuals, revealing patterns in the DNA we inherit and the diseases to which we are vulnerable. This has enormous implications for health management and the field of medicine in the future. Genomics is changing the face of epidemiology. However, scientists increasingly

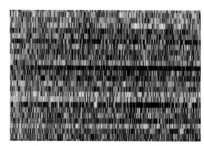

▲ **Figure 1** *The impact of computers on the study of biology is growing all the time. This photo shows data from a gel electrophoresis experiment where a grapevine genome was sequenced*

Synoptic link

You learnt about epidemiology in Topic 12.7, Preventing and treating disease.

recognise, with the exception of a few relatively rare genetic diseases caused by changes in a single gene, that our genes work together with the environment to affect our physical characteristics, our physiology, and our likelihood of developing certain diseases.

Analysing the genomes of pathogens

Sequencing the genomes of pathogens including bacteria, viruses, fungi, and protoctista has become fast and relatively cheap. This enables:

- Doctors to find out the source of an infection, for example bird flu or MRSA in hospitals.

- Doctors to identify antibiotic-resistant strains of bacteria, ensuring antibiotics are only used when they will be effective and helping prevent the spread of antibiotic resistance. For example, the bacteria that cause tuberculosis (TB) are difficult to culture, slow growing, and some strains are resistant to most antibiotics. Whole genome analysis makes it easier to track the spread of transmission and to plan suitable treatment options. This has enormous implications for successful treatment of this potentially fatal disease, especially as TB is spreading fast around the world again, linked to the spread of HIV/AIDS.

- Scientists to track the progress of an outbreak of a potentially serious disease and monitor potential epidemics, for example flu each winter, Ebola virus in 2014/15.

- Scientists to identify regions in the genome of pathogens that may be useful targets in the development of new drugs and to identify genetic markers for use in vaccines.

Identifying species (DNA barcoding)

Using traditional methods of observation, it can be very difficult to determine which species an organism belongs to or if a new species has been discovered. Genome analysis provides scientists with another tool to aid in species identification, by comparison to a standard sequence for the species. The challenge for scientists is to produce stock sequences for all the different species.

One useful technique is to identify particular sections of the genome that are common to all species but vary between them, so comparisons can be made – this technique is referred to as DNA barcoding. In the International Barcode of Life (iBOL) project, scientists identify species using relatively short sections of DNA from a conserved region of the genome. For animals, the region chosen is a 648 base-pair section of the mitochondrial DNA in the gene cytochrome c oxidase, that codes for an enzyme involved in cellular respiration. This section is small enough to be sequenced quickly and cheaply, yet varies enough to give clear differences between species. In land plants, that region of the DNA does not evolve quickly enough to show clear differences between species, but two regions in the DNA of the chloroplasts have been identified that can be used in a similar way to identify species.

Synoptic link

You learnt about how organisms are classified and grouped in Topic 10.1, Classification.

The barcoding system is not perfect – so far scientists have not come up with suitable regions for fungi and bacteria, and they may not be able to do so – but DNA sequencing is nevertheless having a big impact on classification.

Searching for evolutionary relationships

Genome sequencing has given scientists a powerful tool to help them understand the evolutionary relationships between organisms. DNA sequences of different organisms can be compared – because the basic mutation rate of DNA can be calculated scientists can calculate how long ago two species diverged from a common ancestor. DNA sequencing enables scientists to build up evolutionary trees with an accuracy they have never had before.

Genomics and proteomics

Proteomics is the study and amino acid sequencing of an organism's entire protein complement. Traditionally, scientists thought that each gene codes for a particular protein, but we now know that there are 20–25 000 coding genes in the human DNA but a very different number of unique proteins. Estimates range from somewhere between 250 000 and 1 000 000 different proteins to only 17–18 000 so there is still a lot of work to be done. More scientific evidence is emerging that highlights the complexity of the relationship between the genotype and the phenotype of an individual. The DNA sequence of the genome should, in theory, enable you to predict the sequence of the amino acids in all of the proteins it produces. The evidence is that the sequence of the amino acids is not always what would be predicted from the genome sequence alone. Some genes can code for many different proteins.

Spliceosomes

The mRNA transcribed from the DNA in the nucleus includes both the exons and introns. Before it lines up on the ribosomes to be translated, this 'pre-mRNA' is modified in a number of ways. The introns are removed, and in some cases, some of the exons are removed as well. Then the exons to be translated are joined together by enzyme complexes known as spliceosomes to give the mature functional mRNA. The spliceosomes may join the same exons in a variety of ways. As a result, a single gene may produce several versions of functional mRNA, which in turn would code for different arrangements of amino acids, giving different proteins and resulting in several different phenotypes.

Protein modification

Some proteins are modified by other proteins after they are synthesised. A protein that is coded for by a gene may remain intact or it may be shortened or lengthened to give a variety of other proteins.

The study of proteomics is constantly giving us increasing knowledge of the extremely complex relationship between the genotype and the phenotype.

> **Synoptic link**
>
> You learnt about pre-mRNA in Topic 19.2, Control of gene expression.

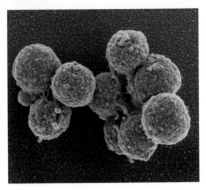

▲ **Figure 2** Mycoplasma mycoides, *the first organism to be controlled by synthetic DNA, was created at the University of California, USA in 2010. Scanning electron micrograph, ×21 000 magnification*

Synoptic link

You will learn more about microorganisms in industrial contexts in Topic 22.4, Microorganisms and biotechnology, Topic 22.6, Culturing microorganisms in the laboratory, and Topic 22.7, Culturing microorganisms on an industrial scale.

Synthetic biology

The ability to sequence the genome of organisms and understand how each sequence is translated into amino acids, along with the ever-increasing ability of computers to store, manipulate, and analyse the data, has led to the development of the new field of science called **synthetic biology**. Synthetic biology is defined by the Royal Society as 'an emerging area of research that can broadly be described as the design and construction of novel artificial biological pathways, organisms or devices, or the redesign of existing natural biological systems.'

Synthetic biology includes many different techniques. These include:

● genetic engineering – this may involve a single change in a biological pathway or relatively major genetic modification of an entire organism (further detail is given in the next topic)

● use of biological systems or parts of biological systems in industrial contexts, for example, the use of fixed or immobilised enzymes and the production of drugs from microorganisms

● the synthesis of new genes to replace faulty genes, for example, in developing treatments for cystic fibrosis (CF), scientists have attempted to synthesise functional genes in the laboratory and use them to replace the faulty genes in the cells of people affected by CF

● the synthesis of an entire new organism. In 2010, scientists announced that they had created an artificial genome for a bacterium and successfully replaced the original genome with this new, functioning genome.

➕ Synthetic life

● Scientists have developed some new nucleotide bases (not adenine, thymine, cytosine, or guanine) which, in a test tube, can be incorporated into a strand of DNA by special enzymes. The bases fit together well – they are not held by hydrogen bonds like the natural bases.

● In 2014, scientists introduced a small section of DNA made with these synthetic bases into bacteria. They found that this unique DNA, including the synthetic nucleotide bases, was replicated time after time as long as they supplied the bacteria with the synthetic bases.

● If these bases can be incorporated into the main DNA of an organism, and then transcribed into RNA, synthetic biologists will have synthetically expanded the genetic code for the very first time.

Suggest a possible practical application for the synthetic expansion of the genetic code.

Infection outbreak – DNA sequencing and clinical intervention

In 2012, there was an outbreak of MRSA (methicillin-resistant *Staphylococcus aureus*) in the Special Care Baby Unit (SCBU) at a UK hospital. The hospital infection control team identified 12 patients carrying MRSA. Researchers at the Wellcome Trust Sanger Institute used DNA sequencing to show that all the bacteria were closely related and this was a hospital-based outbreak.

The sequencing also showed that a number of people living in the community who developed MRSA at the same time all had the same strain as the hospital outbreak. In every case it was found they had a recent link to the hospital.

Two months later, another baby developed MRSA in the same SCBU. Immediate DNA sequencing showed that it was the same strain as the previous outbreak. This suggested that someone working in the hospital was unknowingly carrying MRSA.

Over 150 healthcare workers were screened – and one staff member was found to be carrying MRSA. DNA sequencing confirmed that it was the strain linked to the outbreak. The healthcare worker went through a process to eradicate the MRSA – and the risk of any further infections was removed.

The use of DNA sequencing was critical in identifying that the infections were connected and that a member of staff was a carrier. Without it, this would have been seen as a new outbreak, and many more people could have been infected.

This was the first time DNA sequencing has led to an immediate and successful clinical intervention – but it will certainly not be the last.

▲ **Figure 3** *Babies born early or with health problems are very vulnerable to infection. Sequencing the genome of pathogens so that effective treatment can be introduced as fast as possible and outbreaks halted is a big step forward*

Scientists predict that advances in science and technology may, in a few years, lead to every hospital having its own sequencing machines and hand held devices. These hand held devices would be capable of sequencing the genome, identifying a pathogen within a few minutes from just a drop of blood. Suggest some of the benefits of such developments, and some of the factors that will limit their arrival in our hospitals and local surgeries.

Summary questions

1   Explain the impact of computational biology and bioinformatics on the usefulness of DNA sequencing to scientists.   (*4 marks*)

2   a   Explain how the ability to sequence the genome can be used to identify the source of an outbreak of infectious disease and how this is helpful.   (*4 marks*)

    b   Discuss how DNA sequencing has changed the ways in which we identify species and our understanding of evolutionary relationships.   (*6 marks*)

3   'One-gene-one-polypeptide' is an outdated concept. Discuss how our model of the link between the genotype and phenotype is changing.   (*9 marks*)

21.4 Genetic engineering

Specification reference: 6.1.3

DNA sequencing and proteomics provide us with a detailed understanding of an organism's genetic instructions. Advances in these technologies and molecular biotechnology techniques means it is now possible to manipulate an organism's genome to achieve a desired outcome. This manipulation of the genome is called genetic engineering. The basic principles of genetic engineering involve isolating a gene for a desirable characteristic in one organism and placing it into another organism, using a suitable vector. The two organisms between which the genes are transferred may be the same, similar, or very different species. An organism that carries a gene from another organism is termed 'transgenic' and is often called a genetically modified organism (GMO).

Isolating the desired gene

The first stage of successful genetic modification is to isolate the desirable gene.

The most common technique uses enzymes called restriction endonucleases to cut the required gene from the DNA of an organism. As you learnt in Topic 21.1, DNA profiling, each type of endonuclease is restricted to breaking the DNA strands at specific base sequences within the molecule. Some make a clean, blunt-ended cut in the DNA. However, many restriction endonucleases cut the two DNA strands unevenly, leaving one of the strands of the DNA fragment a few bases longer than the other strand (Figure 2). These regions with unpaired, exposed bases are called sticky ends. The sticky ends make it much easier to insert the desired gene into the DNA of a different organism.

Another technique involves isolating the mRNA for the desired gene and using the enzyme reverse transcriptase to produce a single strand of complementary DNA. The advantage of this technique is that it makes it easier to identify the desired gene, as a particular cell will make some very specific types of mRNA. For example, β cells of the pancreas make insulin, so produce lots of insulin mRNA molecules.

bacterial cell

plasmid

mRNA extracted from human pancreatic cells

mRNA

plasmid obtained from bacteria

cDNA

plasmid cut with restriction enzyme

mRNA treated with reverse transcriptase to make complementary DNA (cDNA)

plasmid and cDNA fused using DNA ligase

recombinant plasmid introduced into host cells

bacteria multiply in a fermenter and produce insulin

separation and purification of human insulin

human insulin can be used by diabetic patients

▲ Figure 1 *The production of human insulin by genetically engineered bacteria was an early success*

a Hpal restriction endonuclease has a recognition site GTTAAC, which produces a straight cut and therefore blunt ends

b HindIII restriction endonuclease has the recognition site AAGCTT, which produces a staggered cut and therefore 'sticky ends'

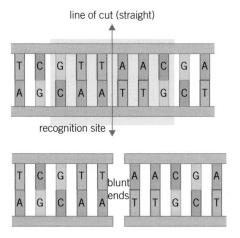

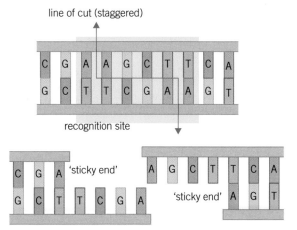

▲ **Figure 2** *Restriction endonucleases are used to isolate genes to be inserted into another organism*

The formation of recombinant DNA

The DNA isolated by restriction endonucleases must be inserted into a vector that can carry it into the host cell.

Vectors

The most commonly used vectors in genetic engineering are bacterial plasmids – small circular molecules of DNA separate from the chromosomal DNA that can replicate independently. Once a plasmid gets into a new host cell it can combine with the host DNA to form what is called **recombinant** DNA. Plasmids are particularly effective in the formation of genetically engineered bacteria used, for example, to make human proteins.

The plasmids that are used as vectors are often chosen because they contain what is known as a marker gene. For example they may have been engineered to have a gene for antibiotic resistance. This gene enables scientists to determine that the bacteria have taken up the plasmid, by growing the bacteria in media containing the antibiotic.

To insert a DNA fragment into a plasmid, first it must be cut open. The same restriction endonuclease as used to isolate the DNA fragment is used to cut the plasmid. This results in the plasmid having complementary sticky ends to the sticky ends of the DNA fragment. Once the complementary bases of the two sticky ends are lined up, the enzyme DNA ligase forms phosphodiester bonds between the sugar and the phosphate groups on the two strands of DNA, joining them together (Figure 2).

The plasmids used as vectors are usually given a second marker gene, which is used to show that the plasmid contains the recombinant gene. This marker gene is itself often placed in the plasmid by genetic engineering methods. The plasmid is then cut by a restriction

Synoptic link

You learnt about bacterial plasmids in Topic 2.6, Prokaryotic and eukaryotic cells.

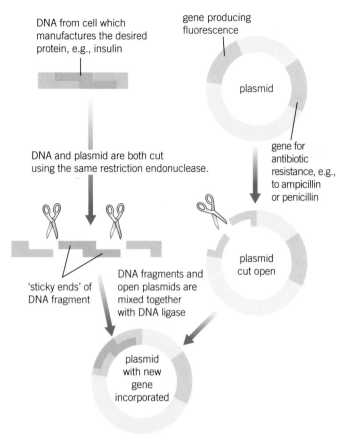

DNA from cell which manufactures the desired protein, e.g., insulin

gene producing fluorescence

plasmid

DNA and plasmid are both cut using the same restriction endonuclease.

gene for antibiotic resistance, e.g., to ampicillin or penicillin

plasmid cut open

'sticky ends' of DNA fragment

DNA fragments and open plasmids are mixed together with DNA ligase

plasmid with new gene incorporated

▲ **Figure 3** *Engineering a desired gene into a plasmid vector*

enzyme within this marker gene to insert the desired gene. If the DNA fragment is inserted successfully, the marker gene will not function. In the early days of genetic engineering, these marker genes were often for antibiotic resistance. There have, however, been many concerns about antibiotic resistance in genetically engineered organisms. As a result, genes producing fluorescence or an enzyme that causes a colour change in a particular medium are now more widely used as marker genes. If a bacterium does not fluoresce, or change the colour of the medium, then it has been engineered successfully and can be grown on (Figure 3).

Transferring the vector

The plasmid with the recombinant DNA must be transferred into the host cell in a process called transformation.

One method is to culture the bacterial cells and plasmids in a calcium-rich solution and increase the temperature. This causes the bacterial membrane to become permeable and the plasmids can enter.

Another method of transformation is **electroporation**. A small electrical current is applied to the bacteria. This makes the membranes very porous and so the plasmids move into the cells.

Electroporation can also be used to get DNA fragments directly into eukaryotic cells. The new DNA will pass through the cell membrane and the nuclear membrane to fuse with the nuclear DNA. Although this technique is effective, the power of the electric current has to be carefully controlled or the membrane is permanently damaged or destroyed, which in turn destroys the whole cell. It is less useful in whole organisms.

Electrofusion

Another way of producing genetically modified (GM) cells is electrofusion. In electrofusion, tiny electric currents are applied to the membranes of two different cells. This fuses the cell and nuclear membranes of the two different cells together to form a hybrid or polyploid cell, containing DNA from both. It is used successfully to produce GM plants.

Electrofusion is used differently in animal cells, which do not fuse as easily and effectively as plant cells. Their membranes have different properties and polyploid animal cells – especially polyploid mammalian cells – do not usually survive in the body of a living organism. However, electrofusion is important in the production

Synoptic link

You learnt about antibodies in Topic 12.6, The specific immune system.

of monoclonal antibodies. A monoclonal antibody is produced by a combination of a cell producing one single type of antibody with a tumour cell, which means it divides rapidly in culture. Monoclonal antibodies are now used to identify pathogens in both animals and plants, and in the treatment of a number of diseases including some forms of cancer.

Engineering in different organisms

The techniques of genetic engineering vary between different types of organisms but the principles are the same. It is much easier to carry out genetic modification of prokaryotes than eukaryotes, and among the eukaryotes, plants are easier to work with than animals.

Engineering prokaryotes

Bacteria and other microorganisms have been genetically modified to produce many different substances that are useful to people. These include hormones, for example insulin and human growth hormone, clotting factors for haemophiliacs, antibiotics, pure vaccines, and many of the enzymes used in industry.

Engineering plants

One method of genetically modifying plants uses *Agrobacterium tumefaciens*, a bacterium that causes tumours in healthy plants. A desired gene – for example, for pesticide production, herbicide-resistance, drought-resistance, or higher yield – is placed in the Ti plasmid of *A. tumefaciens* along with a marker gene, for example, for antibiotic resistance or fluorescence. This is then carried directly into the plant cell DNA. The transgenic plant cells form a callus, which is a mass of GM plant cells, each of which can be grown into a new transgenic plant.

Transgenic plant cells can also be produced by electrofusion. The cells produced have chromosomes from both of the original cells and so are polyploid. The cells that are fused may be from similar species, or very different ones. The main stages in this process involve removal of the plant cell wall by cellulases, electrofusion to form a new polyploid cell, the use of plant hormones to stimulate the growth of a new cell wall, followed by callus formation and the production of many cloned, transgenic plants.

Engineering animals

It is much harder to engineer the DNA of eukaryotic animals, especially mammals, than it is to modify bacteria or plants. This is partly because animal cell membranes are less easy to manipulate than plant cell membranes. However, it is an important technique both to enable animals to produce some medically important proteins and to try and cure human genetic diseases such as CF and Huntington's disease.

Synoptic link

You will learn more about cloning plants in Topic 22.1, Natural cloning in plants.

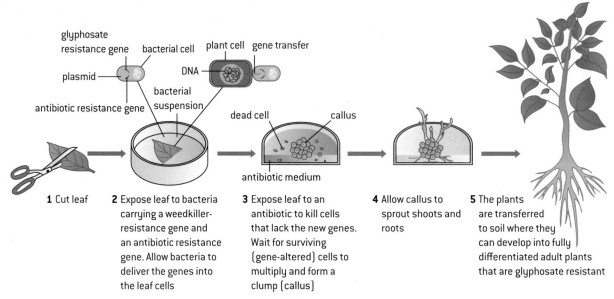

glyphosate resistance gene   bacterial cell   plant cell   gene transfer

plasmid   DNA

antibiotic resistance gene   bacterial suspension

dead cell   callus

antibiotic medium

1 Cut leaf

2 Expose leaf to bacteria carrying a weedkiller-resistance gene and an antibiotic resistance gene. Allow bacteria to deliver the genes into the leaf cells

3 Expose leaf to an antibiotic to kill cells that lack the new genes. Wait for surviving (gene-altered) cells to multiply and form a clump (callus)

4 Allow callus to sprout shoots and roots

5 The plants are transferred to soil where they can develop into fully differentiated adult plants that are glyphosate resistant

▶ **Figure 4** *Top – genetic engineering in plants. Right – A clone of a genetically modified tobacco plant growing in nutrient agar in a petri dish*

Summary questions

1   What is genetic engineering? *(3 marks)*

2   Explain the difference between the way restriction endonucleases and reverse transcriptases produce genes ready for insertion into another organism. *(6 marks)*

3   Describe how a gene is inserted into a bacterial plasmid, which is then taken up by bacteria, and how scientists ensure that they can identify bacteria that have been successfully transformed. *(6 marks)*

4   Produce flow diagrams to show:
    a   The process of genetic engineering a bacterium. *(6 marks)*
    b   The process of engineering a plant. *(6 marks)*

The rapid development of gene technology in recent years has made many amazing scientific advances in the fields of genetic engineering and biotechnology possible – but it has also raised a number of ethical issues.

All scientists have a responsibility to consider the moral and social values, or ethics, of their work. These ethical considerations are important for many reasons including the protection of human rights, human health and safety, animal welfare, and the protection of the environment. Ethical lapses in scientific research can not only cause harm but can also damage the public's trust in scientists and their research. This in turn can have significant implications on the advancement of knowledge and understanding. Clear and open two-way communication between scientists and the public is vital for building trust and regulating research.

Genetic manipulation of microorganisms

Microorganisms, particularly bacteria, have been genetically modified or engineered to produce many different substances that are useful to people, including insulin and vaccines. These substances can be produced in very large quantities in this way.

GM microorganisms are also used to store a living record of the DNA of another organism in DNA libraries. DNA sequencing projects, such as the HGP (Topic 21.2, DNA sequencing and analysis), enable scientists to build a collection of sequenced DNA fragments from one organism that is then stored (and propagated) in microorganisms (usually bacteria or yeast) through the process of genetic engineering (Topic 21.4, Genetic engineering). These libraries serve as a source of DNA fragments for further genetic engineering applications or for further study of their function.

GM microorganisms are a widely used tool in research for developing novel medical treatments and industrial processes, as well as the development of gene technology itself. Genetically engineered pathogens, however, are not widely used in these applications for the obvious health and safety of the researchers and the wider public. In the few cases that genetic modification of pathogens is carried out, this is usually for the purposes of medical and epidemiological research and is strictly regulated. There is, of course, the concern that genetic engineering of pathogens could be used for the purposes of biological warfare. Attempts to modify the genomes of pathogens to be more virulent, or to be resistant to all known treatments, raises serious ethical concerns and is a largely prohibited area of research except in specialised military research facilities.

Learning outcomes

Demonstrate knowledge, understanding, and application of:

→ the ethical issues relating to the genetic modification of organisms

→ the principles of, and potential for, gene therapy in medicine.

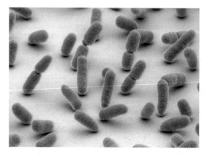

▲ **Figure 1** *Microorganisms such as these gut bacteria (Escherichia coli) are widely used in scientific research into genetic modification. Scanning electron micrograph, ×7500 magnification*

Synoptic link

You will learn more about the use of genetically modified microorganisms and the enzymes they produce in Chapter 22, Cloning and Biotechnology.

Ethical concerns

Initially, some people were uncomfortable with inserting human genes into microorganisms but the pure human medicines, antibiotics, and enzymes produced are now seen as overwhelmingly beneficial. They have been used safely for many years now. As a result, there is relatively little ethical debate about the use of GM microorganisms except for the manipulation of pathogens in biological warfare.

GM plants

Some people think the genetic modification of plants will help feed the ever-growing human population, and overcome environmental issues including excess carbon dioxide production and pollution. Others have major concerns about the process.

Insect resistance in GM soya beans

Soya beans are a major world crop – around 250 million tonnes are produced each year and over half of the plants are from GM strains. In one such modification, scientists have inserted a gene into soya beans so that they produce the Bt protein. The Bt protein is toxic to many of the pest insects that attack the plant and is widely used as a pesticide by organic farmers. One increasingly widely used strain of soya beans has been engineered to be resistant to a common weed killer and to contain Bt protein. This means farmers can spray to get rid of weeds, making all the resources of light, water, and minerals available to the beans, and they do not need to use pesticides. These plants should enable farmers to grow a much higher-yield crop of soya beans with less labour and less expense.

▲ **Figure 2** *Scientists are currently developing vaccines in transgenic plants to many different diseases such as malaria. Tomato, tobacco, and banana plants are among those used*

Benefits and risks of GM crops

Some of the potential benefits and possible problems of GM plant crops are summarised in Table 1. Also note that whenever antibiotic resistance is used as a marker gene to create GM crops (Topic 21.4, Genetic engineering) there is a perceived risk that this resistance could spread to wild populations of plants and into bacteria.

▼ **Table 1** *Pros and cons of GM crops*

Genetically engineered characteristic	Perceived pros of GM crops	Perceived cons of GM crops
Pest resistance	Pest-resistant GM crop varieties reduce the amount of pesticide spraying, protecting the environment and helping poor farmers. Increased yield.	Non-pest insects and insect-eating predators might be damaged by the toxins in the GM plants. Insect pests may become resistant to pesticides in GM crops.
Disease resistance	Crop varieties resistant to common plant diseases can be produced, reducing crop losses/increasing yield.	Transferred genes might spread to wild populations and cause problems, e.g., superweeds.
Herbicide resistance	Herbicides can be used to reduce competing weeds and increase yield.	Biodiversity could be reduced if herbicides are overused to destroy weeds. Fear of superweeds.

▼ Table 1 *Continued*

Genetically engineered characteristic	Perceived pros of GM crops	Perceived cons of GM crops
Extended shelf-life	The extended shelf-life of some GM crops reduces food waste.	Extended shelf-life may reduce the commercial value and demand for the crop.
Growing conditions	Crops can grow in a wider range of conditions/survive adverse conditions, e.g., flood resistance or drought resistance.	
Nutritional value	Nutritional value of crops can be increased, e.g., enhanced levels of vitamins.	People may be allergic to the different proteins made in GM crops.
Medical uses	Plants could be used to produce human medicines and vaccines.	

Patenting and technology transfer

One of the major concerns about GM crops is that people in less economically developed countries will be prevented from using them by patents and issues of technology transfer. When someone discovers a new technique or invents something, they can apply for a legal patent, which means that no-one else can use it without payment. The people who most need the benefits of, for instance, drought- or flood-resistant crops, high yields, and added nutritional value may therefore be unable to afford the GM seed. They also rely on harvesting seed from one year to plant the next – something that patenting may make impossible.

These concerns are based on evidence. The company that developed the herbicide-resistant and pesticide-producing soya beans, have patented them so farmers can buy the beans from them and grow them to use or sell them for food or processing *only* in the year they are bought. They cannot save the seed to grow again the next year – and in 2013 this was upheld in the US Supreme Court.

Some organisations, however, such as the International Rice Research Institute (IRRI), work to develop engineered rice specifically to support farmers in less economically developed countries with whom they share the technological developments without patent constraints on seed harvesting. For example, they have engineered flood-resistant 'scuba' rice, which gives 70–80% of maximum potential yield even if submerged for 2–3 weeks by flooding.

GM animals

It is much harder to produce GM vertebrates, especially mammals, but scientists are researching the use of microinjections – tiny particles of gold covered in DNA – and modified viruses to carry new genes into animal DNA. Such techniques are used with a number of goals in mind, including the transfer of disease resistance from one animal to another, or to modify physiology in farmed animals.

Some examples of GM animals:

● Swine fever-resistant pigs – in 2013 UK scientists successfully inserted a gene from wild African pigs into the early embryos of a European pig strain giving them immunity to otherwise fatal African swine fever.

● Faster-growing salmon – in the USA, GM Atlantic salmon have received genes from faster-growing Chinook salmon. The genes cause them to produce growth hormones all year round. They grow to full adult size in half the time of conventional salmon, making them a very efficient food source.

Pharming

One of the biggest uses of genetic engineering so far in animals is in the production of human medicines – known as **pharming**. There are two aspects to this field of gene technology:

● Creating animal models – the addition or removal of genes so that animals develop certain diseases, acting as models for the development of new therapies, for example, knockout mice have genes deleted so they are more likely to develop cancer.

● Creating human proteins – the introduction of a human gene coding for a medically required protein. Animals are sometimes used because bacteria cannot produce all of the complex proteins made by eukaryotic cells. The human gene can be introduced into the genetic material of a fertilised cow, sheep, or goat egg, along with a promoter sequence so the gene is expressed only in the mammary glands. The fertilised, transgenic female embryo is then returned to the mother. A transgenic animal is born and when it matures and gives birth, it produces milk. The milk will contain the desired human protein and can be harvested.

Ethical issues

There are many potential benefits to people and indeed to animal health of genetic engineering but the process also raises some ethical questions, which include:

● Should animals be genetically engineered to act as models of human disease?

● Is it right to put human genes into animals?

● Is it acceptable to put genes from another species into an animal without being certain it will not cause harm?

● Does genetically modifying animals reduce them to commodities?

● Is welfare compromised during the production of genetically engineered animals?

▲ **Figure 3** *These transgenic sheep have a human gene in their DNA. This gene codes for the protein alpha-1-antitrypsin (AAT). Hereditary deficiency of AAT leads to emphysema. The gene activates AAT production in the sheep's milk which can be isolated and used for therapy*

Gene therapy in humans

Some human diseases such as CF, haemophilia, and severe combined immunodeficiency (SCIDS) are the result of faulty (mutant) genes. Scientists are looking at different ways of replacing the faulty allele with a healthy one. They can remove the desired alleles from healthy cells or synthesise healthy alleles in the laboratory.

Somatic cell gene therapy

This involves replacing the mutant allele with a healthy allele in the affected somatic (body) cells. The potential for helping people with a wide range of diseases is enormous. Until recently there were few success stories as there are problems in getting the healthy alleles into the affected cells, getting the engineered plasmids into the nucleus of the cells, and finally difficulties in starting and maintaining expression of the healthy allele. Viral vectors are often used.

In recent years, **somatic cell gene therapy** is beginning to show signs of fulfilling its potential. Successful treatments have been reported for diseases including retinal disease (people have regained some vision), immune diseases, leukaemias, myelomas, and haemophilia. The first gene therapy has recently (2012) been approved by the European Medicines Agency for lipoprotein lipase deficiency, which can cause severe pancreatitis. However, somatic cell gene therapy is only a temporary solution for the treated individual. The healthy allele will be passed on every time a cell divides by mitosis but somatic cells have a limited life, and are replaced from stem cells, which will have the faulty allele. In addition, a treated individual will still pass the faulty allele on to any children they have.

Germ line cell gene therapy

The alternative to treating the somatic (body) cells of people already affected by a disease is to insert a healthy allele into the germ cells – usually the eggs – or into an embryo immediately after fertilisation (as part of in vitro fertilisation (IVF) treatment). The individual would be born healthy with the normal allele in place – and would pass it on to their own offspring. This is called **germ line cell gene therapy**.

Such therapy has been successfully done with animal embryos but is illegal for human embryos in most countries as a result of various ethical and medical concerns. These concerns include the fact that the potential impact on an individual of an intervention on the germ cells is unknown. Also, the human rights of the unborn individual could be said to be violated because it is, of course, done without consent and once done the process is irrevocable. Another major ethical concern is that the technology might eventually be used to enable people to choose desirable or cosmetic characteristics of their offspring.

Summary questions

1 Suggest why there is relatively little debate about the ethics of genetically engineering microorganisms.
(3 marks)

2 Draw a table to compare somatic cell gene therapy and germ line cell therapy.
(6 marks)

3 a Suggest potential benefits of genetically engineering plants *(6 marks)*

b Discuss the ethical issues *(6 marks)*

4 Describe the process of genetically modifying animals to produce human proteins and discuss some of the ethical issues this raises.
(6 marks)

Practice questions

1 The diagram is a simplified representation of the apparatus used to separate fragments of DNA in genetic engineering.

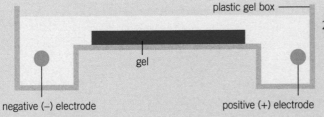

plastic gel box

gel

negative (–) electrode   positive (+) electrode

a (i) State the name of the process.
(*1 mark*)

(ii) Copy the diagram and indicate where the sample of DNA fragments would be placed and the direction the fragments would move when an electric field is turned on. (*2 marks*)

(iii) State the reason why the fragments are placed in the position you have indicated. (*1 mark*)

(iv) Explain the reason for using a gel in the separation process. (*2 marks*)

b Outline how you could identify the position of a fragment you wanted to locate. (*4 marks*)

The fragment obtained, which usually contains a particular gene under investigation, then undergoes another process called the polymerase chain reaction.

c Outline the reason for carrying out the polymerase chain reaction. (*2 marks*)

The diagram shows one step in the polymerase chain reaction.

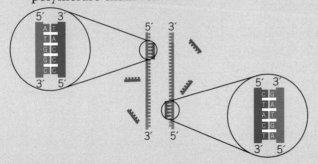

d (i) Describe what is happening in the diagram. (*2 marks*)

(ii) Explain the reason for the step you have described. (*2 marks*)

e (i) Explain why fragments of DNA are denatured during the polymerase chain reaction. (*3 marks*)

(ii) Compare the denaturation of DNA with protein denaturation. (*5 marks*)

2 It has been discovered that genes make up only about 3% of the human genome and over 50% of the genome consists of repetitive nucleotide sequences. Only recently has it been possible to start to process the data produced from the sequencing of these repetitive DNA sequences. Bioinformatics is a field of study that has developed methods and software programs to handle the large quantities of data involved.

Epigenetic refers to processes that alter the activity of genes without changing nucleotide sequences.

a (i) State two techniques that might be used to supply the data used in bioinformatics. (*2 marks*)

(ii) Discuss why the evidence for epigenetics has only been discovered as bioinformatics has developed as a field of study. (*4 marks*)

Synthetic biology refers to the design and synthesis of biological components that do not exist in the natural world or the re-design of existing biological systems.

The guiding principles of synthetic biologists have been stated as below:

- public benefit
- responsible stewardship
- intellectual freedom and responsibility
- democratic deliberation
- justice and fairness

b Discuss what you believe to be the relative importance of each of these principles. (*7 marks*)

3 DNA profiling is used to identify organisms on the basis of their nucleotide sequences. There are long sections of repetitive DNA, called satellite DNA, in the genomes of most organisms. Different individual have different numbers of repeating segments in their satellite DNA.

A sample of DNA to be sequenced is first amplified, then fragmented and analysed using gel electrophoresis.

a (i) Explain why gel electrophoresis is used to compare the DNA fragments in the samples. (*2 marks*)

(ii) State the name of the process used to amplify a DNA sample. (*1 mark*)

(iii) State the name of the enzymes used to produce the fragments of DNA and explain their specificity to certain sequences of nucleotides. (*3 marks*)

(iv) Explain why the degree of match required differs for the samples of DNA being compared in paternity testing and forensic investigation. (*3 marks*)

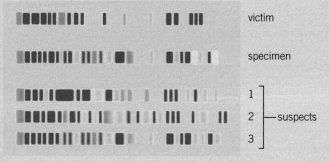

b (i) Using the information in the diagram, identify the most likely suspect. (*1 mark*)

(ii) DNA profiles can also be used to determine if an organism is homozygous or heterozygous at particular loci.

Suggest how this is possible. (*3 marks*)

4 The graph shows the changes in the area of land growing genetically modified crops in different parts of the world.

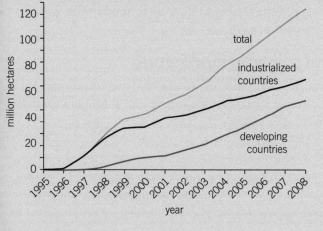

a (i) Describe the changes in the areas of land used for growing GM crops in different parts of the world. (*3 marks*)

(ii) Suggest the reason(s) for the different trends. (*3 marks*)

b Discuss the ethical issues relevant to the genetic modification of plants. (*6 marks*)

5 **a** Outline the differences between somatic and germ line gene therapy. (*4 marks*)

Some diseases are a result of defective mitochondria such as diabetes, nerve degeneration, and some forms of blindness. It has been proposed that mitochondrial diseases could be cured during the process of IVF.

b (i) Explain why tissues with a high metabolic demand such as muscle and nervous tissue are often affected by mitochondrial disease. (*2 marks*)

(ii) Suggest how the process of IVF could be modified to cure mitochondrial disease. (*3 marks*)

(iii) Discuss whether the manipulation of mitochondria to cure genetic disease overcomes the arguments against germline gene therapy. (*5 marks*)

▲ **Figure 1** *Blueberries grow on low bushes – a single clone of plants may stretch for around half a mile*

▲ **Figure 2** *Each baby spider plant is a natural clone*

Asexual reproduction is a form of **cloning** and it results in offspring produced by mitosis and known as **clones.** Clones are usually genetically identical to both the parent organism and to each other.

Natural cloning

Vegetative propagation, or natural cloning, occurs in many species of flowering plants. A structure forms which develops into a fully differentiated new plant, which is genetically identical to the parent. The new plant may be propagated from the stem, leaf, bud, or root of the parent, depending on the type of plant, and it eventually becomes independent from its parent, for example, strawberries and spider plants.

Vegetative propagation often involves perennating organs, which enables plants to survive adverse conditions. These contain stored food from photosynthesis and can remain dormant in the soil. They are often not only a means of asexual reproduction, but also a way of surviving from one growing season to the next. Natural plant cloning occurs in:

● bulbs, for example, daffodil. The leaf bases swell with stored food from photosynthesis. Buds form internally which develop into new shoots and new plants in the next growing season.

● Runners, for example, a strawberry or spider plant. A lateral stem grows away from the parent plant and roots develop where the runner touches the ground. A new plant develops – the runner eventually withers away leaving the new individual independent.

● Rhizomes, for example, marram grass. A rhizome is a specialised horizontal stem running underground, often swollen with stored food. Buds develop and form new vertical shoots which become independent plants.

● Stem tubers, for example, potato. The tip of an underground stem becomes swollen with stored food to form a tuber or storage organ. Buds on the storage organ develop to produce new shoots (e.g., the 'eyes' on a potato).

Using natural clones in horticulture

Natural plant cloning is exploited in horticulture by farmers and gardeners to produce new plants. Splitting up bulbs, removing young plants from runners, and cutting up rhizomes all increase plant numbers cheaply, and the new plants have exactly the same genetic characteristics as their parents.

It is also possible to take cuttings of many plants – short sections of stems are taken and planted either directly in the ground (e.g., sugar cane) or in pots, for example, pelargoniums. Rooting hormone is often applied to the base of a cutting to encourage the growth of new roots. Propagation from cuttings has several advantages over using seeds. It is much faster – the time from planting to cropping is much reduced. It also guarantees the quality of the plants. By taking cuttings from good stock, the offspring will be genetically identical and will therefore crop well. The main disadvantage is the lack of genetic variation in the offspring should any new disease or pest appear or if climate change occurs.

Many of the world's most important food crops are propagated by cloning. Bananas, sugar cane, sweet potatoes, and cassava are all propagated from stem cuttings or rhizomes. Coffee and tea bushes are also propagated from stem cuttings.

Cloning sugar cane

Sugar cane is an internationally important crop used to make sugar and manufacture biofuels. It is one of the fastest growing crop plants in the world – the stems can grow 4–5 metres in 11 months if conditions are good – and it is usually propagated by cloning. Short lengths of cane about 30 cm long, with three nodes, are cut and buried in a clear field in shallow trenches, covered with a thin layer of soil. Per hectare, 10–25 000 lengths of stem are planted.

 ## Practical cloning

Many popular houseplants are propagated by taking cuttings. There are a number of points which increase the success rate of most cuttings:

- Use a non-flowering stem
- Make an oblique cut in the stem
- Use hormone rooting powder
- Reduce leaves to two or four
- Keep cutting well watered
- Cover the cutting with a plastic bag for a few days.

▲ **Figure 3** *Taking cuttings on a commercial scale*

1 How does each of the above points increase the likelihood that a cutting will succeed?
2 Why are cuttings useful for investigating the effect of growing conditions on plants?

Synoptic link

You learnt about the use of hormonal rooting powders in Topic 16.5, The commercial use of plant hormones.

Study tip

Make sure you use terms such as cloning, propagation, and taking cuttings correctly.

Summary questions

1 What are perennating organs and how are they involved in cloning and survival? (*5 marks*)

2 Explain the advantages and disadvantages of propagating crop plants by cuttings over using seeds. (*4 marks*)

3 Suggest why it is important to describe clones as genetically identical to their parent rather than simply identical – and why even this may not always be true. (*5 marks*)

22.2 Artificial cloning in plants

Specification reference: 6.2.1

Learning outcomes

Demonstrate knowledge, understanding, and application of:

→ the production of artificial clones of plants by micropropagation and tissue culture

→ the arguments for and against artificial cloning in plants.

People have propagated plants by cloning for centuries, but there is a limit to how many 'natural' clones you can make from one plant. Many plant cells are totipotent – they can differentiate into all of the different types of cells in the plant. Scientists have developed ways of using this property to produce huge numbers of identical clones from one desirable plant.

Micropropagation using tissue culture

Micropropagation is the process of making large numbers of genetically identical offspring from a single parent plant using tissue culture techniques. This is used to produce plants when a desirable plant:

- does not readily produce seeds
- doesn't respond well to natural cloning
- is very rare
- has been genetically modified or selectively bred with difficulty
- is required to be 'pathogen-free' by growers, for example, strawberries, bananas, and potatoes.

There are a number of ways in which plants can be micropropagated. One protocol, based on work done at the Royal Botanic Garden at Kew, uses sodium dichloroisocyanurate, the sterilising tablets used to make emergency drinking water and babies' bottles safe. This keeps the plant tissues sterile without being in a sterile lab so it is extremely useful for scientists in the field working with rare and endangered plant material – and also for use in schools. Other protocols are more suited to industrial micropropagation where large sterilising units are available.

The basic principles of micropropagation and tissue culture are as follows:

- Take a small sample of tissue from the plant you want to clone – the meristem tissue from shoot tips and axial buds is often dissected out in sterile conditions to avoid contamination by fungi and bacteria. This tissue is usually virus-free.

- The sample is sterilised, usually by immersing it in sterilising agents such as bleach, ethanol, or sodium dichloroisocyanurate. The latter does not need to be rinsed off which means the tissue is more likely to remain sterile. The material removed from the plant is called the explant.

- The explant is placed in a sterile culture medium containing a balance of plant hormones (including auxins and cytokinins) which stimulate mitosis. The cells proliferate, forming a mass of identical cells known as a callus.

- The callus is divided up and individual cells or clumps from the callus are transferred to a new culture medium containing a different mixture of hormones and nutrients which stimulates the development of tiny, genetically identical plantlets.

- The plantlets are potted into compost where they grow into small plants.

- The young plants are planted out to grow and produce a crop.

The scale of micropropagation is increasing. It now takes place in bioreactors, effectively making artificial embryo plants to be packaged in artificial seeds.

▲ **Figure 1** *Micropropagation*

Advantages and disadvantages of micropropagation

The number of common plants that are largely produced by micropropagation is growing constantly and includes potatoes, sugar cane, bananas, cassava, strawberries, grapes, chrysanthemums, Douglas firs, and orchids. Here are some of the points both for and against this process

Arguments for micropropagation

- Micropropagation allows for the rapid production of large numbers of plants with known genetic make-up which will yield good crops.

- Culturing meristem tissue produces disease-free plants.

- It makes it possible to produce viable numbers of plants after genetic modification of plant cells.

- It provides a way of producing very large numbers of new plants which are seedless and therefore sterile to meet consumer tastes (e.g., bananas and grapes).

- It provides a way of growing plants which are naturally relatively infertile or difficult to grow from seed (e.g., orchids).

- It provides a way of reliably increasing the numbers of rare or endangered plants.

▲ **Figure 2** *Micropropagation of orchids means they are no longer only available to very wealthy people*

Arguments against micropropagation

- It produces a monoculture – many plants which are genetically identical – so they are all susceptible to the same diseases or changes in growing conditions.

- It is a relatively expensive process and requires skilled workers.

- The explants and plantlets are vulnerable to infection by moulds and other diseases during the production process.

- If the source material is infected with a virus, all of the clones will also be infected.

- In some cases, large numbers of new plants are lost during the process.

▲ **Figure 3** *Bananas are the fourth most important food worldwide – and the ones we eat are all clones*

Synoptic link

You learnt about Black Sigatoka in Topic 12.2, Animal and plant diseases.

 Yes, we have no bananas…

Bananas are now thought to be one of the oldest crops – and possibly the first to be cloned. A wild banana is full of hard seeds and it is virtually inedible. A mutation made them parthenocarpic which means they produce fruit without fertile seeds – which made them good to eat but also made them sterile. Scientists therefore think that since the dawn of agriculture, people cloned bananas using natural asexual reproduction to propagate the plants producing the seedless, tasty fruit. Sweet bananas are widely eaten in more economically developed countries, whilst plantains (cooking bananas) are a staple food in many less economically developed countries.

In the early 20th century almost all of the sweet bananas eaten were the cultivar Gros Michel. Then fungal Panama disease wiped them out in the major banana growing countries – none of the clones had any resistance and a new cultivar took over. Cavendish bananas, while apparently not as tasty as Gros Michel bananas, are resistant to Panama disease. But Cavendish bananas are also clones. Now another banana disease, Black Sigatoka, is destroying Cavendish plantations, and is also spreading to other cooking varieties of bananas.

New biotechnologies for example genetic engineering and micropropagation offer hope for the future. Genetically engineered strains of bananas with resistance genes from the original wild fruit could be micropropagated and used to restock banana plantains across the whole growing region.

1 Gros Michel bananas have been almost entirely lost. Why?
2 How might new technology enable us to retain the Cavendish strain?
3 Discuss how banana culture has changed over the centuries.

Summary questions

1 Make a flow chart to show the main stages in the micropropagation of plants. *(7 marks)*

2 What do you consider the two main advantages of micropropagation of crop plants? Explain your choices. *(6 marks)*

3 a What is the potential of natural cloning for saving important crops such as the banana against disease? Give arguments for and against the technique. *(6 marks)*

  b What is the potential of micropropagation for saving important crops against disease in contrast to natural cloning? Give arguments for and against the technique. *(6 marks)*

22.3 Cloning in animals
Specification reference: 6.2.1

Cloning is a natural part of the reproductive cycle in many plants. Perhaps surprisingly, it is not uncommon in many animal species and even occurs in human beings.

Natural animal cloning

Natural cloning is common in invertebrate animals. Although it is less common in vertebrates, it still occurs in the form of twinning.

Cloning in invertebrates

Natural cloning in invertebrates can take several forms. Some animals, such as starfish, can regenerate entire animals from fragments of the original if they are damaged. Flatworms and sponges fragment and form new identical animals as part of their normal reproductive process, all clones of the original. *Hydra* produce small 'buds' on the side of their body which develop into genetically identical clones. In some insects, females can produce offspring without mating. Scientists are increasingly finding differences between the mother and daughters, however, suggesting that as a result of high mutation rates the offspring are not true clones.

Cloning in vertebrates

The main form of vertebrate cloning is the formation of **monozygotic twins (identical twins)**. The early embryo splits to form two separate embryos. No one is sure of the trigger which causes this to happen. The frequency at which identical twins occur varies between species. For example, domestic cattle rarely if ever produce identical twins naturally, while the incidence in natural human pregnancies is around 3 per 1000. When monozygotic twins are born, although genetically identical, they may look different as a result of differences in their position and nutrition in the uterus.

Some female amphibians and reptiles will produce offspring when no male is available. The offspring are often male rather than female, so they are not clones of their mother, yet all of their genetic material arises from her.

Artificial clones in animals

It is relatively easy to produce artificial clones of some invertebrates – liquidise a sponge or chop up a starfish and new animals will regenerate from most of the fragments. It is much more difficult to produce artificial clones of vertebrates, especially mammals. However, two methods are now used widely in the production of high-quality farm animals and in the development of genetically engineered animals for pharming.

Learning outcomes

Demonstrate knowledge, understanding, and application of:

→ natural clones in animal species

→ how artificial clones in animals can be produced by artificial embryo twinning or by enucleation and somatic cell nuclear transfer (SCNT)

→ the arguments for and against artificial cloning in animals.

▲ **Figure 1** *The small <u>Hydra</u> is a clone of the parent and will eventually separate and live independently*

Synoptic link

You learnt about pharming in Topic 21.5, Gene technology and ethics.

Artificial twinning

After an egg is fertilised, it divides to form a ball of cells. Each of these individual cells is **totipotent** – it has the potential to form an entire new animal. As the cells continue to divide, the embryo becomes a hollow ball of cells. Soon after this the embryo can no longer divide successfully.

In natural twinning, an early embryo splits and two foetuses go on to develop from the two halves of the divided embryo. In **artificial twinning** the same thing happens, but the split in the early embryo is produced manually. In fact, the early embryo may be split into more than two pieces and results in a number of identical offspring. Artificial twinning, like embryo transfer which preceded it, is used by the farming community to produce the maximum offspring from particularly good dairy or beef cattle or sheep.

The stages of artificial twinning in cattle can be summarised as follows:

- A cow with desirable traits is treated with hormones so she super-ovulates, releasing more mature ova than normal.
- The ova may be fertilised naturally, or by artificial insemination, by a bull with particularly good traits. The early embryos are gently flushed out of the uterus.
- Alternatively, the mature eggs are removed and fertilised by top-quality bull semen in the lab.
- Usually before or around day six, when the cells are still totipotent, the cells of the early embryo are split to produce several smaller embryos, each capable of growing on to form a healthy full-term calf.
- Each of the split embryos is grown in the lab for a few days to ensure all is well before it is implanted into a surrogate mother. Each embryo is implanted into a different mother as single pregnancies carry fewer risks than twin pregnancies.
- The embryos develop into foetuses and are born normally, so a number of identical cloned animals are produced by different mothers.

In pigs, a number of cloned embryos must be introduced into each mother pig. This is because they naturally produce a litter of piglets, and the body may reject and reabsorb a single foetus.

This technology makes it possible to greatly increase the numbers of offspring produced by the animals with the best genetic stock. Some of the embryos may be frozen. This allows the success of a particular animal to be assessed and, if the stock is good, remaining identical embryos can be implanted and brought to term.

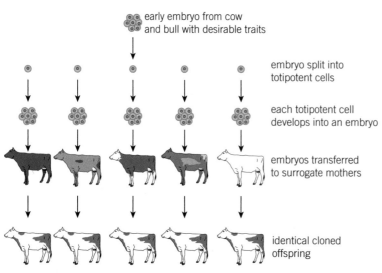

early embryo from cow and bull with desirable traits

embryo split into totipotent cells

each totipotent cell develops into an embryo

embryos transferred to surrogate mothers

identical cloned offspring

▲ **Figure 2** *Artificial twinning in cows*

Somatic cell nuclear transfer

Artificial twinning clones an embryo. However, it is now possible to clone an adult animal, by taking the nucleus from an adult somatic (body) cell and transferring it to an **enucleated** egg cell (an oocyte which has had the nucleus removed). A tiny electric shock is used to fuse the egg and nucleus, stimulate the combined cell to divide, and form an embryo that is a clone of the original adult. This process is known as **somatic cell nuclear transfer** (SCNT). The first adult mammal to be cloned in this way was Dolly the sheep in 1996. Since then scientists have cloned a wide range of species including mice, cows, horses, rabbits, cats, and dogs. SCNT is simple in theory, although in practice there are many difficulties so the technique is still not widely used. As you can see in Figure 3, animals of different breeds are often used as the cell donor, the egg donor, and the surrogate mother to make it easier to identify the original animal at each stage.

1 The nucleus is removed from a somatic cell of an adult animal.

2 The nucleus is removed from a mature ovum harvested from a different female animal of the same species (it is enucleated).

3 The nucleus from the adult somatic cell is placed into the enucleated ovum and given a mild electric shock so it fuses and begins to divide. In some cases, the nucleus from the adult cell is not removed – it is simply placed next to the enucleated ovum and the two cells fuse (electrofusion) and begin to divide under the influence of the electric current.

5 The embryo that develops is transferred into the uterus of a third animal, where it develops to term.

6 The new animal is a clone of the animal from which the original somatic cell is derived, although the mitochondrial DNA will come from the egg cell.

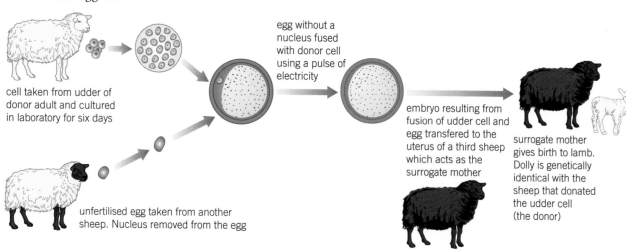

▲ **Figure 3** *Somatic cell nuclear transfer (SCNT)*

This process is also known as reproductive cloning, because live animals are the end result. The cloned embryo can then be split to produce several identical clones.

There have been some problems with the animals produced by SCNT – Dolly the sheep had to be put down when she was only six years old

Synoptic link

You learnt about electrofusion in Topic 21.4, Genetic engineering.

because she suffered from arthritis and lung disease, usually seen in much older sheep. However, scientists have improved the technique and whilst concerns about premature ageing in clones produced by SCNT persist, researchers in Japan have produced 581 clones from one original donor mouse, through 25 generations. The mice in each generation were cloned to produce the next generation. Furthermore, they all had babies naturally to prove they functioned normally. All of the mice had normal lifespans. The same team has also produced SCNT clones from the bodies of mice which had been frozen for 16 years.

SCNT can be used in a number of ways. It is used in pharming – the production of animals which have been genetically engineered to produce therapeutic human proteins in their milk. It can also be used to produce genetically modified (GM) animals which grow organs that have the potential to be used in human transplants.

Pros and cons of animal cloning

Animal cloning is still not widespread, although it is increasingly used in agriculture and the world of animal breeding and medicine. A number of arguments are put forward both for and against the process.

Arguments *for* animal cloning

Artificial twinning enables high-yielding farm animals to produce many more offspring than normal reproduction.

Artificial twinning enables the success of a sire (the male animal) at passing on desirable genes to be determined. If the first cloned embryo results in a successful breeding animal, more identical animals can be reared from the remaining frozen clones. The use of meat from animals born to a cloned parent is now permitted in the US.

SCNT enables GM embryos to be replicated and to develop, giving many embryos from one engineering procedure. It is an important process in pharming – the production of therapeutic human proteins in the milk of genetically engineered farm animals, such as sheep and goats.

SCNT enables scientists to clone specific animals, for example, replacing specific pets or cloning top-class race horses. Pet cats and dogs have been cloned in the US at great expense.

SCNT has the potential to enable rare, endangered, or even extinct animals to be reproduced. In theory, the nucleus from dried or frozen tissue could be transferred to the egg of a similar living species and used to produce clones of species that have been dead for a long time.

Arguments *against* animal cloning

SCNT is a very inefficient process – in most animals it takes many eggs to produce a single cloned offspring.

Many cloned animal embryos fail to develop and miscarry or produce malformed offspring.

Many animals produced by cloning have shortened lifespans, although cloned mice have now been developed which live a normal two years.

SCNT has been relatively unsuccessful so far in increasing the populations of rare organisms or allowing extinct species to be brought back to life. For example, scientists have attempted to clone the gaur and the banteng – both extremely rare breeds of wild cattle. One gaur calf was born in 2001 and died within a couple of days. Two banteng calves were born in 2003 – one was deformed and euthanised, the other grew normally but its natural lifespan was halved. The idea of restoring extinct organisms is exciting but scientists are increasingly unconvinced that it will be possible by this method.

Cloning humans

- Scientists have reproduced clones of primates by artificial twinning but it is proving very difficult to produce a SCNT clone of a primate.

- Part of the problem seems to be that the spindle proteins needed for cell division in primate cells are sited very close to the nucleus, so the removal of the nucleus to produce the enucleated primate ovum also destroys the mechanism by which the cell divides. This is not a problem in the ova of many other mammals because the spindle proteins are more dispersed in the cytoplasm.

- In addition, the synchronisation of the stage of the embryo and the state of the reproductive organs of the mother have to be exactly attuned in primates – there seems to be more flexibility in some other mammals.

- In recent years scientists have finally produced embryonic primate stem cell lines by SCNT. This means it may eventually be possible to develop these potentially important therapeutic cells from human beings.

- In most countries there is strict legislation to prevent reproductive cloning of human beings, even if the technical problems of primate cloning are overcome. A modified version of SCNT has the potential, however, to produce human embryonic stem cells from an adult which could produce cells to be used to grow new tissues for that individual patient. Research in this process is strictly controlled so it cannot be used for reproductive cloning – it is known as therapeutic cloning to make it clear that the end result is not to reproduce a person. However, this form of SCNT can potentially make it possible to grow replacement organs which will not trigger an immune response in a patient and which will enable us to cure many currently life-threatening conditions.

> ### Synoptic link
> You learnt about spindle formation during cell division in Topic 6.1, Cell cycle.

Some people claim to have produced a cloned human baby, although they have never produced the child and the adult it was cloned from.

1 Explain how this could easily be proved if scientists were given access to the individuals

2 Suggest why these claims seem very unlikely to be true.

Summary questions

1 How is artificial twinning different from natural twinning? *(6 marks)*

2 The evidence suggests that monozygotic twins do not occur naturally in cattle. Suggest ways in which this might be investigated. *(2 marks)*

3 Primate clones have been produced by artificial twinning but not, in 2014, by SCNT.
  a Explain the similarities in the two processes. *(3 marks)*
  b Explain the differences between the two processes. *(6 marks)*

4 Should funding continue for projects attempting to recreate extinct animals by SCNT? Include an evaluation of the process in your answer. *(6 marks)*

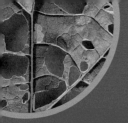

22.4 Microorganisms and biotechnology

Specification reference: 6.2.1

Learning outcomes

Demonstrate knowledge, understanding, and application of:

→ the use of microorganisms in biotechnological processes.

→ the advantages and disadvantages of using microorganisms to make food for human consumption.

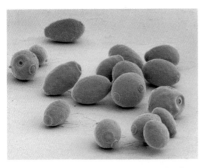

▲ **Figure 1** *Yeasts were probably the first microorganism to be used in biotechnology, and they are still of immense importance worldwide today, scanning electron micrograph approx ×3250 magnification*

The earliest recorded use of microorganisms by people was around 6000 BC when the Sumerians and Babylonians were using yeast to make beer. By 4000 BC the Egyptians were using yeast to make their bread rise. These are all examples of the development and use of biotechnology over several millennia.

Defining biotechnology

Biotechnology involves applying biological organisms or enzymes to the synthesis, breakdown, or transformation of materials in the service of people. It describes a range of processes, from the traditional production of cheese, yoghurt, wine, bread, and beer to the latest molecular technologies using DNA manipulation to produce genetically engineered microorganisms synthesising drugs such as insulin and antibiotics, and the use of biological systems to remove soil and water pollution in processes known as **bioremediation** (Topic 22.5, Microorganisms, medicines, and bioremediation).

The most commonly used organisms in biotechnology processes (bioprocesses) are fungi, particularly the yeasts, and bacteria, which are particularly useful in the newer technologies based around genetic manipulation.

The use of microorganisms

Most biotechnology involves using biological catalysts (enzymes) in a manufacturing process and the most stable, convenient, and effective form of the enzymes is often a whole microorganism. Microorganisms are ideal for a variety of reasons.

● There are no welfare issues to consider – all that is needed is the optimum conditions for growth.

● There is an enormous range of microorganisms capable of carrying out many different chemical syntheses or degradations that can be used.

● Genetic engineering allows us to artificially manipulate microorganisms to carry out synthesis reactions that they would not do naturally, for example, to produce human insulin.

● Microorganisms have a very short life cycle and rapid growth rate. As a result, given the right conditions of food, oxygen, and temperature, huge quantities of microorganisms can be produced in short periods of time.

● The nutrient requirements of microorganisms are often very simple and relatively cheap. Genetic manipulation means we can modify them so that the microorganisms can utilise materials which would otherwise be wasted, making the raw materials for microorganism-controlled syntheses much cheaper than the raw materials needed for most other industrial processes.

- The conditions which most microorganisms need to grow include a relatively low temperature, a supply of oxygen and food, and the removal of waste gases. They provide their own catalysts in the form of enzymes. This makes bioprocesses relatively cheap compared to the high temperatures and pressures and expensive catalysts often needed in non-biological industrial processes.

Indirect food production

Microorganisms are widely used in biotechnological processes to make food such as bread, yogurt, and cheese. The microorganisms have an indirect effect – it is their actions on other food that is important,. When you eat bread you are mainly eating flour, when you eat yoghurt or cheese it is mainly milk.

The advantages of using microorganisms in this way are all of the ones listed previously as advantages of using microorganisms in biotechnology generally. There are few disadvantages to using microorganisms indirectly in the production of human foods. If the conditions are not ideal (e.g., too hot or too cold) the microorganisms do not grow properly and so they do not work efficiently. Conditions that are ideal for the microorganisms can also be ideal for microorganisms that cause the food to go off or cause disease and so the processes have to be sterile. Increasingly the microorganisms used in food production have been genetically engineered, and some people have ethical issues with the use of GM organisms, although this is generally much less the case with microorganisms than with animals and plants.

There are around 900 different types of cheese made around the world. Some are still made by very small-scale, traditional methods and others are produced commercially on a very large scale.

Extra microorganisms

Sometimes microorganisms are used in the same biotechnological process in more than one way. Traditionally bacteria are used in cheesemaking (Table 1) and proteolytic enzymes are also added to the milk to help form curds and whey. Originally these came from rennet, a substance extracted from the stomachs of calves, cows, and pigs containing the enzyme chymosin. In modern cheesemaking, the chymosin used comes mainly from microorganisms – either from fungal sources or GM bacteria.

Suggest three different advantages of using modern sources of chymosin.

▲ **Figure 2** *Different types of cheeses*

▼ **Table 1** *Examples of microorganisms involved in commercial processes*

Process	Microorganism involved	Steps in commercial process
Baking	Yeast – mixed with sugar and water to respire aerobically. Carbon dioxide produced makes bread rise.	• The active yeast mixture is added to flour and other ingredients. Mixed and left in warm environment to rise. • Dough is knocked back (excess air removed), kneaded, shaped, and left to rise again. • Cooked in a hot oven – the carbon dioxide bubbles expand, so the bread rises more. Yeast cells are killed during cooking.
Brewing	Yeast – respires anaerobically to produce ethanol. Traditional yeasts ferment at 20–28 °C. GM yeasts ferment at lower, and therefore cheaper, temperatures, and clump together (flocculate) and sink at the end of the process leaving the beer very clear.	• Malting – barley germinates producing enzymes that break starch molecules down to sugars which yeast can use. Seeds then killed by slow heating but enzyme activity retained to produce malt. • Mashing – the malt is mixed with hot water (55–65 °C) and enzymes break down starches to produce wort. Hops are added for flavour and antiseptic qualities. The wort is sterilised and cooled. • Fermentation – wort is inoculated with yeast. Temperature maintained for optimum anaerobic respiration (fermentation). Eventually yeast is inhibited by falling pH, build up of ethanol, and lack of oxygen. • Maturation – the beer is conditioned for 4–29 days at temperatures of 2–6 °C in tanks • Finishing – the beer is filtered, pasteurised, and then bottled or canned with the addition of carbon dioxide • The alcohol content varies between about 4% and 9%.

Cheese-making	Bacteria – feed on lactose in milk, changing the texture and taste, and inhibiting the growth of bacteria which make milk go off.	• The milk is pasteurised (heated to 95 °C for 20 seconds to kill off most natural bacteria) and homogenised (the fat droplets evenly distributed through the milk). • It is mixed with bacterial cultures and sometimes chymosin enzyme and kept until the milk separates into solid curds and liquid whey. • For cottage cheese, the curds are separated from the whey, packaged, and sold. • For most cheese, the curds are cut and cooked in the whey then strained through draining moulds or cheesecloth. The whey is used for animal feeds. • The curds are put into steel or wooden drums and may be pressed. They are left to dry, mature, and ripen before eating as the bacteria continue to act for anything from a few weeks to several years.
Yoghurt-making	Bacteria – often *Lactobacillus bulgaricus* (forms ethanal) and *Streptococcus thermophilus* (forms lactic acid). Both produce extracellular polymers that give yoghurt its smooth, thick texture.	• Skimmed milk powder is added to milk and the mixture is pasteurised, homogenised, and cooled to about 47 °C. • The milk is mixed with a 1:1 ratio of *Lactobacillus bulgaricus* and *Streptococcus thermophilus* and incubated at around 45 °C for 4–5 hours. • At the end of the fermentation, the yoghurt may be put into cartons at a temperature of around 10 °C as plain yoghurt or mixed with previously sterilised fruit. • Thick-set yoghurts are mixed and ferment in the pot. • Yoghurt has a shelf-life of about 19 days if stored at 2–3 °C.

Direct food production

People have eaten fungi for thousands of years in the form of a wide variety of mushrooms. In recent times, facing potential protein shortages around the world, scientists are developing more ways of using microorganisms to directly produce protein you can eat. It is known as single-cell protein or SCP.

The best known SCP is Quorn™. This is made of the fungus *Fusarium venetatum*, a single celled fungus that is grown in large fermenters using glucose syrup as a food source (You will learn more about fermenters in Topic 22.7, Culturing microorganisms on an industrial scale). The microorganisms are combined with albumen (egg white) and then compressed and formed into meat substitutes. Quorn™ is not only suitable for vegetarians, it is also a healthy choice as it is high in protein and low in fat. People are very conservative in their food choices and when the new food was launched, no mention was made of the fungi used to produce it. Using the term mycoprotein meant most people did not recognise what it was made of. However a combination of good marketing and a good product meant that people tried Quorn™ and liked it, and it has been internationally successful as a novel protein food.

Other attempts to make proteins from microorganisms have not yet been as successful. Yeasts, algae, and bacteria can be used to grow proteins that match animal proteins found in meat as well as plant proteins. They can be grown on almost anything, are relatively cheap and low in fat, yet none of the alternative protein sources has proved successful so far. People have many reservations about eating food grown on waste. Increasingly single celled proteins are being used to feed animals that we prefer to eat – from fish to cattle. If the world protein shortage continues, however, people may yet turn to eating foods made directly from microorganisms. Table 2 shows some of the advantages and disadvantages of using microorganisms directly to make food for human consumption.

▼ Table 2

Advantages of using microorganisms to produce human food	Disadvantages of using microorganisms to produce human food
microorganisms reproduce fast and produce protein faster than animals and plants	some microorganisms can also produce toxins if the conditions are not maintained at the optimum
microorganisms have a high protein content with little fat	the microorganisms have to be separated from the nutrient broth and processed to make the food
microorganisms can use a wide variety of waste materials including human and animal waste, reducing costs	need sterile conditions that are carefully controlled adding to costs
microorganisms can be genetically modified to produce the protein required	often involve GM organisms and many people have concerns about eating GM food
production of microorganisms is not dependant on weather, breeding cycles etc – it takes place constantly and can be increased or decreased to match demand	the protein has to be purified to ensure it contains no toxin or contaminants
no welfare issues when growing microorganisms	many people dislike the thought of eating microorganisms grown on waste
can be made to taste like anything	has little natural flavour – needs additives

Summary questions

1 What are the main advantages of using microorganisms in biotechnological processes? (6 marks)

2 Compare the way yeast is used in the process of baking and brewing. (6 marks)

3 a Why is milk pasteurised before being used commercially to make cheese and yoghurt? (2 marks)
  b Why is milk homogenised before being used commercially to make cheese and yoghurt? (2 marks)
  c Give two important differences between the production processes of cheese and yogurt. (2 marks)

22.5 Microorganisms, medicines, and bioremediation

Specification reference: 6.2.1

Learning outcomes

Demonstrate knowledge, understanding, and application of:

→ the use of microorganisms in biotechnological processes to include penicillin production, insulin production, and bioremediation.

Synoptic link

In Topic 12.7, Preventing and treating disease, you learnt about antibiotics and their importance in the treatment of bacterial diseases. In Topic 14.4, Diabetes and its control, you learnt about the importance of insulin to diabetics.

Since the discovery of penicillin in the 1920s, biotechnology has played a key role in the development of medicines.

Producing penicillin

The first effective antibiotic was penicillin, produced by a mould called *Penicillium notatum*. The yield of penicillin from this mould was very small. Commercial production of the drug in the quantities needed to treat everyone who needed it did not begin until the discovery of *Penicillium chrysogenum* by Mary Hunt on a melon from a market stall.

P. chrysogenum needs relatively high oxygen levels and a rich nutrient medium to grow well. It is sensitive to pH and temperature. This affects the way it is produced commercially. A semi-continuous batch process is used (Topic 22.7, Culturing microorganisms on an industrial scale). In the first stage of the production process the fungus grows. In the second stage it produces penicillin. Finally the drug is extracted from the medium and purified.

- The process uses relatively small fermenters (40–200 dm³) because it is very difficult to maintain high levels of oxygenation in very large bioreactors.
- The mixture is continuously stirred to keep it oxygenated.
- There is a rich nutrient medium.
- The growth medium contains a buffer to maintain pH at around 6.5.
- The bioreactors are maintained at about 25–27 °C.

Making insulin

As you learnt in Topic 21.4, Genetic engineering, biotechnology in the form of genetic engineering is important in the production of human medicines – for example, the production of human insulin. People with type 1 diabetes, and some people with type 2 diabetes, need regular injections of insulin to control their blood sugar levels. In the past, insulin was extracted from the pancreas of animals, usually pigs or cattle, slaughtered for meat. This meant the supply was erratic because it depended on the demand for meat – when fewer animals were killed, less insulin was available but the number of people with diabetes stayed the same. There were a number of other problems. Some people were allergic to the animal insulin as it was often impure, although eventually very pure forms were developed which overcame this problem. The peak activity of animal insulin is several hours after it is injected, which made calculating when to eat meals difficult. For some faith groups, using pig products is not permitted. The development of genetically engineered bacteria which can make

▲ Figure 1 *Penicillin is produced by the large-scale cultivation of* Penicillium *mould followed by processing and purification of the drug*

human insulin revolutionised the supply from the 1970s onwards. The bacteria are grown in a fermenter and downstream processing results in a constant supply of pure human insulin. You will learn about the bioreactors used to produce human insulin in Topic 22.7, Culturing microorganisms on an industrial scale.

Synoptic link

You learnt about genetic engineering and the production of insulin in Topic 21.4, Genetic engineering.

Bioremediation

In bioremediation, microorganisms are used to break down pollutants and contaminants in the soil or in water. There are different approaches to bioremediation:

1   Using natural organisms – many microorganisms naturally break down organic material producing carbon dioxide and water. Soil and water pollutants are often biological, for example, sewage and crude oil. If these naturally occurring microorganisms are supported, they will break down and neutralise many contaminants. For example, in an oil spill, nutrients can be added to the water to encourage microbial growth, and the oil can be dispersed into smaller particles to give the maximum surface area for microbial action.

2   GM organisms – scientists are trying to develop GM bacteria which can break down or accumulate contaminants which they would not naturally encounter, for example, bacteria have been engineered that can remove mercury contamination from water. Mercury is very toxic and accumulates in food chains. The aim is to develop filters containing these bacteria to remove mercury from contaminated sites.

Often, bioremediation takes place on the site of the contamination. Sometimes material is removed for decontamination. In most cases, natural microorganisms outperform GM ones – but as our ability to change the genetic material of microorganisms increases, it may be possible to use bioremediation even more effectively than it is used now.

▲ **Figure 2** *When huge areas of water are contaminated with oil, as in the Deepwater Horizon spill in 2010, bioremediation by microorganisms is the best hope for the environment*

➕ Plants and bioremediation

There are some pollutants which microorganisms cannot, at the moment, remove from the soil. In a number of cases, special plants can be used instead. In the early 1970s, a tree was discovered which produced a blue sap that turned out to contain 26% nickel in its dry mass. Plants which can take up large quantities of metals from the ground are known as hyperaccumulators. The process by which hyperaccumulators take up metals from the ground is known as bioleaching.

Suggest why plants can be used as bioremediators for heavy metal contamination but microorganisms cannot.

Summary questions

1   What is the main difference between the use of fungi to produce penicillin and the use of bacteria to produce human insulin? *(2 marks)*

2   What is bioremediation? Why it is often carried out on the site of contamination? *(6 marks)*

3   With your knowledge of genetic engineering and biotechnology so far, produce a flow diagram to show the stages in the production of human insulin. *(6 marks)*

22.6 Culturing microorganisms in the laboratory

Specification reference: 6.2.1

To investigate microorganisms for the medical diagnosis of disease or for scientific experiments you need to **culture** them. This often involves growing large enough numbers of the microorganisms for us to see them clearly with the naked eye.

Whenever microorganisms are cultured in the laboratory the correct health and safety procedures must be followed because even when the microorganisms are expected to be completely harmless:

● there is always the risk of a mutation taking place making the strain pathogenic

● there may be contamination with pathogenic microorganisms from the environment.

Culturing microorganisms

The microorganisms to be cultured need food as well as the right conditions of temperature, oxygen, and pH. The food provided for microorganisms is known as the nutrient medium. It can be either in liquid form (broth) or in solid form (agar). Nutrients are often added to the agar or the broth to provide a better medium for microbial growth. Some microorganisms need a precise balance of nutrients but often the medium is simply enriched with good protein sources such as blood, yeast extract, or meat. Enriched nutrient media allow samples containing a very small number of organisms to multiply rapidly. The nutrient medium must be kept sterile (free from contamination by microorganisms) until it is ready for use. **Aseptic techniques** are important.

Once the agar or nutrient broth is prepared the bacteria must be added in a process called inoculation.

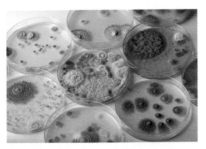

▲ **Figure 1** *Culturing microorganisms like these different types of bacteria allows us to see what we are working with*

Inoculating broth

1 Make a suspension of the bacteria to be grown.

2 Mix a known volume with the sterile nutrient broth in the flask.

3 Stopper the flask with cotton wool to prevent contamination from the air.

4 Incubate at a suitable temperature, shaking regularly to aerate the broth providing oxygen for the growing bacteria.

Inoculating agar

This also involves a suspension of bacteria but the process is slightly more complicated.

1 The wire inoculating loop must be sterilised by holding it in a Bunsen flame until it glows red hot. It must not be allowed to touch any surfaces as it cools to avoid contamination.

▲ **Figure 2** *Cultures grown in broth are usually prepared in flasks or test tubes whilst agar-based cultures are prepared in Petri dishes. The hot sterile liquid agar is poured onto the plates which are immediately resealed and cooled so that they set ready for use*

2   Dip the sterilised loop in the bacterial suspension. Remove the lid of the Petri dish and make a zig-zag streak across the surface of the agar. Avoid the loop digging into the agar by holding it almost horizontal. However many streaks are applied, the surface of the agar must be kept intact.

3   Replace the lid of the Petri dish. It should be held down with tape but not sealed completely so oxygen can get in, preventing the growth of anaerobic bacteria. Incubate at a suitable temperature.

▲ **Figure 3** *Inoculating an agar plate with bacteria*

The growth of bacterial colonies

Bacteria can reproduce very rapidly, undergoing asexual reproduction every 20 minutes in optimum conditions. If a single bacterium had unlimited space and nutrients, and if all its offspring continued to divide at the same rate, then at the end of 48 hours there would be 2.2×10^{43} bacteria, weighing 4000 times the weight of the Earth. Fortunately, in a closed system limited nutrients and a build-up of waste products always acts as a brake on reproduction and growth. Logarithmic numbers (logs) are mainly used to represent the bacterial population because the difference in numbers from the initial organism to the billions of descendants is sometimes too great to represent using standard numbers.

There are four stages to this growth curve:

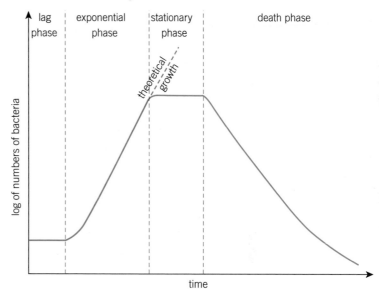

▲ **Figure 4** *The exponential growth theoretically possible for a bacterial population (red dashed line) is prevented by a variety of limiting factors which, fortunately, result in the typical growth curve of a bacterial population in a closed system*

- the lag phase when bacteria are adapting to their new environment. They are growing, synthesising the enzymes they need, and are not yet reproducing at their maximum rate.

- the log or exponential phase is when the rate of bacterial reproduction is close to or at its theoretical maximum.

- the stationary phase occurs when the total growth rate is zero – the number of new cells formed by binary fission is cancelled out by the number of cells dying.

- the decline or death stage comes when reproduction has almost ceased and the death rate of cells is increasing.

There are several limiting factors which prevent exponential growth in a culture of bacteria. These include:

- Nutrients available – initially there is plenty of food, but as the numbers of microorganisms multiply exponentially it is used up. The nutrient level will become insufficient to support further growth and reproduction unless more nutrients are added.

- Oxygen levels – as the population rises, so does the demand for respiratory oxygen so oxygen levels can become limiting.

- Temperature – the enzyme-controlled reactions within microorganisms are affected by the temperature of the culture medium. For most bacteria, a low temperature slows down growth and reproduction, and a higher temperature speeds it up. If the temperature gets too high it will denature the enzymes, killing the microorganisms – even thermophiles have a maximum temperature they can withstand.

- Build-up of waste – as bacterial numbers rise, the build-up of toxic material may inhibit further growth and can even poison and kill the culture.

- Change in pH – as carbon dioxide produced by the respiration of the bacterial cells increases, the pH of the culture falls until a point where the low pH affects enzyme activity and inhibits population growth.

Investigating factors which affect the growth of microorganisms – serial dilutions and bacterial counting

You can investigate the factors which affect the growth of bacterial colonies in a number of ways. For example, you can:

- set up identical colonies in different conditions of temperature

- set up serial dilutions of nutrients or pH, at a set temperature.

It is essential when carrying out these experiments that precautions are taken to ensure aseptic conditions (free from contamination). For example, using sterile equipment and a fresh pipette after each dilution.

To see the effect of the conditions you need to be able to measure the number of microorganisms at the beginning and end of your investigations. One method is to use another application of serial dilutions (Figure 5). The assumption is made that each of the colonies on an agar plate grows from a single, viable microorganism. If two bacterial colonies can be seen after culturing, then there were two living bacteria on the plate, and if 50 patches form there were 50 bacteria on the plate when it was inoculated. However, in most cases when a plate is inoculated a solid mass of microbial growth is present after culturing – you cannot count the individual colonies. This

is overcome by carrying out a serial dilution of the original culture broth until, when you culture a given volume of the broth on an agar plate, you can count the number of colonies. Multiply the number of colonies by the dilution factor to give you a total viable cell count per volume for the original colony. As long as you can count the number of colonies on two or more plates, you can calculate the mean of the number of organisms in a particular culture.

A student made up serial dilutions of a bacterial culture from 10^{-1} to 10^{-10}. A $0.1cm^3$ sample was cultured from each of these dilutions. From the original dilutions the numbers of bacterial colonies counted were:
A Original dilution 10^{-5-} – 500 colonies
B Original dilution 10^{-6} – 52 colonies
C Original dilution 10^{-7} – 4 colonies
Calculate (a) the actual dilution of each sample grown on the plate
(b) the number of bacteria cm^{-3} in the original sample as shown on each plate
(c) the mean number of bacteria cm^{-3} in the original sample.

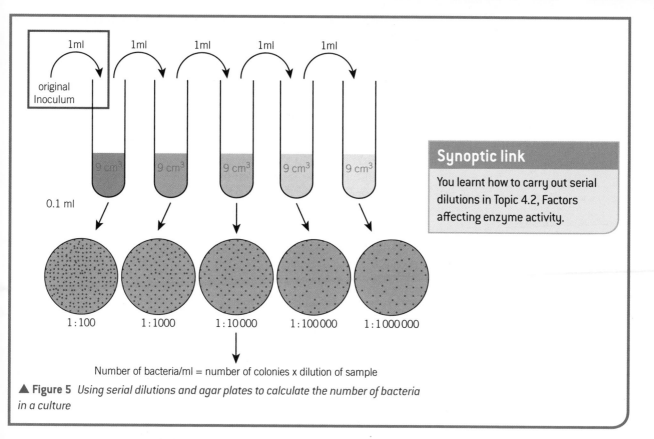

Synoptic link

You learnt how to carry out serial dilutions in Topic 4.2, Factors affecting enzyme activity.

Number of bacteria/ml = number of colonies x dilution of sample

▲ **Figure 5** *Using serial dilutions and agar plates to calculate the number of bacteria in a culture*

Summary questions

1 Compare the processes of culturing bacteria in broth and on agar. (*6 marks*)

2 Why are there such clear differences between the theoretical growth curve of a bacterial colony and the actual growth curve in a closed culture? (*5 marks*)

3 Explain the following statements in terms of factors affecting bacterial growth:
   a Vinegar is a very good preservative. (*2 marks*)
   b Food eventually goes bad in a fridge. (*2 marks*)
   c In the Northern hemisphere, material placed in a compost heap rots down much faster in August than it does in December. (*2 marks*)

4 Make a flow chart to show how you would calculate the affect of a factor on bacterial growth using serial dilutions and agar plating. (*6 marks*)

22.7 Culturing microorganisms on an industrial scale

Specification reference: 6.2.1

In any bioprocess the microorganism involved must be able to synthesise or break down the chemical required, work reasonably fast, give a good yield of the product, use relatively cheap nutrients, and not require extreme (and therefore expensive) conditions. It must not produce any poisons that contaminate the product or mutate easily into non-functioning forms.

Primary and secondary metabolites

What is wanted from the microorganisms varies from one bioprocess to another.

Sometimes, you would want as much microorganism as possible, because the microorganism itself is the product to be sold, for example, single-celled protein such as Quorn, or baker's yeast.

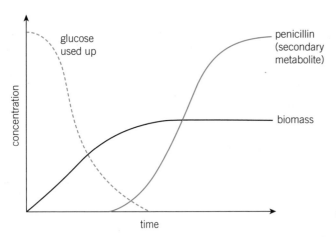

▲ **Figure 1** *Primary metabolites are usually formed in the period of active growth, whilst secondary metabolites tend to be formed during the stationary phase of the life of the culture once the cell mass has reached its maximum. The time at which the culture is harvested will depend on what we actually want from the microorganisms*

Sometimes, substances are wanted which are formed as an essential part of the normal functioning of a microorganism, for example, ethanol (a product of anaerobic respiration in yeast), ethanoic acid, and a range of amino acids and enzymes. They are known as primary metabolites.

In some circumstances, organisms produce substances which are not essential for normal growth, but are still used by the cells. Examples include many pigments, and the toxic chemicals plants produce to protect themselves against attack by herbivores. The organism would not suffer, at least in the short term, without them. These chemicals are known as secondary metabolites, and they are often the required product in a bioprocess, for example, penicillin and many other antibiotics.

Types of bioprocess

Once a microorganism has been chosen, and the ideal size and shape of the bioreactor (reaction vessel) decided, the organisation of the commercial production has to be decided. Two of the main ways of growing microorganisms are **batch fermentation** and **continuous fermentation**.

Batch fermentation

- The microorganisms are inoculated into a fixed volume of medium.
- As growth takes place, nutrients are used up and both new biomass and waste products build up.

- As the culture reaches the stationary phase, overall growth ceases but during this phase the microorganisms often carry out biochemical changes to form the desired end products (such as antibiotics and enzymes).

- The process is stopped before the death phase and the products harvested. The whole system is then cleaned and sterilised and a new batch culture started up.

Continuous culture

- Microorganisms are inoculated into sterile nutrient medium and start to grow.

- Sterile nutrient medium is added continually to the culture once it reaches the exponential point of growth.

- Culture broth is continually removed – the medium, waste products, microorganisms, and product – keeping the culture volume in the bioreactor constant.

Continuous culture enables continuous balanced growth, with levels of nutrients, pH, and metabolic products kept more or less constant.

Both methods of operating a bioreactor can be adjusted to ensure either the maximum production of biomass or the maximum production of the primary or secondary metabolites. Most systems are adapted for maximum yield of metabolites. The majority of industrial processes use batch or semi-continuous cultivation. Continuous cultivation is largely used for the production of single-celled protein and in some waste water treatment processes.

All bioreactors produce a mixture of unused nutrient broth, microorganisms, primary metabolites, possibly secondary metabolites, and waste products. The useful part of the mixture has to be separated out by downstream processing. This is one of the most difficult and expensive parts of the whole bioprocess – the percentage of the total cost of a product which is due to downstream processing costs varies from 15–40%.

Controlling bioreactors

Whether a bioreactor is simply a container containing microbial broth or a complex aseptic fermenter, it is very important to control and manipulate the growing conditions to maximise the yield of product required. Factors which need to be controlled include:

Temperature

If the temperature is too low the microorganisms will not grow quickly enough. If the temperature gets too high, enzymes start to denature and the microorganisms are inhibited or destroyed. Bioreactors often have a heating and/or a cooling system linked to temperature sensors and a negative feedback system to maintain optimum conditions.

Nutrients and oxygen

Oxygen and nutrient medium can be added in controlled amounts to the broth when probes or sample tests indicate that levels are dropping.

▲ **Figure 2** *Bioreactors range from around a 10 litre capacity to hundreds of litres and are widely used in the pharmaceutical industry*

Mixing things up

Inside a bioreactor there are large volumes of liquid, which may be quite thick and viscous due to the growth of microorganisms. Simple diffusion is not enough to ensure that all the microorganisms receive enough food and oxygen or that the whole mixture is kept at the right temperature, so most bioreactors have a mixing mechanism and many are stirred continuously.

Asepsis

If a bioprocess is contaminated by microorganisms from the air, or from workers, it can seriously affect the yield. To solve this problem most bioreactors are sealed, aseptic units. If the process involves genetically engineered organisms, it is a legal requirement that they should be contained within the bioreactor and not be released into the environment.

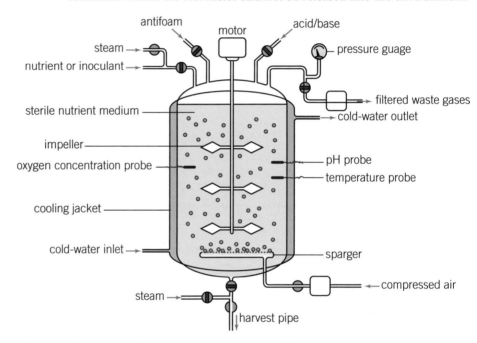

▲ **Figure 3** *The design of bioreactors is moving forward all the time as new and more sensitive ways of controlling conditions during fermentation processes become available, making it easier to get the maximum yield from any process*

22.8 Using immobilised enzymes

Specification reference: 6.2.1

Enzymes, mainly from microorganisms, have been used indirectly for thousands of years in biotechnologies including brewing, baking, and making cheese and yoghurt. Many biotechnological processes still use whole microorganisms as their enzyme source. More recently they are also being used in isolation.

Isolated enzymes

Using isolated enzymes instead of whole organisms has some clear advantages.

- Less wasteful – whole microorganisms use up substrate growing and reproducing, producing biomass rather than product. Isolated enzymes do not.

- More efficient – isolated enzymes work at much higher concentrations than is possible when they are part of the whole microorganism.

- More specific – no unwanted enzymes present, so no wasteful side reactions take place.

- Maximise efficiency – isolated enzymes can be given ideal conditions for maximum product formation, which may differ from those needed for the growth of the whole microorganism.

- Less downstream processing – pure product is produced by isolated enzymes. Whole microorganisms give a variety of products in the final broth, making isolation of the desired product more difficult and therefore expensive.

Most of the isolated enzymes used in industrial processes are extracellular enzymes produced by microorganisms. They are generally easier and therefore cheaper to use than intracellular enzymes.

- Extracellular enzymes are secreted, making them easy to isolate and use.

- Each microorganism produces relatively few extracellular enzymes, making it easy to identify and isolate the required enzyme. In comparison, each microorganism produces hundreds of intracellular enzymes which would need extracting from the cell and separating.

- Extracellular enzymes tend to be much more robust than intracellular enzymes. Conditions outside a cell are less tightly controlled than conditions in the cytoplasm, so extracellular enzymes are adapted to cope with greater variations in temperature and pH than intracellular enzymes.

However, in spite of the advantages of using extracellular enzymes, intracellular enzymes are still sometimes used as isolated enzymes in manufacturing processes. This is because there is a bigger range of intracellular enzymes (bullet 2) so in some cases they provide the ideal enzyme for a process. In these cases, the benefits of using a

— active site

▲ **Figure 1** *The shape of the active site gives an enzyme great specificity and makes it an invaluable tool to the biotechnologist*

very specific intracellular enzyme outweigh the disadvantages of the more expensive extraction and isolation process and the need for more tightly controlled conditions. Examples of intracellular enzymes used as isolated enzymes in industry include glucose oxidase for food preservation, asparaginase for cancer treatment, and **penicillin acylase** for converting natural penicillin into semi-synthetic drugs which are more effective.

Immobilised enzymes

Isolated enzymes are more efficient than whole organisms, but using free enzymes is often very wasteful. Enzymes are not cheap to produce, but at the end of the process they cannot usually be recovered and so they are simply lost.

Increasingly enzymes used in industrial processes are immobilised – attached to an inert support system over which the substrate passes and is converted to product. This is a case of technology mimicking nature – enzymes in cells are usually bound to membranes to carry out their repeated cycles of catalysis. Because immobilised enzymes are held stationary during the catalytic process, they can be recovered from the reaction mixture and reused time after time. The enzymes do not contaminate the end product, so less downstream processing is needed. These things all make the process more economical.

Advantages of using immobilised enzymes

- Immobilised enzymes can be reused – which is cheaper.
- Easily separated from the reactants and products of the reaction they are catalysing so reduced downstream processing – which is cheaper.
- More reliable – there is a high degree of control over the process as the insoluble support provides a stable microenvironment for the immobilised enzymes.
- Greater temperature tolerance – immobilised enzymes are less easily denatured by heat and work at optimum levels over a much wider range of temperatures, making the bioreactor less expensive to run.
- Ease of manipulation – the catalytic properties of immobilised enzymes can be altered to fit a particular process more easily than those of free enzymes – for example, immobilised **glucose isomerase** can be used continuously for over 1000 hours at temperatures of 60–65 °C. The ability to keep bioreactors running continuously for long periods without emptying and cleaning helps to keep running costs low.

Disadvantages of using immobilised enzymes

- Reduced efficiency – the process of immobilising an enzyme may reduce its activity rate.
- Higher initial costs of materials – immobilised enzymes are more expensive than free enzymes or microorganisms. However, the immobilised enzymes, unlike free enzymes, do not need to be replaced frequently.

- Higher initial costs of bioreactor – the system needed to use immobilised enzymes is different from traditional fermenters so there is an initial investment cost.

- More technical issues – reactors which use immobilised enzymes are more complex than simple fermenters – they have more things which can go wrong.

How are enzymes immobilised?

Enzymes can be immobilised in a number of ways. They may be bound to the surface of insoluble supporting materials either by adsorption onto the surface or by covalent or ionic bonds. They may be entrapped in a matrix, encapsulated in a microcapsule (like proteases for detergent use), or behind a semi-permeable membrane. Each of these methods has advantages and disadvantages as summarised in Table 1.

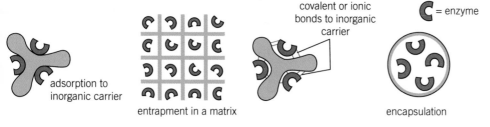

▲ **Figure 2** *Four commonly used ways of immobilising enzymes*

▼ **Table 1**

Method of immobilising the enzymes	Advantages	Disadvantages
surface immobilisation – adsorption to inorganic carriers, e.g., cellulose, silica, carbon nanotubes, and polyacrylamide gel	simple and cheap to do can be used with many different processes enzymes very accessible to substrate and their activity is virtually unchanged	enzymes can be lost from matrix relatively easily
surface immobilisation – covalent or ionic bonding to inorganic carriers covalent bonding, e.g., carriers with amino, hydroxyl, carboxyl groups ionic bonding, e.g., polysaccharides such as cellulose, synthetic polymers	cost varies enzymes strongly bound and therefore unlikely to be lost enzymes very accessible to substrate pH and substrate concentration often have little effect on enzyme activity	cost varies active site of the enzyme may be modified in process, making it less effective
entrapment – in matrix, e.g., polysaccharides, gelatin, activated carbon	widely applicable to different processes	may be expensive can be difficult to entrap diffusion of the substrate to and product from the active site can be slow and hold up the reaction effect of entrapment on enzyme activity very variable, depending on matrix

entrapment – membrane entrapment in microcapsules (encapsulation) or behind a semi-permeable membrane, e.g., polymer-based semi-permeable membranes	relatively simple to do relatively small effect on enzyme activity widely applicable to different processes	relatively expensive diffusion of the substrate to and product from the active site can be slow and hold up the reaction

In some cases whole microorganisms rather than just the enzymes are immobilised. This has many of the same advantages but avoids the time-consuming and expensive process of extracting the pure enzyme and immobilising it before the process starts. On the other hand, the organisms need food, oxygen, and a carefully controlled environment to work at their optimum rate.

Using immobilised enzymes

Immobilised enzymes are very useful when large quantities of product are wanted, because they allow continuous production. Examples include:

- Immobilised penicillin acylase used to make semi-synthetic penicillins from naturally produced penicillins. Many types of bacteria have developed resistance to naturally occurring penicillins so they are no longer very effective drugs. Fortunately, many bacteria are still vulnerable to the semi-synthetic penicillins produced by penicillin acylase so they are very important in treating infections caused by bacteria resistant to the original penicillin. Hundreds of tonnes of these medicines are made every year by immobilised penicillin acylase.

- Immobilised glucose isomerase used to produce fructose from glucose. Fructose is much sweeter than sucrose or glucose and is widely used as a sweetener in the food industries. Glucose is produced from cheap, starch-rich plant material. Glucose isomerise is then used to turn the cheap glucose into very marketable fructose.

- Immobilised **lactase** used to produce lactose-free milk. Some people, and cats, are intolerant of lactose (milk sugar). Immobilised lactase hydrolyses lactose to glucose and galactose, giving lactose-free milk.

- Immobilised aminoacylase used to produce pure samples of L-amino acids used in the production of pharmaceuticals, organic chemicals, cosmetics, and food.

- Immobilised glucoamylase, which can be used to complete the breakdown of starch to glucose syrup. Amylase enzymes break starch down into short chain polymers called dextrins. The final breakdown of dextrins to glucose is catalysed by immobilised glucoamylase.

- Immobilised nitrile hydratase, an enzyme which is playing an increasing role in the plastics industry. Acrylamide is a very important compound which is used in the production of many plastics. It is made by the hydration of acrylonitrile. Traditionally the hydration of acrylonitrile to acrylamide was done using sulphuric acid with a reduced copper catalyst, but the conditions

Study tip

Nitrile hydratase is the enzyme which catalyses the conversion of acrylonitrile to acrylamide in a hydration reaction.

Nitrilases are a group of enzymes that catalyse the hydrolysis of nitriles, for example, acrylonitrile to *carboxylic* acids and ammonia.

needed are extreme and therefore expensive. Furthermore, unwanted by-products form and the yield is poor. Using immobilised nitrile hydratase the conversion takes place under moderate conditions so the process is cheaper and it also gives a 99% yield and no unwanted by-products.

Immobilised enzymes in medicine

Immobilised enzymes and microbial cells are increasingly important both as diagnostic tools in medicine – the manufacture of drugs – for example, the fungus *Rhizopus arrhizus* is immobilised and used in the production of the steroid drug cortisone.

Biosensors are used in accurate monitoring of blood and urine levels of substances such as glucose, urea, amino acids, ethanol, and lactic acid, as well as in the monitoring of waste treatment, water analysis, and the control of complex chemical processes. They are based on an electrochemical sensor in close proximity to an immobilised enzyme membrane. The enzymes react with a specific substrate and the chemicals produced

are detected by the sensor. Because the size of the response is related to the concentration of the substrate, these sensors can be very sensitive and accurate in their measurements and so can be used, for example, by people with diabetes to determine their blood glucose levels and therefore judge the insulin dose they need.

1 Give one clear advantage of using immobilised enzymes in biosensors for medical use and one possible disadvantage, both linked to what you know of the properties of enzymes.
2 Suggest how immobilising the enzyme might help overcome your suggested disadvantage.

Summary questions

1 a What is meant by an immobilised enzyme? (*2 marks*)
   b What are the main advantages of immobilised enzymes over whole microorganisms? (*4 marks*)
   c What are the main advantages of immobilised enzymes over free enzymes? (*4 marks*)

2 Summarise the main ways in which enzymes are immobilised. (*4 marks*)

3 How can immobilisation:
   a increase the effectiveness of an enzyme (*3 marks*)
   b decrease the effectiveness of an enzyme? (*4 marks*)

4 Investigate one reaction catalysed by immobilised enzymes. Give the reaction catalysed, the way the enzyme is immobilised, the importance of the process, and the advantages of using immobilised enzymes over any other way of catalysing the reaction. (*6 marks*)

Practice questions

1   **a**   Outline the differences between reproductive and therapeutic cloning.
         (3 marks)

   b   Evaluate the ethical concerns regarding the reproductive cloning of human beings.
         (4 marks)

   c   (i)   Describe what is meant by the term somatic cell nuclear transfer (SCNT).
         (3 marks)

       (ii)   Suggest why clones produced by SCNT are not exact genetic clones.
         (2 marks)

   d   Describe the reasons for embryo splitting.
         (3 marks)

2   Micropropagation is the name of a process used to produce artificially cloned plants.

   a   Explain the meaning of the following terms and outline their roles in the micropropagation.
         (4 marks)

      •   Explant

      •   Callus

   b   Describe how plant hormones are used to stimulate the development of cells obtained from the callus.   *(5 marks)*

3   The diagram shows a type of fermentation.

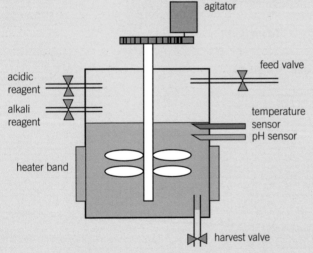

   a   State, giving your reasons, the type of fermentation being conducted using the equipment shown in the diagram
         (3 marks)

   b   Outline the roles of the valves and sensors shown in the diagram.   *(4 marks)*

   c   Discuss the advantages and disadvantages of this type of fermentation.   *(4 marks)*

4   The cloning of animals is a controversial subject due to the shortened life expectancies and health problems of cloned animals.

The graph shows the data collected from an investigation into the health problems faced by cloned and conventional cows.

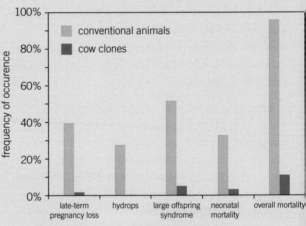

Calves with large offspring syndrome usually have breathing and cardiovascular problems as well difficult births. Hydrops is an abnormal build-up of fluid in new-born calves and is an indication of more serious health problems.

A survey carried out by the European Commission investigating European Attitudes Towards Animal Cloning in October 2008 produced the results shown in the graph.

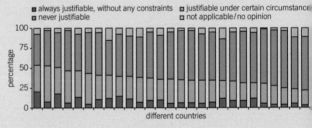

Discuss, using information from both graphs, the arguments for and against the cloning of animals in the production of food.   *(6 marks)*

5   Microorganisms, such as the fungus Fusarium, can be grown and then purified to produce mycoprotein, This mycoprotein can be used as a food source for humans.

The table compares mycoprotein with beef.

Food	Content per 100g					
	Energy (kJ)	Protein (g)	Carbohydrate (g)	Total fat (g)	Saturated fat	Iron (mg)
mycoprotein	357	12	9	2.9	0.6	0.1
beef	1163	26	0	18.2	7.0	2.6

Use the data in the table to describe and explain the advantages and disadvantages of using microorganisms to produce food for human consumption. *(7 marks)*

OCR F212 2010

6 A number of different methods can be used to determine the number of bacteria in a sample.

a (i) Outline the differences between using a graticule and dilution plating to count bacteria. *(4 marks)*

(ii) State the name of another, more simple technique that uses the same principle as graticule counts. *(1 mark)*

b Suggest why it is necessary to maintain aseptic conditions when working with bacteria. *(3 mark)*

A 1 ml sample containing bacteria was mixed with an equal volume of the dye methylene blue. A drop the resulting mixture was placed on a graticule and observed under a microscope.

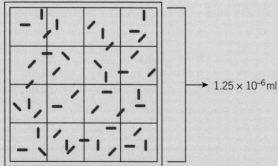

The graticule held 1.25×10^{-6} ml of sample.

c Calculate the number of bacteria per ml in the original sample. *(3 marks)*

A serial dilution was carried out on a sample containing *E. coli*. Samples from the last three dilutions were plated and incubated. The number of colonies that grew on each plate is shown in the diagram.

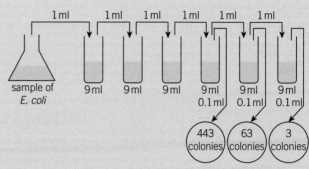

d Calculate the number of bacteria in the original sample. *(3 marks)*

7 Starch phosphorylase catalyses the conversion of starch and inorganic phosphate into glucose-1-phosphate. It has an important role in the metabolism of starch in plants. It has been used in the production of glucose-1-phosphate for use in the treatment of some heart conditions.

When used for the production of glucose-1-phosphate starch phosphorylase is normally immobilized.

a Outline the different ways that enzymes are immobilized. *(4 marks)*

The diagram shows the changing activities of free and immobilized starch phosphorylase at different temperatures.

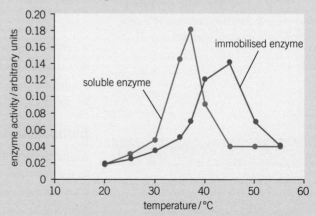

b Explain the reasons for the differences in optimum temperature and maximum activity of the free and immobilized enzymes. *(5 marks)*

23 ECOSYSTEMS
23.1 Ecosystems

Specification reference: 6.3.1

Learning outcomes

Demonstrate knowledge, understanding, and application of:

→ ecosystems and the influence of biotic and abiotic factors.

Synoptic link

Look back at Chapter 11, Biodiversity to remind yourself of the importance of maintaining biodiversity.

▲ **Figure 1** *A biome is a large naturally occurring community of flora (plants) and fauna (animals) occupying a major habitat such as a rainforest*

Synoptic link

You will learn more about competition in Chapter 24, Populations and sustainability.

Ecology is the name given to the study of the relationships between organisms and their environment. It is essential that scientists study the natural world and the vast variety of life that exists, to help us understand the interdependence of living organisms and to help ensure the survival of as much of Earth's biodiversity as possible.

Defining ecosystems

An **ecosystem** is made up of all the living organisms that interact with one another in a defined area, and also the physical factors present in that region. Ecosystems can vary dramatically in size – from a tiny bacterial colony to the entire biosphere of Earth.

The boundaries of a particular ecosystem being studied are defined by the person or team carrying out the study. For example, individual habitats may be studied such as a rock pool or a large oak tree, or small areas of land such as a playing field or a particular stretch of a river.

Factors that affect ecosystems

All ecosystems are dynamic, meaning that they are constantly changing. This is a result of the living organisms present and the environmental conditions.

A large number of factors affect an ecosystem. The factors can be divided into two groups:

- **biotic factors** – the living factors. For example, in a forest ecosystem, the presence of shrews and hedgehogs are biotic factors, as is the size of their populations – the competition between these two animal populations for a food source (e.g., insects) is also a biotic factor.
- **abiotic factors** – the non-living or physical factors. Within the forest ecosystem, abiotic factors include the amount of rainfall received and the yearly temperature range of the ecosystem.

Biotic factors

Biotic factors often refer to the interactions between organisms that are living, or have once lived. These interactions often involve competition, either within a population or between different populations. Examples of things for which animals compete include food, space (territory), and breeding partners.

Abiotic factors

Light

Most plants are directly affected by light availability as light is required for photosynthesis. In general the greater the availability of light, the greater the success of a plant species.

Plants develop strategies to cope with different light intensities. For example, in areas of low light they may have larger leaves. They may also develop photosynthetic pigments that require less light, or reproductive systems that operate only when light availability is at an optimum.

Temperature

The greatest effect of temperature is on the enzymes controlling metabolic reactions. Plants will develop more rapidly in warmer temperatures, as will ectothermic animals. (Endothermic animals control their internal temperature, and so are less affected by the external environment.) Changes in the temperature of an ecosystem, for example, due to the changing seasons, can trigger migration in some animal species, and hibernation in others. In plant species it can trigger leaf-fall, dormancy, and flowering.

Water availability

In most plant and animal populations, a lack of water leads to water stress, which, if severe, will lead to death.

A lack of water will cause most plants to wilt, as water is required to keep cells turgid and so keep the plant upright. It is also required for photosynthesis. Cacti are an example of xerophytes, plants that have developed successful strategies to cope with water stress.

Oxygen availability

In aquatic ecosystems, it is beneficial to have fast-flowing cold water as it contains high concentrations of oxygen. If water becomes too warm, or the flow rate too slow, the resulting drop in oxygen concentration can lead to the suffocation of aquatic organisms.

In waterlogged soil, the air spaces between the soil particles are filled with water. This reduces the oxygen available for plants.

Edaphic (soil) factors

Different soil types have different particle sizes. This has an effect on the organisms that are able to survive in them. There are three main soil types:

- clay – this has fine particles, is easily waterlogged, and forms clumps when wet
- loam – this has different-sized particles, it retains water but does not become waterlogged
- sandy – this has coarse, well-separated particles that allow free draining – sandy soil does not retain water and is easily eroded.

▲ Figure 2 *Bromeliads, mosses, and small trees growing on the upper branches of a large tree in the cloud forest in Ecuador. This strategy allows small plants to receive sufficient light for photosynthesis*

Synoptic link

You learnt about xerophytes in Topic 9.5, Plant adaptations to water availability.

▲ Figure 3 *The scimitar Oryx (*Oryx dammah*) has adapted to survive for 9–10 months without water. Their kidneys are specially adapted to produce the barest minimum of urine. They also sweat very little and so lose hardly any of their overall water intake*

Summary questions

1 Explain what is meant by the term 'dynamic ecosystem'. To which ecosystems does the term apply? *(2 marks)*

2 State two biotic and two abiotic factors in a pond ecosystem. *(4 marks)*

3 Explain why abiotic factors often have a greater effect on plant species than on animal species in an ecosystem. *(4 marks)*

23.2 Biomass transfer through an ecosystem

Specification reference: 6.3.1

Learning outcomes

Demonstrate knowledge, understanding, and application of:

→ biomass transfers through ecosystems.

Synoptic link

You learnt about photosynthesis in Chapter 17, Energy for biological processes.

All organisms found within an ecosystem require a source of energy to perform the functions needed to survive. Ultimately, the Sun is the source of energy for almost all ecosystems on Earth. Through the process of photosynthesis, the Sun's light energy is converted into chemical energy in plants and other photosynthetic organisms. This chemical energy is then transferred to other non-photosynthetic organisms as food.

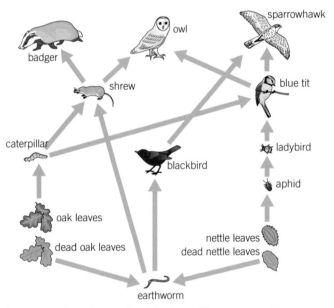

▲ **Figure 1** *Part of a woodland food web. The arrows on the diagram show the direction of energy flow.*

Trophic levels

You should be familiar with food chains and food webs (systems of interlinked food chains). These are diagrams that scientists use to show the transfer of biomass (mass of living material), and therefore energy, through the organisms in an ecosystem. Each stage in the chain is known as a **trophic level**.

The first trophic level is always a **producer** – an organism that converts light energy into chemical energy by the process of photosynthesis. The subsequent trophic levels are all **consumers** – organisms that obtain their energy by feeding on other organisms. The second trophic level is occupied by a primary consumer – an animal that eats a producer.

The following trophic levels are labelled successively – secondary consumer (an animal that eats a primary consumer), tertiary consumer (an animal that eats a secondary consumer), and a quaternary consumer (an animal that eats a tertiary consumer). Food chains rarely have more trophic levels than this as there is not sufficient biomass and stored energy left to support any further organisms.

▲ **Figure 2** *Food chains can be presented diagrammatically as a pyramid of numbers*

tertiary consumers

secondary consumers

primary consumers

producers

Decomposers are also important components of food webs – they break down dead organisms releasing nutrients back into the ecosystem. You will learn more about decomposers in Topic 23.3, Recycling within ecosystems.

Food chains can be presented diagrammatically as a pyramid of numbers, with each level representing the number of organisms at each trophic level (Figure 2). In a pyramid the producers are always placed at the bottom of the diagram with subsequent trophic levels added above.

Measuring biomass

Biomass is the mass of living material present in a particular place or in particular organisms. It is an important measure in the study of food chains and food webs as it can be equated to energy content.

To calculate biomass at each trophic level, you multiply the biomass present in each organism by the total number of organisms in that trophic level. This information can be presented diagrammatically as a pyramid of biomass (Figure 3). This represents the biomass present at a particular moment in time – it does not take into account seasonal changes.

The easiest way to measure biomass is to measure the mass of fresh material present. However, water content must be discounted and the presence of varying amounts of water in different organisms makes this technique unreliable unless very large samples are used. Scientists therefore usually calculate the 'dry mass' of organisms present. This is not without problems. Organisms have to be killed in order to be dried. The organisms are placed in an oven at 80 °C until all water has evaporated – this point is indicated by at least two identical mass readings. To minimise the destruction of organisms (particularly animals) only a small sample is taken. However, this sample may not be representative of the population as a whole.

Biomass is measured in grams per square metre ($g\,m^{-2}$) for areas of land, or grams per cubic metre ($g\,m^{-3}$) for areas of water.

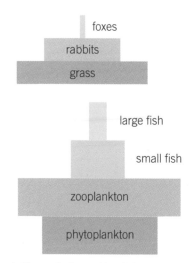

▲ **Figure 3** *Pyramids of biomass are virtually always a pyramid shape (top). An exception is a pyramid of biomass for a marine ecosystem (bottom). Phytoplankton are microorganisms that can photosynthesise. The mass of phytoplankton at any given time is often quite small, but they reproduce very quickly so over a period of time there is always more phytoplankton than zooplankton*

Efficiency of biomass and energy transfer between trophic levels

The biomass in each trophic level is nearly always less than the trophic level below. This is because biomass consists of all the cells and tissues of the organisms present, including the carbohydrates and other carbon compounds the organisms contain. As carbon compounds are a store of energy, biomass can be equated to energy content. When animals eat, only a small proportion of the food they ingest is converted into new tissue. It is only this part of the biomass (and hence energy) which is available for the next trophic level to eat.

The energy available at each trophic level is measured in kilojoules per metre squared per year ($kJ\,m^{-2}\,yr^{-1}$), to allow for changes in photosynthetic production and consumer feeding patterns throughout the year. As biomass is transferred between trophic levels, so the energy contained is

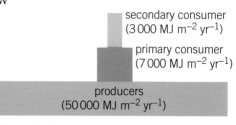

▲ **Figure 4** *A pyramid of energy for grassland*

transferred. The efficiency with which biomass or energy is transferred from one trophic level to the next is called the **ecological efficiency**.

The amount of biomass or energy converted to new biomass by each trophic level in a food chain can be represented by a pyramid of energy (Figure 4).

Efficiency at producer level

Producers only convert 1–3% of the sunlight (solar or light energy) they receive into chemical energy and hence biomass. This is because:

- not all of the solar energy available is used for photosynthesis – approximately 90% is reflected, some is transmitted through the leaf, and some is of unusable wavelength
- other factors may limit photosynthesis, such as water availability
- a proportion of the energy is 'lost', as it is used for photosynthetic reactions.

The total solar energy that plants convert to organic matter is called the gross production. However, plants use 20–50% of this energy in respiration. The remaining energy is converted into biomass. This is the energy available to the next trophic level and is known as the net production.

The energy available to the next trophic level can be calculated using the following formula:

Net production = gross production – respiratory losses

Note that this calculation can be applied equally to the biomass or energy production within an organism. The generation of biomass in a producer is referred to as primary production – in a consumer, it is known as secondary production.

 Worked example 1: Calculating net production (biomass)

A group of scientists measured the gross production of a grassland area as $60\,g\,m^{-2}\,year^{-1}$. If the respiration loss was $20\,g\,m^{-2}\,year^{-1}$, calculate the annual net production of this area of grassland.

Net production = gross production – respiratory losses

$$= 60 - 20$$

$$= 40\,g\,m^{-2}\,year^{-1}$$

This is an example of primary production.

 Worked example 2: Calculating net production (energy)

Sheep in a grassland digest and absorb 15 000 $kJ\,m^{-2}\,yr^{-1}$ from the biomass they take in. Of this, $8000\,kJ\,m^{-2}\,yr^{-1}$ is used in respiration. How much energy is available to humans, the next organism in the food chain?

Net production = gross production – respiratory losses

$$= 15\,000 - 8000$$

$$= 7000\,kJ\,m^{-2}\,yr^{-1}$$

This is an example of secondary production.

Efficiency at consumer levels

Consumers at each trophic level convert at most 10% of the biomass in their food to their own organic tissue. This is because:

- not all of the biomass of an organism is eaten, for example, plant roots or animal bones may not be consumed

- some energy is transferred to the environment as metabolic heat, as a result of movement and respiration

- some parts of an organism are eaten but are indigestible – these parts (and their energy content) are egested as faeces

- some energy is lost from the animal in excretory materials such as urine.

Only around 0.001% of the total energy originally present in the incident sunlight is finally embodied as biomass in a tertiary consumer.

You can use the following formula to calculate the efficiency of the energy transfer (approximately equivalent to biomass transfer) between each trophic level of a food chain:

$$\text{Ecological efficiency} = \frac{\text{energy or biomass available after the transfer}}{\text{energy or biomass available before the transfer}} \times 100$$

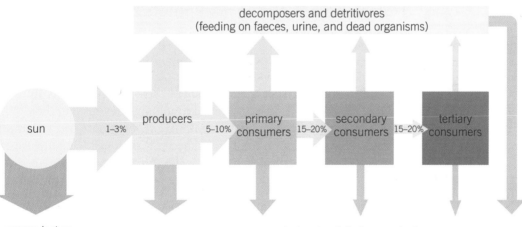

▲ **Figure 5** *This diagram shows the energy flow through different levels of a food chain. The figures stated for percentage energy transfer are approximate – energy transfers vary considerably between different plants, animals, and ecosystems*

Worked example: Calculating ecological efficiency

Look at Figure 6. This diagram shows the energy available in a food chain in Cayuga Lake.

Calculate the ecological efficiency of the energy transfer between smelt and trout.

$$\text{Ecological efficiency} = \frac{\text{energy or biomass available after the transfer}}{\text{energy or biomass available before the transfer}} \times 100$$

$$= \frac{250}{1250} \times 100 = 20\%$$

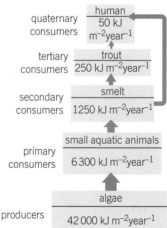

▲ **Figure 6** *This diagram shows the energy available at each trophic level in a food chain found in Cayuga Lake, New York State, USA*

Study tip

Remember that energy cannot be created or destroyed. However, it can be transferred, including to less useful forms such as heat, which is dissipated to the environment. This energy cannot be used to create biomass, and is therefore not available to the next trophic level.

▲ **Figure 7** *Intensively farmed animals are kept in a strictly-controlled environment. The conditions are chosen to ensure as much energy from their food as possible is turned into biomass. For example, small heated cages reduce energy loss through movement and maintaining body temperature*

Worked example: Efficiency of biomass transfer

A region of grassland has a net production of $40\,\text{g}\,\text{m}^{-2}\,\text{year}^{-1}$. A goat grazes an area of $20\,\text{m} \times 20\,\text{m}$ of this grassland. Assume that the goat consumes all of the biomass in this area.

1 Calculate the total biomass consumed by the goat each year:

Biomass consumed = mass (per metre squared per year) × area of land

$$= 40 \times (20 \times 20)$$
$$= 16\,000\,\text{g}$$
$$= 16\,\text{kg}$$

2 The mass of the goat increases in this time by $2.4\,\text{kg}$. Calculate the efficiency of biomass transfer between the grass and the goat.

$$\text{Efficiency of transfer} = \frac{\text{biomass available after transfer}}{\text{biomass available before transfer}} \times 100$$
$$= \frac{2.4}{16} = 15\%$$

Human activities can manipulate biomass through ecosystems

Human civilisation depends on agriculture. Agriculture involves manipulating the environment to favour plant species that we can eat (crops) and to rear animals for food or their produce. Plants and animals are provided with the abiotic conditions they need to thrive such as adequate watering and warmth (e.g., greenhouse use, stabling of animals). Competition from other species is removed (e.g., the use of chemicals such as pesticides) as well as the threat of predators (e.g., by creating barriers such as fences to exclude wild herbivores or predators).

In a natural ecosystem, humans would occupy the second, third or even fourth trophic level. As you have learnt, at each trophic level there are considerable energy losses, therefore only a tiny proportion of the energy available at the start of the food chain is turned into biomass for consumption at these third and fourth levels.

Agriculture creates very simple food chains. In farming animals or animal produce for human consumption, only three trophic levels are present – producers (animal feed), primary consumers (livestock), and secondary consumers (humans). In cultivating plants for human consumption, there are just two trophic levels – producers (crops) and primary consumers (humans). This means that the minimum energy is lost since there are fewer trophic levels present than in the natural ecosystem. This ensures that as much energy as possible is transferred into biomass that can be eaten by humans.

Summary questions

1 Describe how the biomass of a trophic level is measured. *(3 marks)*

2 Explain how human activities can manipulate the transfer of biomass through ecosystems. *(2 marks)*

3 Using Figure 6, calculate the ecological efficiency of energy transfer between algae and small aquatic animals. *(2 marks)*

4 Explain why biomass decreases at each level in a food chain. *(4 marks)*

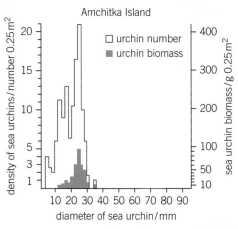

Monitoring biomass during conservation

Sea urchins (*Strongylocentrotus* spp) are marine invertebrates that feed on kelp (a type of seaweed). In regions where sea urchins are abundant, kelp forest ecosystems can be disrupted. The urchins eat the kelps' holdfasts, these are strong structures which anchor the kelp to the sea bed. The remainder of the plant floats away resulting in an ecosystem known as an 'urchin barren' ecosystem, which contains so little biomass of seaweeds that few species are able to live in this region. The presence or absence of kelp beds therefore has a major influence on the structure of the marine community.

▲ **Figure 8** *Forests of kelp are felled by sea urchins which eat the kelp's holdfasts using their five self-sharpening teeth*

In many areas, sea otters (*Enhydra lutris*) feed on urchins, keeping their levels low and therefore the kelp forests intact. During the 19th century, this ecological balance was destroyed when populations of sea otters were virtually wiped out by excessive hunting for otter fur. As a result, urchin numbers grew rapidly and kelp forests were destroyed. This balance has since been restored by the cessation of the hunting of sea otters, allowing them to again control the abundance of the urchins. In turn, the productive kelp forests have been able to redevelop.

As part of a conservation management project, scientists studied sea urchin populations around two of the Aleutian Islands off the coast of Alaska. Data on sea urchin size, density, and biomass were recorded per 0.25 m² from samples collected from Amchitka Island (which has sea otters) and Sheyma Island (which has no sea otters) The results are shown in Figure 9.

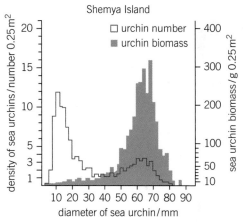

▲ **Figure 9**

1. State which trophic level of this marine community sea otters belong to.

2. a State the diameter of the largest sea urchin studied on Sheyma Island.

   b State the diameter of sea urchins with the most frequent biomass on Amchitka Island.

   c Suggest with reasons which island has the oldest sea urchins.

3. Oil spills can result in the death of sea urchins. Explain how scientists can use the presence of kelp biomass to determine how an area is recovering from oil damage.

23.3 Recycling within ecosystems

Specification reference: 6.3.1

Energy has a linear flow through an ecosystem. It enters the ecosystem from the Sun, and is ultimately transferred to the atmosphere as heat. As long as the Sun continues to supply Earth with energy, life will continue. In contrast, nutrients constantly have to be recycled throughout ecosystems in order for plants and animals to grow. This is because they are used up by living organisms and there is no large external source constantly replenishing nutrients in the way the Sun supplies energy.

Decomposition

Decomposition is a chemical process in which a compound is broken down into smaller molecules, or its constituent elements. Often an essential element, such as nitrogen or carbon, cannot be used directly by an organism in the organic form it is in, in dead or waste matter. This organic material must be processed into inorganic elements and compounds, which are a more usable form, and returned to the environment.

A **decomposer** is an organism that feeds on and breaks down dead plant or animal matter, thus turning organic compounds into inorganic ones (nutrients) available to photosynthetic producers in the ecosystem. Decomposers are primarily microscopic fungi and bacteria, but also include larger fungi such as toadstools and bracket fungi.

▲ **Figure 1** *Oyster mushrooms. Fungi are the main decomposers in forests. Only wood-decay fungi have evolved the enzymes necessary to decompose lignin, a chemically complex substance found in wood*

Decomposers are saprotrophs because they obtain their energy from dead or waste organic material (saprobiotic nutrition). They digest their food externally by secreting enzymes onto dead organisms or organic waste matter. The enzymes break down complex organic molecules into simpler soluble molecules – the decomposers then absorb these molecules. Through this process, decomposers release stored inorganic compounds and elements back into the environment.

Detritivores are another class of organism involved in decomposition. They help to speed up the decay process by feeding on detritus – dead and decaying material. They break it down into smaller pieces of organic material, which increases the surface area for the decomposers to work on. Examples of detritivores include woodlice that break down wood, and earthworms that help break down dead leaves. Detritivores perform internal digestion.

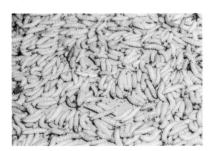

▲ **Figure 2** *Maggots are an example of a detritivore that helps to break down animal tissue*

Recycling nitrogen

Nitrogen is an essential element for making amino acids (and consequently proteins) and nucleic acids in both plants and animals. Animals obtain the nitrogen they need from the food they eat, but plants have to take in nitrogen from their environment.

Nitrogen is abundant in the atmosphere, 78% of air is nitrogen gas (N_2). However, in this form nitrogen cannot be taken up by plants. To be used by living organisms, nitrogen needs to be combined with other elements such as oxygen or hydrogen. Bacteria play a very important role in converting nitrogen into a form useable by plants. Without bacteria, nitrogen would quickly become a limiting factor in ecosystems.

Nitrogen fixation

Nitrogen-fixing bacteria such as *Azotobacter* and *Rhizobium* contain the enzyme nitrogenase, which combines atmospheric nitrogen (N_2) with hydrogen (H_2) to produce ammonia (NH_3) – a form of nitrogen that can be absorbed and used by plants. This process is known as **nitrogen fixation**.

Azotobacter is an example of a free-living soil bacterium. However, many nitrogen-fixing bacteria such as *Rhizobium* live inside root nodules. These are growths on the roots of leguminous plants such as peas, beans, and clover. The bacteria have a symbiotic mutualistic relationship with the plant, as both organisms benefit:

- the plant gains amino acids from *Rhizobium*, which are produced by fixing nitrogen gas in the air into ammonia in the bacteria
- the bacteria gain carbohydrates produced by the plant during photosynthesis, which they use as an energy source.

Other bacteria then convert the ammonia that is produced by nitrogen fixation into other organic compounds that can be absorbed by plants.

+ Reward and punishment

Recent work suggests that legumes 'select' the *Rhizobium* colonies which provide them with the most nitrates. Careful measurements show that the plants reward the nodules which make lots of nitrates with extra carbohydrates – but the punishment for nodules containing less-productive bacteria is swift and unforgiving. The plant cuts off the supply of carbohydrates and starves the nodule to death. This is a form of natural selection which maximises the benefit to the plant – and to those bacteria which deliver the goods.

> Discuss the nitrogen-fixation/carbon-fixation mutualism of legumes and *Rhizobium*, including the pattern of the relationship and its importance to living organisms.

Nitrification

Nitrification is the process by which ammonium compounds in the soil are converted into nitrogen-containing molecules that can be used by plants. Free-living bacteria in the soil called nitrifying bacteria are involved.

Study tip

Take care to use the correct terminology when referring to the various nitrogen compounds involved in the nitrogen cycle.

Nitrogen (N_2) is present in the atmosphere as a gas, but cannot be used directly by plants.

Ammonia (NH_3), or ammonium ions (NH_4^+), are produced by nitrogen-fixing bacteria and by decomposers.

Nitrites (NO_2^-) are formed by nitrifying bacteria from ammonia. These cannot be absorbed by plants.

Nitrates (NO_3^-) are produced from nitrites by other nitrifying bacteria. Plants are able to absorb nitrates.

Any reference to ammonia, nitrates, or nitrites should not be written as 'nitrogen' – rather, use the term 'nitrogen-containing molecules'.

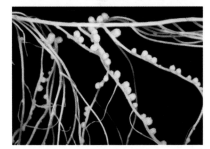

▲ **Figure 3** *Pink nodules of the nitrogen-fixing bacteria* <u>Rhizobium leguminosarum</u> *on the roots of a pea plant. The nodules are the sites where the bacteria fix atmospheric nitrogen, approx ×1.5 magnification*

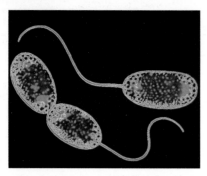

▲ **Figure 4** *Coloured transmission electron micrograph of* <u>Nitrobacter</u> *sp. They oxidise nitrogen-containing molecules for energy, converting nitrites into nitrates. Magnification × 6000*

Nitrification is an oxidation reaction, and so only occurs in well-aerated soil. It takes place in two steps:

1 Nitrifying bacteria (such as *Nitrosomonas*) oxidise ammonium compounds into nitrites (NO_2^-).

2 *Nitrobacter* (another genus of nitrifying bacteria) oxidise nitrites into nitrates (NO_3^-).

Nitrate ions are highly soluble, and are therefore the form in which most nitrogen enters a plant.

Denitrification

In the absence of oxygen, for example, in waterlogged soils, denitrifying bacteria convert nitrates in the soil back to nitrogen gas. This process is known as **denitrification** – it only happens under anaerobic conditions. The bacteria use the nitrates as a source of energy for respiration and nitrogen gas is released.

Ammonification

Ammonification is the name given to the process by which decomposers convert nitrogen-containing molecules in dead organisms, faeces, and urine into ammonium compounds.

Nitrogen cycle

The processes of nitrogen fixation, nitrification, denitrification, and ammonification all form part of the nitrogen cycle. Their place in the cycle can be seen in Figure 5.

▶ **Figure 5** *The nitrogen cycle. This shows how nitrogen is converted into a useable form and then passed on between organisms and the environment. Artificial processes, such as the Haber process used to make ammonia for fertiliser, have an effect on this natural cycle*

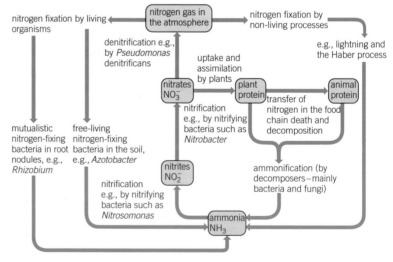

Recycling carbon

Carbon is a component of all the major organic molecules present in living organisms such as fats, carbohydrates, and proteins. The main source of carbon for land-living organisms is the atmosphere. Although carbon dioxide (CO_2) only makes up 0.04% of the atmosphere, there is a constant cycling of carbon between the atmosphere, the land, and living organisms. Figure 6 summarises the key points of the carbon cycle.

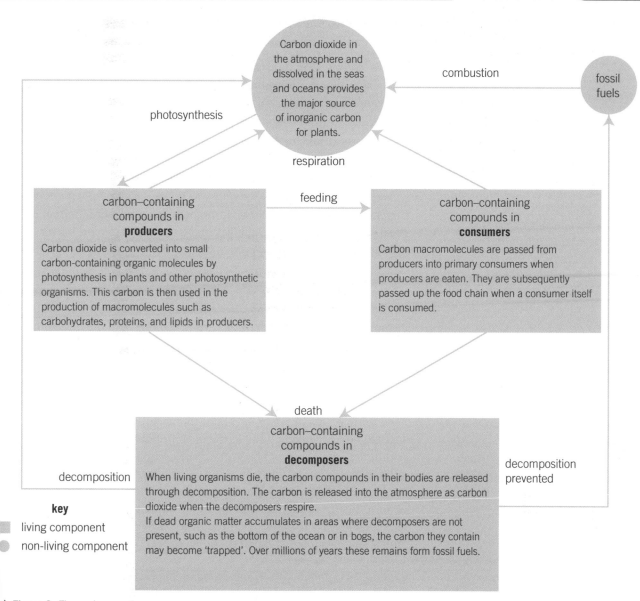

The following describes the carbon cycle diagram:

Carbon dioxide in the atmosphere and dissolved in the seas and oceans provides the major source of inorganic carbon for plants.

fossil fuels

combustion

photosynthesis

respiration

carbon–containing compounds in producers

Carbon dioxide is converted into small carbon-containing organic molecules by photosynthesis in plants and other photosynthetic organisms. This carbon is then used in the production of macromolecules such as carbohydrates, proteins, and lipids in producers.

feeding

carbon–containing compounds in consumers

Carbon macromolecules are passed from producers into primary consumers when producers are eaten. They are subsequently passed up the food chain when a consumer itself is consumed.

death

carbon–containing compounds in decomposers

When living organisms die, the carbon compounds in their bodies are released through decomposition. The carbon is released into the atmosphere as carbon dioxide when the decomposers respire.

If dead organic matter accumulates in areas where decomposers are not present, such as the bottom of the ocean or in bogs, the carbon they contain may become 'trapped'. Over millions of years these remains form fossil fuels.

decomposition

decomposition prevented

key
- living component
- non-living component

▲ **Figure 6** *The carbon cycle*

Fluctuations in atmospheric carbon dioxide

Carbon dioxide levels fluctuate throughout the day. Photosynthesis only takes place in the light, and so during the day photosynthesis removes carbon dioxide from the atmosphere. Respiration, however, is carried out by all living organisms throughout the day and night, releasing carbon dioxide at a relatively constant rate into the atmosphere. Therefore, atmospheric carbon dioxide levels are higher at night than during the day.

Localised carbon dioxide levels also fluctuate seasonally. Carbon dioxide levels are lower on a summer's day than a winter's day, as photosynthesis rates are higher.

▲ **Figure 7** *This is limestone. Although the bodies of marine animals decompose quickly, their shells and bones sink to the ocean floor. Over millions of years these are turned into carbon-containing sedimentary rock such as limestone and chalk. Eventually this carbon is returned to the atmosphere as the rock is weathered*

Over the past 200 years, global atmospheric carbon dioxide levels have increased significantly. This is mainly due to:

- the combustion of fossil fuels – which has released carbon dioxide back into the atmosphere from carbon that had previously been trapped for millions of years below the Earth's surface
- deforestation – which has removed significant quantities of photosynthesising biomass from Earth. As a result, less carbon dioxide is removed from the atmosphere. In many cases the cleared forest is burnt, therefore releasing more carbon dioxide into the atmosphere.

Increased levels of atmospheric carbon dioxide trap more thermal energy (heat) in the atmosphere – it is called a greenhouse gas for this reason. Its production through human activities is contributing to global warming.

The amount of carbon dioxide dissolved in seas and oceans is affected by the temperature of the water (the higher the temperature, the less gas is dissolved). Therefore, global warming reduces the carbon bank in the oceans and releases more carbon dioxide into the atmosphere – further contributing to the process in a positive feedback loop.

Atmospheric carbon dioxide levels have varied significantly over million-year timescales. To gain information about how the atmosphere has changed over time, samples are taken from deep within a glacier. For example, at a depth of 3.6 km in the Antarctic glacier the ice is 420 000 years old. When the ice formed air bubbles were trapped within the ice – these bubbles reflect the composition of the atmosphere at this point in time. Analysis of the gases present within these bubbles therefore reveals the composition of the atmosphere at this point in history. The graphs in Figure 9 show the variations in carbon dioxide levels which have occurred over time. The temperature of the atmosphere is directly related to the level of carbon dioxide present.

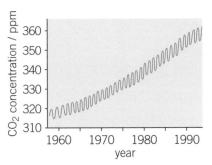

▲ **Figure 8** *Atmospheric levels of carbon dioxide across a 30-year period measured at Mauna Loa, a volcanic mountain in Hawaii. The increase is a result of human activities – primarily burning fossil fuels, and through changed land usage*

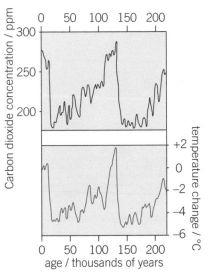

▲ **Figure 9** *Comparison of changes in atmospheric carbon dioxide (from analysis of bubbles in glacial ice) and temperature changes (from oxygen isotope studies)*

Summary questions

1   State the main differences between a decomposer and a detritivore.   *(2 marks)*

2   State and explain three ways in which atmospheric carbon dioxide levels increase.   *(3 marks)*

3   Describe what is meant by saprobiotic nutrition.   *(2 marks)*

4   ⚙️ Explain how the scientific community have produced evidence that atmospheric carbon dioxide levels have varied naturally over time.   *(2 marks)*

5   Explain the role of microorganisms, giving named examples where possible, in the recycling of nitrogen in an ecosystem.   *(8 marks)*

23.4 Succession

Specification reference: 6.3.1

As you have learnt, ecosystems are dynamic – they are constantly changing. One process by which ecosystems change over time is called **succession**. Have you noticed how the types of plants present change as you move inland from a beach? In the sand very few species exist – the further you move away from the sea, the more biodiverse the ecosystem becomes. This is an example of succession.

Succession occurs as a result of changes to the environment (the abiotic factors), causing the plant and animal species present to change.

There are two types of succession:

1 Primary succession – this occurs on an area of land that has been newly formed or exposed such as bare rock. There is no soil or organic material present to begin with.

2 Secondary succession – this occurs on areas of land where soil is present, but it contains no plant or animal species. An example would be the bare earth that remains after a forest fire.

Although much of the natural landscape has taken hundreds of years to reach its existing form, primary succession is still taking place. Primary succession occurs when:

- volcanoes erupt, depositing lava – when lava cools and solidifies igneous rock is created
- sand is blown by the wind or deposited by the sea to create new sand dunes
- silt and mud are deposited at river estuaries
- glaciers retreat depositing rubble and exposing rock.

Stages of succession

Succession takes place in a number of steps, each one known as a **seral stage** (or sere). At each seral stage key species can be identified that change the abiotic factors, especially the soil, to make it more suitable for the subsequent existence of other species.

The main seral stages are pioneer community, intermediate community, and climax community, these are summarised in Figure 1.

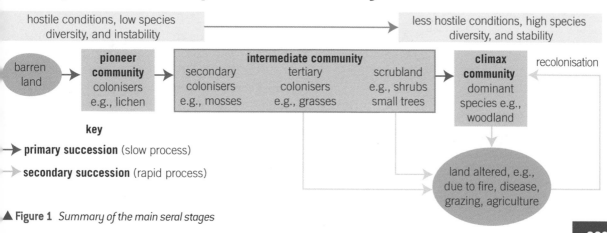

▲ Figure 1 *Summary of the main seral stages*

▲ **Figure 2** *Lichens are an example of a pioneer species. A lichen is not a single organism – it is a stable symbiotic (and mutualistic) association between a fungus and algae and/or cyanobacteria. They are often the first organisms to colonise bare rock. They are able to survive without soil, taking up rainwater and mineral salts through the whole of their body surface*

Pioneer community

Primary succession begins by the colonisation of an inhospitable environment, by organisms known as **pioneer species** (or pioneer colonisers). This represents the first seral stage. These species arrive as spores or seeds carried by the wind from nearby land masses or sometimes by the droppings of birds or animals passing through. Examples of pioneer species include algae and lichen.

Pioneer species have a number of adaptations that enable them to colonise this bare environment, including:

● the ability to produce large quantities of seeds or spores, which are blown by the wind and deposited on the 'new land'

● seeds that germinate rapidly

● the ability to photosynthesise to produce their own energy – light, rainfall, and air (and so carbon dioxide) are often the only abiotic factors present

● tolerance to extreme environments

● the ability to fix nitrogen from the atmosphere, so adding to the mineral content of the soil.

Intermediate community

Over time weathering of the bare rock produces particles that form the basis of a soil. On its own this cannot support other species. However, when organisms of the pioneer species die and decompose small organic products are released into the soil. This organic component of soil is known as **humus**. The soil becomes able to support the growth of new species of plant, known as secondary colonisers, as it contains minerals including nitrates and has an ability to retain some water. These secondary colonisers arrive as spores or seeds. Mosses are an example of a secondary coloniser species. In some cases, pioneer species also provide a food source for consumers, so some animal species will start to colonise the area.

As the environmental conditions continue to improve, new species of plant arrive such as ferns. These are known as tertiary colonisers. These plants have a waxy cuticle that protects them from water loss. These species can survive in conditions without an abundance of water – however, they need to obtain most of their water and mineral salts from the soil.

At each stage the rock continues to be eroded and the mass of organic matter increases. When organisms decompose they contribute to a deeper, more nutrient-rich soil, which retains more water. This makes the abiotic conditions more favourable initially for small flowering plants such as grasses, later shrubs, then finally small trees.

This period of succession is known as the intermediate community and in many cases multiple seral stages evolve during this period until climax conditions are attained.

At each seral stage different plant and animal species are better adapted to the current conditions in the ecosystem. These organisms outcompete many of the species that were previously present and become the dominant species. These are the most abundant species (by mass) present in the ecosystem at a given time.

Climax community

The final seral stage is called the **climax community**. The community is then in a stable state – it will show very little change over time. There are normally a few dominant plant and animal species. Which species make up the climax community depends on the climate. For example, in a temperate climate where the temperatures are mild and there is plenty of water, large trees will normally form the climax community. By comparison in a sub-arctic climate, herbs or shrubs make up the climax community as temperature and water availability are low.

Although biodiversity generally increases as succession takes place, the climax community is often not the most biodiverse. Biodiversity tends to reach a peak in mid-succession. It then tends to decrease due to the dominant species out-competing pioneer and other species, resulting in their elimination. The more successful the dominant species, the less the biodiversity in a given ecosystem.

▲ **Figure 3** *Oak woodland is a common climax community in the UK*

Animal succession

Alongside the succession of plant species, animal species undergo similar progression. Primary consumers such as insects and worms are first to colonise an area as they consume and shelter in the mosses and lichens present. They must move in from neighbouring areas so animal succession is usually much slower than plant succession, especially if the 'new land' is geographically isolated, for example, a new volcanic island.

Secondary consumers will arrive once a suitable food source has been established and the existing plant cover will provide them with suitable habitats. Again these species must move in from neighbouring areas. Eventually larger organisms such as mammals and reptiles will colonise the area when the biotic conditions are favourable.

Synoptic link

You learnt about the importance of maintaining biodiversity and the techniques involved in Topic 11.7, Reasons for maintaining biodiversity and Topic 11.8, Methods of maintaining biodiversity.

Deflected succession

Human activities can halt the natural flow of succession and prevent the ecosystem from reaching a climax community. When succession is stopped artificially, the final stage that is formed is known as a **plagioclimax**. Agriculture is one of the main reasons deflected succession occurs. For example:

- grazing and trampling of vegetation by domesticated animals – this results in large areas remaining as grassland

- removing existing vegetation (such as shrub land) to plant crops – the crop becomes the final community

- burning as a means of forest clearance – this often leads to an increase in biodiversity as it provides space and nutrient-rich ash for other species to grow, such as shrubs.

▲ **Figure 4** *Mowing your garden prevents succession occurring. The growing shoots of woody plants are cut off by the lawnmower, so larger plants cannot establish themselves. A grassland plagioclimax is formed*

Conservation

Deflected succession is an important conservation technique. To ensure the survival of certain species, it is important to preserve their habitat in its current form. This may require ecological land management to prevent further succession from occurring. An example of this is at the National Trust's Studland Heath nature reserve in Dorset, England.

Studland Heath is home to a number of British reptiles, including the UK's rarest reptile, the smooth snake. This region is heathland – if succession were allowed to occur, woodland would develop as the climax community. Other animal species would then inhabit the area, ultimately leading to the replacement of the smooth snake by other animal species. This would risk the elimination of this reptile species from the UK. So the heathland is managed by the National Trust to ensure that the precious ecosystem is preserved.

A range of conservation techniques are being used:

* Physical removal of established bracken, and saplings such as birch and pine. Gorse, which is a legume, adds to the nutrient value of the soil. Heathlands are traditionally nutrient-poor areas of land, and so areas of gorse are also removed through physical means (cutting, crushing, and controlled burning). These techniques are used to restore the heathland to its former state.

* Mimicking the controlled grazing to limit the spread of bracken, gorse, and tree saplings. Livestock crush and nibble new growth, therefore limiting the spread of these species. This technique is used to maintain the heathland in its current state.

Through using a combination of these techniques, low nutrient levels are maintained in the soil. This produces a varied vegetation structure, thus supporting a biodiverse heathland community.

> A herd of Red Devon cows are used to graze Studland Heath.
> Explain their role in maintaining the structure of the heathland.

▲ **Figure 5** *Britain's rarest native reptile, the smooth snake*

Synoptic link

Look back at 11.8, Methods of maintaining biodiversity, to remind yourself of the various techniques scientists use to maintain biodiversity

Succession on a sand dune

One of the few examples where all the stages of succession can often be seen clearly in one place is when a series of sand dunes form on a beach. The youngest dunes will be found closest to the shore and the oldest furthest away. Figure 6 shows a simplified version of what occurs:

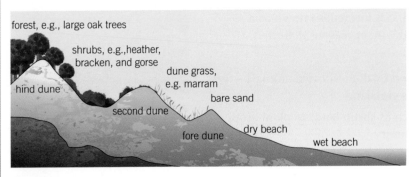

forest, e.g., large oak trees

shrubs, e.g.,heather, bracken, and gorse

dune grass, e.g. marram

hind dune

bare sand

second dune

fore dune

dry beach

wet beach

◀ **Figure 6** *Simplified diagram of succession*

Seeds are blown onto the dunes or washed onto the sand by the sea. At this stage the rooting conditions are poor due to drought, strong winds, and salty sea-water immersion. The presence of a large number of sea shells also makes the environment very alkaline. As the wind blows across the dunes it moves the sand, constantly changing the profile of the dunes and allowing rainwater to soak through rapidly. However, some species such as sea couch grass are able to survive these harsh conditions.

Marram grass becomes the next dominant species. As these plants grow sand is trapped between the roots helping to stabilise the dunes, and the formation of soil begins. Minerals released from decaying pioneer plants create more fertile growing conditions, and the soil becomes less alkaline as pioneer plants grow and trap rainwater. The grasses also reduce salt spray into the hind dunes, and act as a partial wind barrier. Less hardy plants can now grow – over time, these species start to shade out the pioneers. As more species of plants colonise the dunes, the sand disappears and the dunes change colour – from yellow to grey.

Finally taller plants, such as sea buckthorn, or more complex plant species like moorland heathers can grow. Plants from earlier stages are no longer present.

When the water table reaches or nearly reaches the surface, dune slacks can occur. Plants that are specially adapted to be water-tolerant grow here adding to the biodiversity of the sand dunes.

1 State what type of succession occurs on a sand dune.
2 a Name an example of a pioneer species.
   b Suggest and explain two features of the species you named in question 2a.
3 Describe the role marram grass plays in succession.
4 Describe how the abiotic conditions on a sand dune change over time.
5 Explain why fewer pioneer species are present in a climax community of trees or heathers.

Summary questions

1 State the difference between a climax community and a plagioclimax community. *(1 mark)*

2 Describe the difference between primary and secondary succession. *(1 mark)*

3 State and explain the common developments that occur in succession in any ecosystem. *(6 marks)*

4 Evaluate the use of deflected succession as a conservation technique to help maintain the population of an endangered animal species. *(6 marks)*

Synoptic link

You learnt about the adaptations of marram grass in Topic 9.5, Plant adaptations to water availability and Topic 10.7, Adaptations.

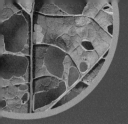

23.5 Measuring the distribution and abundance of organisms

Specification reference: 6.3.1

Learning outcomes

Demonstrate knowledge, understanding, and application of:

→ how the distribution and abundance of organisms in an ecosystem can be measured

→ the use of sampling and recording methods to determine the distribution and abundance of organisms in a variety of ecosystems.

Synoptic link

You learnt about sampling techniques in Topic 11.2, Sampling and Topic 11.3, Sampling techniques. This topic reminds you briefly of the techniques, but you should refer to those topics also.

Study tip

Remember that two other types of non-random sampling could also be used – opportunistic and stratified sampling. See Topic 11.2, Sampling.

Synoptic link

You will learn more about how population size changes in Topic 24.1, Population size.

Scientists use a number of techniques to study the distribution and abundance of organisms within an ecosystem. This is a way of measuring and observing the biodiversity present within an ecosystem. These techniques can also be used to study how the organisms present change during succession. Many of these techniques you met in Chapter 11, Biodiversity.

Distribution of organisms

The distribution of organisms refers to *where* individual organisms are found within an ecosystem. The distribution of organisms is usually uneven throughout an ecosystem. Organisms are generally found where abiotic and biotic factors favour them, therefore their survival rate is high as all the resources they need to live are available and predation/pressure from consumers is low.

Measuring distribution

To measure the distribution of organisms within an ecosystem, a line or belt transect is normally used. A line transect involves laying a line or surveyor's tape along the ground and taking samples at regular intervals. A belt transect provides more information – two parallel lines are marked, and samples are taken of the area between these specified points. Belt and line transects are forms of systematic sampling, a type of non-random sampling.

In systematic sampling different areas within an overall habitat are identified, which are then sampled separately. This can have advantages over random sampling as it allows scientists to study how the differing abiotic factors in different areas of the habitat affect the distribution of a species. For example, systematic sampling may be used to study how plant species change as you move inland from the sea. This would therefore be used to study the successional changes that take place along a series of sand dunes.

Abundance of organisms

The abundance of organisms refers to the *number* of individuals of a species present in an area at any given time. This number may fluctuate daily:

- Immigration and births will increase the numbers of individuals.
- Emigration and deaths will decrease the number of individuals.

Measuring abundance

A population is a group of similar organisms living in a given area at a given time. Populations can rarely be counted accurately – for example, some animals elude capture, it may be too time-consuming to count all members of a population, or the counting process could damage the environment. Populations are therefore estimated using sampling techniques.

A sample, however, is never entirely representative of the organisms present in a habitat. To increase its accuracy you should use as large a sample size as possible. The greater the number of individuals studied, the lower the probability that chance will influence the result. You should also use random sampling to reduce the effects of sampling bias.

Measuring plant abundance

To measure the abundance of plants, quadrats are placed randomly in an area. The abundance of the organisms in that area is measured by counting the number of individual plants contained within the quadrat. Using the following formula, the abundance can be estimated:

$$\text{Estimated number in population (m}^{-2}) = \frac{\text{Number of individuals in sample}}{\text{Area of sample (m}^2)}$$

▲ **Figure 1** *This student is conducting a survey using a quadrat. A quadrat is a frame of standard size, often a square divided into equal sections by string, which is used to isolate an area from which a sample is to be taken. Quadrats are often used for measuring the abundance of different species within a given area*

 Worked example: Estimating plant abundance

A sample was taken using five quadrats with a combined area of $5\,\text{m}^2$. If the quadrats contained a total of 40 buttercup plants, what is the estimated abundance of buttercups?

Step 1: Identify the equation needed

$$\text{Estimated number in population (m}^{-2}) = \frac{\text{Number of individuals in sample}}{\text{Area of sample (m}^2)}$$

Step 2: Substitute the values into the equation and calculate the answer $= \dfrac{40}{5} = 8\,\text{m}^{-2}$

Meaning eight buttercups per square metre.

Study tip

Individual plants in a quadrat can sometimes be difficult to count so scientists also make estimates using frequency or percentage cover. Look back at Topic 11.3, Sampling techniques to remind yourself of these different techniques.

Measuring animal abundance

Quadrats cannot be used to measure the abundance of animals (unless they are very slow moving, such as barnacles and mussels on a sea shore) so the capture–mark–release–recapture technique is often used to estimate population size.

The technique can be carried out as follows:

1 Capture as many individuals as possible in a sample area.

2 Mark or tag each individual.

3 Release the marked animals back into the sample area and allow time for them to redistribute themselves throughout the habitat.

4 Recapture as many individuals as possible in the original sample area.

5 Record the number of marked and unmarked individuals present in the sample. (Release all individuals back into their habitat.)

6 Use the Lincoln index to estimate the population size:

$$\text{Estimated population size} = \frac{\substack{\text{number of individuals in first sample} \times \\ \text{number of individuals in second sample}}}{\text{number of recaptured marked individuals}}$$

▲ **Figure 2** *The animals sampled can be collected in a number of ways including using pooters, sweep nets, and pitfall traps. (Topic 11.3, Sampling techniques). This student is collecting crawling animals off a tree using a pooter*

Worked example: Capture–mark–release–recapture technique

In a random sample of a school playing field, students captured 12 snails. They marked the underside of each snail's shell with a dot of non-toxic paint.

The students then released the snails back onto the field. A week later, another sample of snails from the same location was taken. This time 15 snails were collected, of which three were found to have a paint dot.

Estimate the total snail population.

Step 1: Sum the total the number of individuals in the first sample = 12

Step 2: Sum the total number of individuals in the second sample = 15

Step 3: Estimate the population size by calculating the Lincoln index:

Estimated population size =

$$\frac{\text{number of individuals in first sample} \times \text{number of individuals in second sample}}{\text{number of recaptured marked individuals}} = \frac{12 \times 15}{3} = 60$$

The estimated total snail population of the playing field is 60 snails.

Once the abundance of all the organisms present in a habitat has been determined, scientists will often mathematically calculate the biodiversity present in a habitat. This can be done using **Simpson's Index of Diversity (D)**:

D = diversity index

N = total number of organisms in the ecosystem

n = number of individuals of each species

$$D = 1 - \Sigma\left(\frac{n}{N}\right)^2$$

Simpson's Index of Diversity always results in a value between 0 and 1, where 0 represents no diversity and a value of 1 represents infinite diversity. The higher the value of Simpson's Index of Diversity, the more diverse the habitat.

Synoptic Link

To remind yourself how to calculate Simpson's Index of Diversity, look back at Topic 11.4, Calculating biodiversity.

Monitoring biodiversity in the Sonoran desert

The Sonoran desert is a region of desert in the south-western part of the United States, including parts of Arizona and California. The environment is harsh – summer temperatures regularly exceed 40 °C. Rainfall is rare, often taking the form of intense, violent storms.

A student carried out a study to test the following hypothesis – 'the greater the abundance of plant species in the Sonoran desert, the greater the abundance of animals'. A selection of the data collected is shown in Table 1.

▼ Table 1

Mean plant abundance / 50 m⁻²	Mean animal abundance / 50 m⁻²
0	0.05
1	0.06
2	0.20
3	0.19
4	0.11
5	0.44
6	0.92
7	1.19
8	1.44
9	1.45
10	1.78
11	2.00
12	2.06
13	2.33
14	2.78

1 Suggest how the data could have been collected.
2 Explain how you can tell that the figures stated for animal abundance were a mean of several observations.
3 Plot the data as a scatter diagram.
4 Using your diagram from Q3, state the correlation between the variables, if any.
5 Using the formula $r_s = 1 - \dfrac{6\sum d^2}{n(n^2 - 1)}$, calculate the Spearman's rank correlation coefficient for this data.
6 Use the Spearman's ranked correlation coefficient critical values in the appendix to determine the statistical significance of the correlation, and hence form a conclusion from the data.

Summary questions

1 State the difference between the abundance and distribution of organisms. *(1 mark)*

2 Twenty woodlice were captured in a sample area and marked then released back into the sample area. A week later, 15 woodlice were found in the sample area, of which only two were marked.
   a Suggest which piece of apparatus should be used to collect woodlice. *(1 mark)*
   b Use the Lincoln index to estimate the woodlouse population. *(2 marks)*

3 State and explain one advantage of the following techniques used to measure the distribution of organisms in an ecosystem. *(2 marks)*
   a random sampling   b non-random sampling

4 Discuss the limitations of using the capture–mark–release–recapture technique to estimate population size. *(6 marks)*

Synoptic link

Remind yourself how to calculate the Spearman's rank correlation coefficient by looking back to Topic 10.6, Representing variation graphically.

Study tip

The correlation test is used to see if two different variables are correlated in a linear fashion in the context of a scatter-graph. There are several different types of statistical test that can be used, the OCR GCE Biology specifications will only cover Spearman's rank correlation coefficient. You can find a full table of values in the appendix.

Practice questions

1 Ecological succession describes the change in species within a community over time.

time

| agricultural land is kept in an artificial seral stage | after land is abandoned wild grasses start to grow from wind blown over | over time shrubs start to colonise the grassland | trees become established and climax community develops |

a (i) State, giving your reasons, the type of succession shown in the diagram. *(1 mark)*

(ii) Describe the *ecological role* of the different species involved in the early stages of succession. *(2 marks)*

The graph shows the changing mass of species in a standard example of succession.

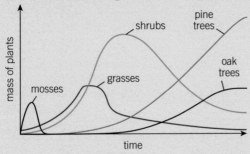

b (i) State the name given to mosses describing their role in this example of succession. *(2 marks)*

(ii) Explain why mosses, grasses, and shrubs all eventually decrease in mass. *(2 marks)*

(iii) State the name given to the community composed primarily of Pine and Oak trees. *(1 mark)*

The quantity of carbon sink changes throughout a period of succession. The graph summarises these changes.

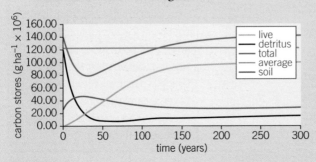

c (i) Describe the changes in carbon content of the components of the ecosystem shown in the diagram. *(4 marks)*

(ii) Suggest how these changes would differ if the succession started without the presence of any soil. *(4 marks)*

As succession progresses the net primary productivity changes as shown in the graph, this is the net rate at which an ecosystem accumulates energy.

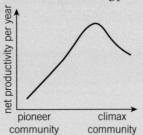

d (i) Describe the change in net primary productivity shown in the graph. *(3 marks)*

(ii) Suggest the reasons for the changes you have described in **d** (ii). *(3 marks)*

2 Ecological pyramids summarise the feeding relationships within an ecosystem in terms of number, biomass or energy.

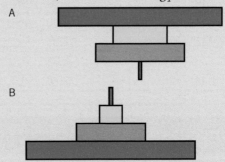

a (i) State the definition of the term trophic level. *(1 mark)*

(ii) Identify which of the lists below represents each of the ecological pyramids shown in the diagram. *(2 marks)*

1 lettuce plant, snail, thrush, sparrowhawk

2 tree, aphid, caterpillar, parasite

(iii) Suggest, giving your reason(s), whether the ecological pyramids in the diagram represent number or biomass. *(2 marks)*

(iv) Describe how you would represent decomposers on pyramid A, or B. *(2 marks)*

3 **a** Outline the factors that can affect primary productivity. (*5 marks*)

It is usually impossible to count all of the members of a population in a particular habitat so animal populations often have to be estimated.

One method to estimate population size uses the Lincoln Index.

b **(i)** Outline the principles that underlie the Lincoln Index. (*3 marks*)

(ii) Discuss the limitations of using the Lincoln Index to estimate population size. (*5 marks*)

In different locations of Utah Lake in the US, 24 064 carp were captured, tagged, and released over a 15 day period.

After the tagging was finished, another 10 357 carp at random locations of the lake were caught and the number of tagged fish counted. Only 208 of the recaptured carp had tags.

c Calculate, using the formula, the estimated population number of carp in the lake. (*2 marks*)

$$\frac{\text{Number marked in second sample}}{\text{Total caught in second sample}} = \frac{\text{Number marked in first sample}}{\text{Size of whole population (N)}}$$

d State, giving your reason(s), if the estimate become more accurate if more individuals are captured and marked. (*2 marks*)

e State one adjustment you would have to make to the investigation if you were estimating a population of snails rather than fish. (*1 mark*)

4 Total plant growth within an ecosystem depends on the light intensity, temperature, and the supply of water and inorganic minerals to the ecosystem.

The table shows the net primary production by plants in four different ecosystems.

Ecosystem	Net primary production (k Jm^{-2} year^{-1})
temperate grassland	9 240
temperate woodland	11 340
tropical grassland	13 340
tropical rainforest	36 160

a Discuss possible reasons for the differences in net primary production in these ecosystems. (*4 marks*)

b To calculate the net primary production figures in the table in k Jm^{-2} year^{-1}, it is necessary to measure the energy content of the primary producers. Outline how the energy content, in kJ, of a primary producer such as grass can be measured in the laboratory. (*2 marks*)

c The efficiency with which consumers convert the food they eat into their own biomass is generally low. The table compares the energy egested, absorbed, and respired in four types of animal.

animal	Percentage of energy consumed that is:			
	egested	absorbed	respired	converted to biomass
grasshopper (herbivorous insect)	63	37	24	13
perch (carnivorous fish)	17	83	61	
cow (herbivorous mammal)	60	40	39	
bobcat (carnivorous mammal)	17	83	77	6

(i) Copy and complete the table to show the percentage of energy consumed that is converted into biomass in the perch and the cow. (*2 marks*)

(ii) Describe and explain, using the data from the table, how the trophic level of a mammal affects the percentage of its food energy that it is able to convert to biomass. (*3 marks*)

(iii) Using the data from the table and your knowledge of energy flow through food chains, suggest which of these four animals could be farmed to provide the maximum amount of food energy in k Jm^{-2} year^{-1} for humans. Explain the reason for your choice. (*4 marks*)

OCR F215 2011

The human population is increasing at a significant rate. The global population has grown from one billion in 1800 to seven billion in 2012, and is predicted to reach 11 billion by the end of this century. Population growth like this cannot be sustained indefinitely as **limiting factors**, such as the availability of food, will prevent the population rising above a certain level. A limiting factor is an environmental resource or constraint that limits population growth.

Population growth curve

If the growth of a new population over time is plotted on a graph, regardless of the organism, most natural populations will show the same characteristics. This is known as a population growth curve.

The graph can be divided into three main phases:

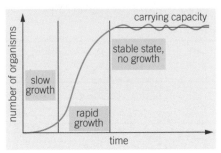

▲ **Figure 1** *Growth curve of most natural populations. This is often referred to as a sigmoid population growth curve*

- Phase 1 – a period of *slow growth*. The small numbers of individuals that are initially present reproduce increasing the total population. As the birth rate is higher than the death rate, the population increases in size.

- Phase 2 – a period of *rapid growth*. As the number of breeding individuals increases, the total population multiplies exponentially. No constraints act to limit the population explosion.

- Phase 3 – a *stable state*. Further population growth is prevented by external constraints. During this time the population size fluctuates, but overall its size remains relatively stable. Birth rates and death rates are approximately equal. Slight increases and decreases can be accounted for by fluctuations in limiting factors, such the presence of predators.

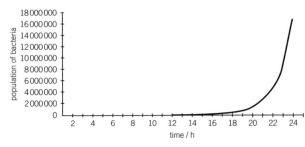

▲ **Figure 2a** *Exponential growth curve. The size of the population doubles each time a fixed time period elapses. The populations of many organisms follow this trend during an initial expansion of population size*

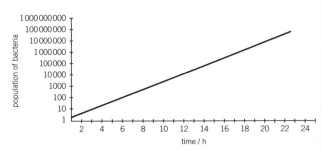

▲ **Figure 2b** *Exponential growth represented on a logarithmic scale. Where data span several orders of magnitude, it is often more appropriate to plot the data using a logarithmic scale on the relevant axis. In this case, population size covers a large range of values—therefore this data (y-axis) is plotted on a logarithmic scale, whereas time (x-axis) is plotted on a linear scale*

Limiting factors

In theory, if all resources were in plentiful supply a population would continue to grow exponentially. However, this is rarely seen in nature. Instead, a short period of exponential growth occurs when conditions are ideal and the maximum growth rate is achieved.

Limiting factors prevent further growth of a population and in some cases cause it to decline. Examples of limiting factors include competition between the organisms for resources, the build-up of the toxic by-products of metabolism, or disease.

Limiting factors can be divided into abiotic and biotic factors:

- Abiotic factors – these non-living factors include temperature, light, pH, the availability of water or oxygen, and humidity.

- Biotic factors – these living factors include predators, disease, and competition.

The maximum population size that an environment can support is known as its **carrying capacity** (Figure 1), although individual years can show slight increases or decreases in population size. The population size remains stable as the number of births and deaths are approximately equal.

Migration

Another important variable which affects population size is migration:

- Immigration – the movement of individual organisms into a particular area increases population size. For example, millions of Christmas Island red crabs (*Gecarcoidea natalis*) migrate each year from forest to coast to reproduce, dramatically increasing the coastal population of red crabs.

- Emigration – the movement of individual organisms away from a particular area decreases population size. For example, the Norway Lemming (*Lemmus lemmus*) emigrates away from areas of high population density or poor habitat.

Density independent factors

Density independent factors are factors that have an effect on the whole population regardless of its size. These can dramatically change population size. These factors include earthquakes, fires, volcanic eruptions, and storms. In some cases, these factors can remove whole populations of a species from a region.

> ## Synoptic link
>
> Look back at Topic 23.1, Ecosystems to remind yourself of the effect of abiotic and biotic factors on organisms.

> ## Study tip
>
> Competition is often considered to be the most important biotic factor controlling population density. You will find out about competition in more detail in 24.2, Competition.

▲ **Figure 3** *Snow Geese breed in the Arctic Tundra, and winter in farmlands on the American south coast. They make an annual round trip journey of more than 8000 km at speeds of more than 80 km/h, changing the population density of geese in each place dramatically as they arrive and depart*

 ## Human population growth

For many years the human population remained fairly stable, as is the case for other natural populations. The population was kept in check by limiting factors.

1 Suggest two limiting factors that historically kept the human population at a constant level.

The development of agriculture, the industrial revolution, and advances in medicine have led to a population explosion. Like other species, the growth of the human population is a result of the imbalance between birth rate and death rate.

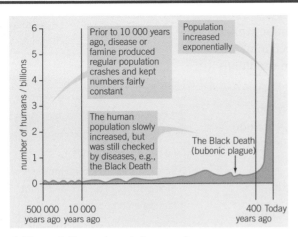

▲ **Figure 4** *Human population growth curve*

Scientists use a measure to calculate how much a population has grown in a certain period of time. This is calculated using the formula:

$$\text{Population growth (\%)} = \frac{\text{population change during the period}}{\text{population at the start of the period}} \times 100$$

If the result is positive, the population has grown. If it is negative, the population has decreased.

> **2** The human population was approximately 3 billion (3×10^9) in 1960. This had risen to 6.8 billion (6.8×10^9) by 2010. Calculate the percentage population growth during this period.

A number of factors affect the birth rate of the human population. In addition to 'natural' limiting factors, economic conditions, cultural and religious backgrounds, and social pressures can limit – or encourage – the birth rate.

Different factors affect the death rate in a population. These include the age profile of the population (in general, the greater the proportion of elderly people, the higher the death rate), the quality of medical care, food availability and quality, and the effects of natural disaster or war.

> **3** Suggest and explain a social, cultural, or religious pressure which may affect the size of a population.

The future size of a population depends upon the number of women of child-bearing age. The age and gender demographic of a population can be represented using an age population pyramid, as shown in Figure 5.

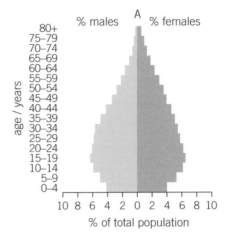

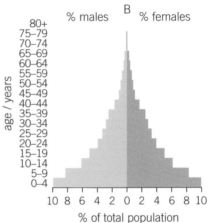

▲ **Figure 5** *Age population pyramids for two different countries*

> **4 a** Explain how the population of each of the countries represented in Figure 5 will change over time.
>
> **b** Sketch or suggest the shape of an age population pyramid for a stable population.

Summary questions

1 State three factors that would cause an increase in population size. *(3 marks)*

2 Describe what happens during a period of exponential population growth. *(2 marks)*

3 Sketch and annotate a graph to show what would happen to the population of duckweed (*Lemna minor*) if a few individuals were introduced into a new pond habitat. *(6 marks)*

24.2 Competition

In the previous topic, you learnt that population size is influenced by limiting factors. Organisms compete for resources including food, shelter, space, and light. As a result, competition between organisms has a significant effect on the number of organisms present in a particular area.

Types of competition

Competition is an example of a *biotic* limiting factor – it is a result of the interactions between living organisms. There are two main types of competition:

1 **Interspecific competition** – competition between different species.

2 **Intraspecific competition** – competition between members of the same species.

Interspecific competition

Interspecific competition occurs when two or more different species of organism compete for the same resource. This interaction results in a reduction of the resource available to both populations. For example, if both species compete for the same food source, there will usually be less available for organisms of each species. As a result of less food, organisms will have less energy for growth and reproduction, resulting in smaller populations than if only one of the species had been present.

If two species of organism, however, are both competing for the same food source but one is better adapted, the less well adapted species is likely to be *outcompeted*. If conditions remain the same, the less well adapted species will decline in number until it can no longer exist in the habitat alongside the better adapted species. This is known as the *competitive exclusion principle* – where two species are competing for limited resources, the one that uses the resources more effectively will ultimately eliminate the other.

Red and grey squirrels in the UK

An example of interspecific competition is the competition between red and grey squirrels for food and territory in the UK. In the 1870s the grey squirrel, a native of North America, was introduced into the wild in the UK. Its population quickly increased in numbers and resulted in the native red squirrel disappearing from many areas. This is primarily because the grey squirrel can eat a wider range of food than the red squirrel and as it is larger it can store more fat. This increases its chances of survival and therefore its ability to reproduce thus increasing its population. An increasing population of grey squirrels further reduces the food supply available to the red squirrels, reducing their ability to survive and reproduce.

> ### Learning outcomes
> Demonstrate knowledge, understanding, and application of:
> → interspecific and intraspecific competition.

▲ **Figure 1** *The bog pondweed* (Potamogeton polygonifolius, *broad leaves*) *and duckweed* (Lemna minor, *small floating leaves*) *growing in this pond are both competing for light. This is an example of interspecific competition. In the absence of the pond weed, the duckweed would multiply to cover the whole surface of the pond*

> ### Synoptic link
> You learnt about biotic and abiotic factors in Topic 23.1, Ecosystems.

▲ **Figure 2** *The red squirrel* (Sciurus vulgaris) *is native to the UK. The introduction of the grey squirrel* (Sciurus carolinensis) *from North America has led to a significant decline in red squirrel numbers*

Intraspecific competition

Intraspecific competition occurs when members of the same species compete for the same resource. The availability of the resource determines the population size – the greater the availability the larger the population that can be supported. This results in fluctuations in the number of organisms present in a particular population over time.

The effects of intraspecific competition on a population are represented in Figure 3.

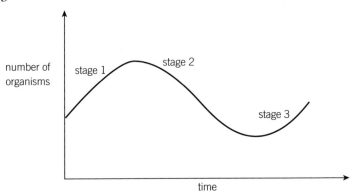

▲ **Figure 3** *The effects of intraspecific competition over time*

▲ **Figure 4** *Robins compete for breeding territory. Each territory provides adequate food for one family of birds. If food is scarce, territories have to become larger, therefore there are fewer breeding pairs and a smaller population*

- Stage 1 – When a resource is plentiful in a habitat (such as food or space), all organisms have enough of the resource to survive and reproduce. This results in an increase in population size.

- Stage 2 – As a result of the increased population, there are many more individuals that share the food or space available. Resources are now limited; not enough is available for all organisms to survive. The population will decrease in size.

- Stage 3 – Less competition exists as the smaller population means less organisms are competing for the same resources. This means more organisms survive and reproduce, resulting in population growth.

This cycle of events will then repeat.

Summary questions

1 State the difference between interspecific and intraspecific competition. *(1 mark)*

2 Describe the cycle of population size changes that occur following intraspecific competition. *(3 marks)*

3 Oak tree saplings compete with each other for light, water, and minerals.
   a State the type of competition that exists between oak tree saplings. *(1 mark)*
   b Suggest how the population of oak trees will vary over time. *(3 marks)*

24.3 Predator–prey relationships

Specification reference: 6.3.2

In the previous topic, you learnt about the effect of competition on population size. Another major biotic factor that has an influence on population size is the role of **predation**. This is where an organism (the predator) kills and eats another organism (the prey). For example, tigers prey on water buffalo and deer. Predation is a type of interspecific competition, operating between prey and predator species.

Predators have evolved to become highly efficient at capturing prey, for example, through sudden bursts of speed, stealth, and fast reactions. Likewise, prey organisms have evolved to avoid capture through camouflage, mimicry, or defence mechanisms such as spines. Prey organisms have had to evolve alongside their predators (and vice versa) – if evolution had not occurred, one or both of the species may have become extinct.

Predator–prey relationships

The sizes of the predator and prey populations are interlinked. As the population of one organism changes, it causes a change in the size of the other population. This results in fluctuations in the size of both populations.

Predator–prey relationships can be represented on a graph (Figure 2).

In general all predator–prey relationships follow the same pattern. The peaks and troughs in the size of the prey population are mirrored by peaks and troughs in the size of the predator population after a time delay.

- Stage one – An increase in the prey population provides more food for the predators, allowing more to survive and reproduce. This in turn results in an increase in the predator population.

- Stage two – The increased predator population eats more prey organisms, causing a decline in the prey population. The death rate of the prey population is greater than its birth rate.

- Stage three – The reduced prey population can no longer support the large predator population. Intraspecific competition for food increases, resulting in a decrease in the size of the predator population.

- Stage four – Reduced predator numbers result in less of the prey population being killed. More prey organisms survive and reproduce, increasing the prey population – the cycle begins again.

Rarely in the wild is the link between the predator and prey population as simple as this. Other factors will also influence the population size – for example, the availability of the food plants of the prey or the presence of other predators. Fluctuations in numbers may also result from seasonal changes in abiotic factors.

Learning outcomes

Demonstrate knowledge, understanding, and application of:

→ predator–prey relationships

▲ **Figure 1** *This predator–prey relationship is often exploited by organic farmers. In a technique known as biological pest control, farmers use the natural predators (in this case ladybirds) to destroy pest populations (aphids) and prevent them damaging crops without the need to use pesticides*

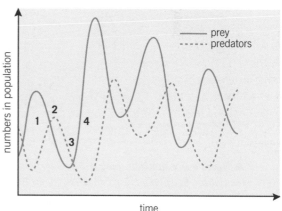

▲ **Figure 2** *A general predator–prey graph*

Canadian lynx and snowshoe hare

When a predator feeds on just one type of prey, there is an interdependence between the predator and prey populations. This means that changes in the population of one animal directly affect the population of the other.

The Canadian lynx and snowshoe hare have an interdependent relationship. This relationship has been widely studied – data exist for over 200 years, as records were kept for the number of lynx furs that were traded in Canada. The data collected is shown in Figure 3.

1 Suggest what assumption was made by scientists when estimating population size from the number of furs traded.
2 a State in which year the hare population was at its highest.
  b Suggest a reason for this.
3 a State in which year the lynx population is at its lowest.
  b Suggest a reason for this.
4 State the approximate time delay between a peak and trough occurring in the snowshoe hare population.
5 Explain the changes in the populations of the lynx and hare that occurred over time.

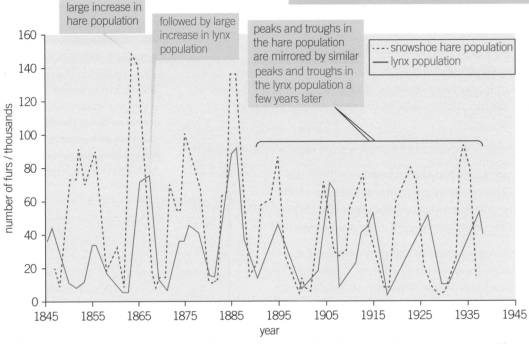

▲ **Figure 3** *The predator–prey relationship illustrated by the number of snowshoe hare and lynx trapped for the Hudson Bay Company in Canada between 1845 and 1940*

Summary questions

1 Describe what would happen to the numbers of a predator population if a fatal disease spread amongst its only prey. *(1 mark)*

2 Suggest two reasons why the populations of an interlinked predator and prey population don't produce an identical graph to the one shown in Figure 2. *(4 marks)*

3 In a controlled environment, an experiment was carried out to study the predator–prey relationship between the unicellular organisms *Didinium* and *Paramecium*. A suitable *Paramecium* population prey was established before a small population of *Didinium* was added. Sketch and annotate a graph to show how the populations of these organisms may vary over time. A constant supply of nutrient material was provided for the *Paramecium*. *(4 marks)*

Specification reference: 6.3.2

You will often hear reports in the media that steps need to be taken to conserve the environment for future generations. This is true not only to preserve the beauty of the natural world, but to try to ensure that the biodiversity of world ecosystems is not lost. There are many reasons for maintaining biodiversity, including economics, preserving genetic variety, and retaining species that might be useful to people. People tend to use the terms **conservation** and **preservation** interchangeably, but in biology they have quite distinct meanings.

Conservation

Conservation means the maintenance of biodiversity through human action or management. This includes maintaining diversity between species, maintaining genetic diversity within a species, and the maintenance of habitats.

Conservation involves the management of ecosystems so that the natural resources in them can be used without running out. This is known as sustainable development. For example, the Forest Stewardship Council ensures that forests are managed so that they provide a sustainable source of timber. Their mission is to promote socially beneficial, environmentally appropriate, and economically viable management of forests across the world. You will find out more about sustainable production in Topic 24.5, Sustainability.

Conservation approaches also include *reclamation* – this is the process of restoring ecosystems that have been damaged or destroyed. For example, a habitat may be destroyed by floods, or as a result of a new building project. Reclamation also involves techniques such as controlled burning of areas of a forest, which can halt succession and increase biodiversity. Conservation is dynamic and needs to adapt to constant change.

Preservation

Preservation is the protection of an area by restricting or banning human interference, so that the ecosystem is kept in its original state. It is most commonly used when preserving ecologically, archaeologically, or palaeontologically sensitive resources, which can easily be damaged or destroyed by disturbances. When lands are preserved, visitation (along with most other activities) is not allowed, except by those who manage and monitor such areas.

Newly discovered caves, called virgin caves, are pristine. These may contain very sensitive geological formations or unique ecosystems – walking from one cave to another can cause irreparable damage, for example, through direct crushing or by the movement of dirt around the cave system. Such damage can be avoided by barring entrance to caves altogether thus preserving these unique habitats. Only through preservation can the integrity of these ecosystems be

Learning outcomes

Demonstrate knowledge, understanding, and application of:

→ the reasons for, and differences between, conservation and preservation.

▲ **Figure 1** *Controlled grazing of fens is a valuable fen management conservation tool. It helps to maintain open, species-rich fen communities by reducing plant biomass and controlling scrub invasion*

Synoptic link

Look back at Topic 11.8, Methods of maintaining biodiversity to remind yourself of the range of in situ and ex situ techniques that scientists use to conserve biodiversity and the importance of conservation.

▲ **Figure 2** *Endangered Gray bats (Myotis grisescens) in a cave in Tennessee*

Conservation of gray bats

The gray bat is an example of a species which has been successfully supported through a series of conservation measures. Inhabiting a small number of caves across the south-eastern US, the gray bat became endangered following the human exploitation of its habitat, and through pesticide bioaccumulation. Following the collection of data about bat numbers and particular habitat requirements, a series of measures were introduced to support the remaining population, which is currently estimated to be around one million:

- In 1982, the gray bat was placed on the International Union for Conservation of Nature (IUCN) endangered species list.

- Caves supporting gray bat populations were gated, preventing human access but allowing bats to freely enter and leave the caves.

- The exploitation of land around the caves, which provide the bats with their food sources, was strictly controlled. This included limiting the use of pesticides in these regions.

- A programme of education was launched for those who inhabited the areas surrounding regions which supported a bat population.

> 1   State one conservation measure and one preservation measure taken to support the population of gray bats.
>
> 2   Suggest and explain the advantages and disadvantages of placing the gray bat on the IUCN endangered species list.

guaranteed. However, this can result in no one being able to enjoy the caves and some argue that there is no point in having a resource that cannot be used.

In reality it is objects and buildings that are more commonly preserved, whereas the natural environment is conserved. Examples of preserved habitats include areas set aside in nature reserves and marine conservation zones where human interference is prohibited.

Importance of conservation

Conservation is important for many reasons (look back at Topic 11.7), which can be broadly divided into three categories: economic, social, and ethical reasons.

- Economic – to provide resources that humans need to survive and to provide an income. For example, rainforest species provide medicinal drugs, clothes, and food that can be traded. Other forests are used for the production of timber and paper.

- Social – many people enjoy the natural beauty of wild ecosystems as well as using them for activities which are beneficial to health by providing a means of relaxation and exercise. Examples of these activities include bird watching, walking, cycling, and climbing.

- Ethical – all organisms have a right to exist, and most play an important role within their ecosystem. Many people believe that we should not have the right to decide which organisms can survive, and which we could live without. We also have a moral responsibility for future generations to conserve the wide variety of existing natural ecosystems.

Summary questions

1   State the difference between conservation and preservation.   *(1 mark)*

2   There are many reasons why natural habitats may be conserved. State one example of why a habitat may be conserved for:
   a   economic reasons   *(1 mark)*
   b   social reasons   *(1 mark)*
   c   ethical reasons.   *(1 mark)*

3   Suggest and explain one conservation and one preservation technique that could be used to maintain biodiversity in an area.   *(4 marks)*

24.5 Sustainability

Specification reference: 6.3.2

An increasing global human population results in the need for more resources. To survive at a basic level you require uncontaminated food and water supplies, shelter, clothing, and access to medical care when the need arises. To live in the manner to which you are accustomed, many more resources are required.

To cope with the increased human demand for resources, intensive methods have been used to exploit environmental resources. This can result in the destruction of ecosystems, a reduction in biodiversity, and the depletion of resources.

Sustainable use of resources

The world's natural resources have conflicting demands placed upon them. To conserve natural resources for future generations, sustainable management of the natural environment is necessary. A **sustainable resource** is a renewable resource that is being economically exploited in such a way that it will not diminish or run out.

The aims of sustainability are to:

- preserve the environment
- ensure resources are available for future generations
- allow humans in all societies to live comfortably
- enable less economically developed countries (LEDCs) to develop, through exploiting their natural resources
- create a more even balance in the consumption of these resources between more economically developed countries (MEDCs) and LEDCs.

Alongside the sustainable management of resources, existing resources should be used more efficiently. This helps to prevent finite resources being used up so quickly. For example, many products can be reused – others, such as aluminium cans, can be recycled into new products.

As technology improves, alternatives may be developed that could ease the strain on current finite resources. However, these new resources may take many years to develop, be more costly, and have negative environmental effects of their own.

Sustainable timber production

The sustainable management of forests is possible. This allows for the maintenance of a forest's biodiversity, while sustaining both our supply of wood to meet demands and the economic viability of timber production. The techniques used depend on the scale of timber production.

Small-scale timber production

To produce sustainable timber on a small scale, a technique known as *coppicing* is often used. This is a technique where a tree trunk is cut close to the ground. New shoots form from the cut surface and mature.

▲ **Figure 1** *The UK government has a campaign to 'Reduce, Reuse, Recycle'. Reduce means lowering your consumption of physical objects and natural resources. Reuse refers to reusing objects in their current form. Recycle means to break an item down into its raw materials, to be used for the manufacture of new items*

▲ **Figure 2** *A hazel coppiced woodland. Some ash trees are left to grow to maturity – the remaining coppice trees are cut to ground level every 7–10 years to provide a constant supply of small wood. The woodland is thus divided into a series of areas cut in different years, ensuring a continuous supply of wood*

un-coppiced ———→ newly coppiced ———→ mature coppice

| before tree to be coppiced | cut close to base in winter | following spring shoots rapidly regrow from stool | 7–20 yrs later coppice ready for harvest |

▲ **Figure 3** *Stages of coppicing in a woodland*

▲ **Figure 4** *Pollarded alder tree (*Alnus glutinosa*) at a walled field boundary. This pollarding was recent. Pollarding is a wood-management technique in which all growth is removed from the tree above 2–3 metres at intervals of several years. This provides a regular supply of wood, and has the incidental benefit of prolonging the lifespan of the tree*

▲ **Figure 5** *Fishermen sorting out their catch ready for market. Fishing quotas vary depending on the species of fish. Scientists study different species to determine how big their population needs to be for the species to maintain its numbers*

Eventually these shoots are cut and in their place more are produced. These shoots have many uses, including fencing.

In most managed woodlands, rotational coppicing takes place. The woodland is divided into sections and trees are only cut in a particular section until all have been coppiced. Coppicing then begins in another area allowing time for the newly coppiced trees to grow. This process continues until you reach the trees that were first coppiced. These will now have grown to mature-sized trees, and the cycle begins again.

Rotational coppicing maintains biodiversity as the trees never grow enough to block out the light. Hence, succession cannot occur and so more species can survive.

An alternative technique to coppicing which may be used is *pollarding*. The technique is similar to coppicing, but the trunk is cut higher up so deer and other animals cannot eat the new shoots as they appear.

Large-scale timber production

Sustainable timber production on a large scale is based around the technique of felling large areas of forest. The felled trees are destroyed and will not regrow.

To ensure that production is sustainable, timber companies:

- Practise selective cutting, which involves removing only the largest trees.
- Replace trees through replanting rather than waiting for natural regeneration. This also helps to ensure that the biodiversity and mineral and water cycles are maintained.
- Plant trees an optimal distance apart to reduce competition. This results in higher yields as more wood is produced per tree.
- Manage pests and pathogens to maximise yields.
- Ensure that areas of forest remain for indigenous people.

The major disadvantage of this technique is that habitats are destroyed, soil minerals are reduced, and the bare soil which is left is susceptible to erosion. Trees are important for binding soil together, removing water from soil, and maintaining nutrient levels through their role in the carbon and nitrogen cycles.

Sustainable fishing

As well as the increased demand for fuel and buildings created as a result of population growth, the demand for food is ever-increasing. Fish provide a valuable source of protein within the human diet. However, overfishing has led to the populations of some species of fish decreasing significantly. These fish populations are then unable to regenerate, meaning that they will no longer be able to provide us with a food source in the future.

To try to overcome this problem international agreements are made about the number of fish that can be caught. An example of this is the Common

Fisheries Policy in the EU. Fishing quotas provide limits on the numbers of certain species of fish that are allowed to be caught in a particular area. The aim is to maintain a natural population of these species that allows the fish to reproduce sufficiently to maintain their population.

Other techniques that have been used include:

- The use of nets with different mesh sizes. For example, mesh sizes can be made sufficiently large enough that immature fish can escape. Only mature fish are caught, thus allowing breeding to continue.

- Allowing commercial and recreational fishing only at certain times of the year. This protects the breeding season of some fish species and allows the fish levels to increase back to a sustainable level. For example fisherman are only allowed to catch red snapper in the Gulf of Mexico between May and July.

- The introduction of fish farming to maintain the supply of protein food, whilst preventing the loss of wild species. For example, tilapia are among the easiest and most profitable fish to farm due to their diet, tolerance of high stocking densities, and rapid growth. In some regions, the fish are placed in rice fields at planting time where they grow to edible size when the rice is ready for harvest.

Study tip

Be careful with the use of the word 'extinct'. Extinct means no organisms of a particular species are found anywhere in the world. Overfishing can result in the removal of a particular species from an area, but this does not mean the species is extinct.

Overfishing of North Sea cod

As a result of increasing boat numbers and improvements in technology, ever larger numbers of fish have been removed from the sea. In recent years, the number of cod caught in the North Sea has declined as the cod population has fallen.

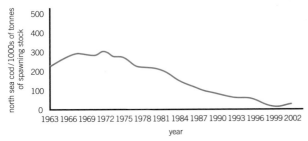

▲ **Figure 6** *The decline of North Sea cod stocks since 1963*

1 State the year the population of cod was at its highest.
2 Suggest why the population of North Sea cod has fallen.

As a result of the near collapse of some fish populations, the European Union introduced fishing quotas to conserve fish stocks.

3 Explain how fishing quotas conserve fish stocks.

These regulations also placed limits on the mesh size of the nets. By increasing the size of the holes in nets, only mature, full-sized fish can be caught. Immature fish can escape.

4 Suggest how increasing mesh size allows the fish population to recover.
5 Evaluate how society uses data on fish populations to inform decisions.

Summary questions

1 State what is meant by a sustainable resource. *(1 mark)*
2 Suggest two reasons why producing resources in a sustainable way is important. *(2 marks)*
3 State and explain two ways that large-scale timber production can be managed sustainably. *(2 marks)*
4 Explain how coppicing is used to produce a sustainable supply of timber. *(4 marks)*

24.6 Ecosystem management – Masai Mara

Specification reference: 6.3.2

Learning outcomes

Demonstrate knowledge, understanding, and application of:

→ how ecosystems can be managed to balance the conflict between conservation/preservation and human needs.

The Masai Mara National Reserve (MMNR) in southern Kenya is an example of an ecosystem that is actively managed to balance the needs of humans and the need for conservation. The reserve was established as a wildlife sanctuary in 1961, and covers around 1500 km^2 – it is situated approximately 1500–2000 m above sea level.

Ecosystem

The MMNR is primarily a savannah ecosystem, divided by the main Mara river. The fertile regions close to the river are a combination of rich grasslands and woodland – further from the river are open plains with scattered shrubs and trees. The region is famous for its annual zebra and wildebeest migrations and is home to a wide range of large mammals, including buffalo, elephants, leopards, lions, and black rhinos.

In the past, the region was dominated by the acacia bush. This provided a habitat for the tsetse fly which is a carrier of African trypanosomiasis (sleeping sickness). To attempt to reduce incidences of the disease, government workers and indigenous communities have cleared major tracts of acacia over the last 50 years. Elephants, fire, and cattle grazing have further reduced the presence of acacia and other woody plants.

▲ **Figure 1** *Wildebeest running in the grassland of the Masai Mara*

Farming

Grazing

Traditionally the Masai Mara has been used by local tribes for livestock grazing. In the past, the Masai practised a traditional method of farming known as semi-nomadic farming. Tribes frequently moved depending on climate variation and the presence of tsetse flies. This allowed vegetation time to recover from animal grazing whenever the farmers moved on to another area.

Grazing is now limited to areas on the edge of the reserve, as local tribes are prevented from entering the park. Populations have grown in these marginal areas. Larger herds graze the grassland areas, and more trees are removed for fuel. As the vegetation is removed, the risk of soil erosion increases.

▲ **Figure 2** *The Reserve is named after the local Masai tribes, whose land surrounds the Reserve. Their homes were set up in groups of huts surrounding an enclosed area in which cattle were housed. Each night, the families would bring their livestock into the enclosure in order to prevent predation and cattle theft*

Cultivation

The level of cultivation around the region of the Masai Mara has increased in recent years. As grassland has been converted into cropland, natural vegetation is removed, and nutrients in the soil are used up. Over time this leads to a reliance on fertilisers for effective crop growth.

Ecotourism

The Masai Mara relies on tourism for most of its economic input. Thousands of people travel to the region each year, eager to see for themselves the vast numbers of animals present in their natural habitat. **Ecotourism** is tourism directed towards natural environments, to support conservation efforts and observe wildlife. It is a type of sustainable development which aims to reduce the impact that tourism has on naturally beautiful environments. This is usually seen as a less invasive use of land than agriculture.

The key principles of ecotourism are to:

- ensure that tourism does not exploit the natural environment or local communities
- consult and engage with local communities on planned developments
- ensure that infrastructure improvements benefit local people as well as visitors.

Ecotourism, however, can have negative impacts on the ecosystem. There is evidence that tourist movements such as the repeated use of hiking trails, or the use of mechanised transport, may contribute to soil erosion and other habitat changes.

Conservation and research

The nature reserve plays an important role in the conservation of endangered species. Some of the most popular large mammals have experienced population declines in recent years – beyond those expected from climate or natural variation.

Black rhinos are one of the most endangered animals in Africa, and appear on the IUCN critically endangered list. Despite the trade being illegal, rhino horn is in huge demand, particularly for use in traditional medicine in south-east Asia. People are lured into poaching by the vast sums of money offered to trade in this material.

In 1972, over 100 rhinos lived in the Masai Mara. By 1982, illegal poaching meant that only a handful remained. An active conservation and protection programme was established to encourage a balance between the needs of local communities and those of the wildlife. The programme included the employment of reserve rangers, and the provision of communication equipment, vehicles, and other necessary equipment and infrastructure. These measures have helped to deter poachers from entering the reserve and by the mid-1990s rhino numbers had increased significantly, although it will be some time before population levels return to those seen in the early 1970s.

▲ **Figure 3** *Increasing numbers of tourists who have come to see the wildlife also visit local Masai tribes to observe their way of life and their traditional dances*

▲ **Figure 4** *Most tourists take part in organised safari tours to try and spot the 'big five' – the African lion, African elephant, Cape buffalo, African leopard, and rhinoceros*

Synoptic link

Look back at Topic 11.8, Methods for maintaining biodiversity for more information on the IUCN.

▲ **Figure 5** *The black rhinoceros* (Diceros bicornis)

A number of scientific research projects have been (or are currently being) undertaken in the Masai Mara. These include:

- Michigan State University, studying the behaviour and physiology of the predator spotted hyena.
- Subalusky and Dutton, completing a flow assessment for the Mara River Basin. The aim of this research is to identify the river flows needed to provide for both the basic human needs of the million people who depend on the water, and to sustain the ecosystem in its current form.
- The Mara Predator Project, which catalogues and monitors lion populations throughout the region. The project aims to identify population trends and responses to changes in land management, human settlements, livestock movements, and tourism.
- The Mara-Meru Cheetah Project, which aims to monitor the cheetah population and evaluate the impact of human activity on cheetah behaviour and survival.

Striking a balance

Some human land uses in Masai Mara are incompatible with wildlife survival – increasing wildlife density also threatens pastoral and cultivation lifestyles. A constant balance has to be maintained between the human and animal populations. For example:

- Elephants, in particular, threaten cultivation. Large elephant populations are often responsible for crop trampling and damage to homesteads. Other grazing animals may also eat the crops. To prevent these problems land may be fenced, but this has a negative effect on natural migration.
- Legal hunting is used to cull excess animals. This can successfully maintain population numbers and bring in considerable amounts of money for conservation work. However, numbers must be constantly monitored to ensure that levels are sufficient to maintain the natural balance within the ecosystem.
- Livestock also faces threats from migratory wildlife. For example during the annual wildebeest migration, the wildebeest outcompete cattle for grass. Diseases are introduced to the domesticated animal populations. Equally, the domesticated cattle eat vegetation that could be used by migrating zebras and wildebeest, and diseases can spread from the domestic to the wild animals.
- As the human population expands more homes are required as well as land for cattle and agriculture. Evidence suggests that wildlife density declines significantly as the density of the built environment rises.

Summary questions

1 Describe the ecosystem of the Masai Mara.   (*1 mark*)

2 State two ways in which humans use the lands of the Masai Mara   (*1 mark*)

3 Explain how and why local Masai tribes have changed their style of farming in recent years.   (*4 marks*)

4 The Masai Mara region receives around 300 000 visitors each year. State and explain the positive and negative impacts of this influx of people on the region.   (*6 marks*)

24.7 Ecosystem management – Terai region of Nepal

Specification reference: 6.3.2

The southern part of Nepal contains a rich agricultural area known as the Terai region. It stretches along Nepal's southern border with India, with a width of around 25–30 km. The Terai lowlands are defined by a belt of well-watered floodplains stretching from the Indian border in the south to the slopes of the Bhabhar and Siwalik mountain ranges to the north.

The land of the Terai region is fertile, and is the main agricultural region of the country. Alongside farming, people are engaged in a range of trades, industries, and services. As a result of the high population density, natural resources are at risk of being overused.

Learning outcomes

Demonstrate knowledge, understanding, and application of:

→ how ecosystems can be managed to balance the conflict between conservation/preservation and human needs.

Ecosystem

The region is hot and humid in the summer months, and is composed of a fertile alluvial soil which is rich in plant nutrients. The Terai is an area of extreme biodiversity – many subtropical plants are found in this region including pipal and bamboo. There are also large areas of thick forest where animals such as the Bengal tiger, the sloth bear, and the Indian rhinoceros can be found.

▲ **Figure 1** *Bengal tigers can be found roaming the forests of Nepal*

Millions of people depend on the Terai forests for their livelihoods. They are also an important source of national income. Primarily as a result of poverty and corruption, large areas of forest have been cleared for agriculture or to sell the timber.

The removal of large parts of the forest has also exacerbated the effects of monsoon flooding, causing severe disruption to communities downstream. If deforestation of the region continues unabated, the communities of the Terai would be left with only small, isolated pockets of forest. This would be devastating not only for the wildlife of the region, but also for the local population who rely on the forests for income through tourism, and through harvesting wood for building products and for burning as fuel.

▲ **Figure 2** *Royal Chitwan National Park, which is located in the Terai region of Nepal*

Sustainable forest management in Nepal

The aim of sustainable forest management is to provide a livelihood for local people, ensure the conservation of forests, and provide the Nepali state with considerable income for general development. This is being achieved through supportive national legislation, and through the development of local community forestry groups.

The local groups develop their own operational plans, set harvesting rules, set rates and prices for products, and determine how surplus income is distributed or spent. Through the creation of cooperative networks,

▲ **Figure 3** *Map of Nepal showing Terai region*

▲ **Figure 4** *Community forestry has contributed to restoring forest resources in Nepal. Forests account for almost 40% of the land in Nepal – however, this area was decreasing at an annual rate of 1.9% during the 1990s. Since 2000 this decline has been reversed, and forest coverage is now increasing*

small forestry businesses can work effectively together – for example, to gain Forestry Stewardship Council (FSC) certification, an international standard which rewards sustainable forestry.

There have been several successes for the community forestry groups:

- Significant improvement in the conservation of the forested regions, both in terms of increased area and improved density.
- Improved soil and water management across the region.
- An increase in the retail price of forestry products, and so a greater economic input to the region.
- Employment and income generation through forest protection, as well as through the production of non-timber forest products.
- Sustainable wood fuel sources, which contribute three-quarters of the local household energy needs.
- Securing the biodiversity of the forested areas.

Promoting sustainable agriculture

The Terai requires a range of management strategies for sustainable land use, to prevent damage of the ecosystem including the further degradation of the Terai forests. These include:

- promoting the production of fruits and vegetables in the hills and mountain regions to avoid further intensification of the Terai
- improving irrigation facilities to enhance crop production
- multiple cropping, where more than one crop is grown on a piece of land each growing season
- the growth of nitrogen-fixing crops such as pulses and legumes to enhance the fertility of the soil
- growing crop varieties resistant to various soil, climatic, and biotic challenges through the use of modern biotechnology and genetic engineering
- improving fertilisation techniques to enhance crop yields – for example, using manure to improve the nutrient content of the soil.

Through the implementation of sustainable forestry and agricultural practices, the Terai region is now being managed in a manner that will secure both the biodiversity of the region, and the economic welfare of its residents for the future.

Summary questions

1   Describe the ecosystem of the Terai region of Nepal.   *(2 marks)*

2   State two ways humans use the Terai region of Nepal.   *(1 mark)*

3   Explain how sustainable forestry and agricultural practices are being used in the Terai region to maintain biodiversity, while also meeting the needs of the local population.   *(4 marks)*

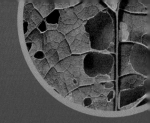

24.8 Ecosystem management – peat bogs

Specification reference: 6.3.2

A peat bog is a region of wet, spongy ground that contains decomposing vegetation. Undisturbed peatland is a 'carbon sink', meaning that it is a store of carbon dioxide. However, once dried, peat can be used as a fuel. As well as releasing thermal energy, burning peat releases carbon dioxide into the atmosphere. It takes many thousands of years for peat bogs to form – the preservation of existing peat bogs is therefore an important component in preventing further climate change.

As well as being used as a fuel, peat is also important for farmers and gardeners, who mix it with soil to improve soil structure and to increase acidity. Peat has very favourable moisture-retaining properties when soil is dry, and prevents excess water killing roots when soil is wet. Although peat can store nutrients, it is not fertile in itself. Commercial peat extraction to supply gardeners and nursery growers is a major threat to this ecosystem.

Ecosystem

Peat forms when plant material is inhibited from fully decaying by acidic and anaerobic conditions. This normally occurs in wet or boggy areas, and therefore peat is mainly composed of wetland vegetation including mosses, sedges, and shrubs.

The plants that grow on peatlands, such as sphagnum mosses (*Sphagnum* spp.), bog cotton or cottonsedge (*Eriophorum angustifolium*), and heathers (typically *Calluna vulgaris*), have adapted to grow and thrive in wet conditions with few nutrients. Bogs also support a wide range of insects such as butterflies, moths, dragonflies, and damselflies. The lack of predators and human disturbance makes some peatlands ideal for birds to nest and bring up their chicks. The abundance of insects, spiders, and frogs, plus nutritious vegetation and berries, provides food for many species. The large areas of open ground provide ideal hunting grounds for birds of prey.

Loss of ecosystem

Lowland raised bogs are one example of a peatland ecosystem. They are a rare and threatened habitat. In the UK, the area of relatively undisturbed lowland raised bog is estimated to have diminished by over 90%, from around 950 km² to only 60 in the last 100 years. It is essential that the remaining areas are conserved to maintain biodiversity – their maintenance will also contribute to flood management, erosion control downstream, and carbon storage.

Learning outcomes

Demonstrate knowledge, understanding, and application of:

→ how ecosystems can be managed to balance the conflict between conservation/preservation and human needs.

Synoptic link

Look back at Topic 23.3, Recycling within ecosystems to remind yourself of how carbon is recycled in the carbon cycle.

▲ **Figure 1** *Peat beds on the Somerset Levels. The peat from here is extracted and sold*

▲ **Figure 2** *This peat is drying to be used as fuel. As peat dries out it releases greenhouse gases into the atmosphere*

▲ **Figure 3** *Traditional peat cutting in a bog in south-west Ireland. Slabs of peat are cut from a shallow trench and stacked in small heaps to dry . Hand-cutting peat is a slow, labour-intensive process that can allow the bog to partially recover. It is very different from industrialised, mechanical extraction. The peat companies deep-drain peatlands and strip all vegetation from vast expanses of bog surface, wiping out whole bogs. The Wildlife Trusts are active in attempting to stop this destruction and working to conserve and protect the few remaining areas*

Historically, the greatest decline has occurred through afforestation (the establishment of a forest or stand of trees in an area where there was no forest), peat extraction, and agricultural intensification, including land drainage. These activities have all contributed to the drying out of the bogs.

Conserving lowland bogs

The key to conserving lowland bogs is to maintain or restore appropriate water levels. Steps which are taken to conserve areas of lowland bog include:

- Ensuring that the peat and vegetation of the bog surface is as undisturbed and as wet as possible. Most bogs are surrounded by ditches to allow water to run off, preventing flooding of nearby land. In restoring a bog, ditch blocking may be required for a period of time to raise the water table to the bog surface.

- Removal of seedling trees from the area. Trees have a high water requirement due to transpiration. Therefore, any tree seedling that has the potential to remove water from an area of peatland, or to reduce its ability to support bog vegetation, should be removed to maintain water levels in the area.

- Using controlled grazing to maintain the biodiversity of peatland. Grazing ensures a diverse wetland surface in terms of structure and species composition. This in turn provides a wide range of habitats for many rare insect species.

Continuing intensive land use threatens the existence of much of the remaining peatlands in the UK. Although around 10% of the UK is classified as peat bog of one form or another, around 80% of these areas are in a poor condition. A number of organisations are working to preserve these important ecosystems, including The Wildlife Trusts, Natural England, and the Royal Society for the Protection of Birds (RSPB). It is hoped that, through the conservation work of these and other organisations, an appropriate balance can be met between our need to exploit the land, and the maintenance of a sustainable, biodiverse ecosystem.

Summary questions

1 State what is meant by a peat bog. *(1 mark)*

2 Describe two conservation measures which can be taken to conserve a peat bog ecosystem. *(2 marks)*

3 State and explain why the UK has lost such a large proportion of its peatlands over the past century. *(4 marks)*

24.9 Environmentally sensitive ecosystems

Specification reference: 6.3.2

Although all ecosystems are, to some extent, fragile, some regions are less resistant to changes than others. These regions are often referred to as *environmentally sensitive ecosystems*. For example, although mass tourism can bring economic vitality to an area, it can also bring attendant changes to an environment that are not always positive, such as the overdevelopment of a coastline. This level of human activity can lead to losses in the biodiversity of the region.

In many environmentally sensitive areas the same types of management techniques are used. These include:

- limiting the areas tourists can visit
- controlling the movement of livestock
- introducing anti-poaching measures
- replanting of forests and native plants
- limiting hunting through quotas and seasonal bans.

This topic looks at examples of some of the world's environmentally sensitive ecosystems, which without control of human activities will be lost.

The Galapagos Islands, Pacific Ocean

The Galapagos is an archipelago of volcanic islands that rise up from the bed of the Pacific Ocean 1000 km west of Ecuador. They are of special interest because they have never been connected to the mainland. The original flora and fauna that reached the islands' shores had to survive a crossing of hundreds of kilometres of ocean. Darwin used evidence he gained from his visits to these islands to develop his theory of evolution by natural selection.

Animals present

The majority of land animals living on the islands are reptiles – there is only one species of land mammal, the Galapagos rice rat. These species arrived here by being washed away from mainland river banks, floating on rafts of vegetation.

Over millions of years these animals, and many of the marine birds that also arrived on the islands, have adapted to their environment in isolation, resulting in many species that are unique to the islands. These include:

- The Galapagos giant tortoise (*Chelonoidis nigra*), which grows to over 150 cm in length.
- The flightless cormorant (*Phalacrocorax harrisi*), whose reduced wings were better for fishing underwater, when flight was not needed to escape mainland predators.

Learning outcomes

Demonstrate knowledge, understanding, and application of:

→ the effects of human activities on the animal and plant populations in environmentally sensitive ecosystems and how these human activities are controlled.

▲ **Figure 1** *The Galapagos giant tortoise (*Chelonoidis nigra*)*

- The marine iguana (*Amblyrhynchus cristatus*), which contains the advantageous mutation, over a land iguana, of the ability to swim effectively. Unless they are trying to attract a mate, the iguana appears black or dark grey, this allows these ectotherms to bask in the sun and raise raise their body temperatures to approximately 36 °C before swimming in the cold sea where they forage for food. The higher the temperature the longer they can forage.

Plants present

On the larger islands three distinct regions exist, each of which supports particular plant species. These regions are:

- the coastal zone, which contains salt-tolerant species such as mangrove and saltbush
- the arid zone, which contains drought-tolerant species such as cacti and the carob tree
- the humid zone, which contains dense cloud-forest. These trees support populations of mosses and liverworts.

Control of human activities

Until the 19th century the islands were hardly visited by humans. However, as a result of the whaling trade this all changed. The whalers disrupted the islands' fragile ecosystems by allowing domestic animals to roam loose, chopping forests for fires to render down whale fat, and removing tens of thousands of live giant tortoises, whose meat would sustain the whalers on their long sea voyage due to their ability to survive for long periods without food or water. The goats they introduced have also outcompeted giant tortoises on many islands.

The Galapagos National Park was established in 1959. Since then, measures have been taken to protect the living and non-living parts of this unique ecosystem, including:

- introduction of park rangers across the islands
- limiting human access to particular islands, or specific parts of islands
- controlling migration to and from the islands
- strict controls over movement of introduced animals such as pigs (the presence of these were noted by Darwin).

It is hoped that through these and many other measures, the islands' flora and fauna can be protected for future generations.

Antarctica

Antarctica is the coldest, highest, driest, windiest, and emptiest continent. It is almost entirely covered by an ice sheet, which averages 2 km thick. This ice sheet contains around 70% of the world's fresh water.

The average temperature in Antarctica in the winter is below −30 °C. Unlike most parts of the Earth Antarctica has just two seasons, summer and winter – during the summer, many parts of the continent experience 24-hour sunlight – in the winter many parts of the continent experience 24-hour darkness.

▲ **Figure 2** *The* Scalesia *tree, which dominates the humid zone of the Galapagos Islands. This tree originates from the same family as the diminutive daisy, but grows to heights of over 10 metres*

Animals present

All endothermic animals living on and around Antarctica rely on thick layers of blubber to insulate them from the cold. These include whales, seals, and penguins. For example, the blubber layer on a Weddell seal (*Leptonychotes weddellii*) can be 10 cm thick.

The emperor penguin (*Aptenodytes forsteri*) is the only warm-blooded animal that remains on the Antarctic continent during the winter. Females lay one egg in June (mid-winter) and leave to spend the winter at sea. The male penguins stay on land, surviving the most extreme winter conditions for up to nine weeks (with no food), keeping their egg warm by balancing it on their feet and covering it with a flap of abdominal skin.

A few invertebrates live on the continent all year. The largest is a wingless midge called *Beligica antarctica*, although at a body size of around 5 mm it can only be viewed properly under a microscope.

Plants present

Plants can only grow in the ice-free regions (around 2% of the continent). Lichens and moss grow in any favourable niche such as in sand, soil, rock, and on the weathered bones and feathers of dead animals. Algae are also able to grow in many sheltered areas.

Control of human activities

With the exception of a few specialised scientific settlements, Antarctica is too cold for people to live. However, in the last 100 years increasing numbers of tourists have been visiting Antarctica. Most visit the Antarctic Peninsula, accessible from Chile, where the climate is mild in comparison with the rest of the continent, and there is much wildlife.

Although strictly controlled in Antarctica (only allowing visits for a few hours to selected area), human activity has had a number of effects on the continent:

- Planet-wide impacts such as global warming (causing some parts of the coastal ice sheet to break up) and ozone depletion, caused by human activities elsewhere.
- Hunting of whales and seals, and fishing of some Antarctic species, has depleted stocks of these organisms.
- Soil contamination, particularly around scientific research stations.
- Discharging of waste into the sea, including human sewage.

The Antarctic treaty was established in 1961 to protect the unique nature of the Antarctic continent. This treaty remains in force indefinitely. Some of its provisions include:

- scientific cooperation between nations
- protection of the Antarctic environment
- conservation of plants and animals
- designation and management of protected areas
- management of tourism.

▲ **Figure 3** *There are many different species of penguins in Antarctica, including the large, colourful emperor penguin shown here with its chick*

▲ **Figure 4** *These are humpback whales (*Megaptera novaeangliae*) exhaling in calm waters off the Antarctic Peninsula. The International Whaling Commission (IWC) was established in 1949 to protect whales such as the blue and humpback whales whose numbers were diminished by whaling. Indicators show that whale populations are beginning to recover, but such long-lived species with low reproductive rates are incapable of rebuilding their numbers in just a few years*

▲ Figure 5 *Snowdonia National Park*

Snowdonia National Park, Wales

Snowdonia National Park covers 2000 square kilometres of countryside in north-west Wales. It contains the highest mountain range in England and Wales, with four peaks over 1000 metres (including Mount Snowdon at 1085 m). The rugged terrain includes lakes and fast-flowing rivers, and wide tracts of ancient woodland and heath.

Animals present

The rich diversity of habitats in the region provides homes for a wide range of birds. These include:

- Coast and estuary birds such as choughs, cormorants, and oystercatchers.
- Forest birds such as pied flycatcher, redstart, and wood warblers.
- Moorland and mountain birds such as ospreys, buzzards, and sparrowhawks.

There are also over 40 species of land mammal present in Snowdonia including badgers, voles, deer, and hedgehogs.

Plants present

Snowdonia supports an equally diverse range of plant species. For example, if you climb to the top of Mount Snowdon you may come across the Snowdon lily (*Gagea serotina*) and other hardy arctic-alpine plants that have evolved to cope with extreme conditions. Lower down the slopes, the mountain is fringed by woodlands of oak, alder, and wych elm.

Control of human activities

Snowdonia National Park was created in 1951 to conserve the biodiversity of the region. The Park is home to over 25 000 people, many of whom work on the land. It also attracts several million visitors each year. Climbing, walking, cycling, and watersports are some of the most popular activities.

The key purposes of the Park authority are to:

- Conserve and enhance the natural beauty, wildlife, and cultural heritage of the area.
- Promote opportunities for the understanding and enjoyment of the special qualities of the Park.
- Enhance the economic and social well-being of communities within the Park.

▲ Figure 6 *Walkers are asked to keep to designated footpaths in many areas of the park to prevent plants being trampled. The areas in which mountain bikes can be used is even more limited as large tyres with deep treads damage vegetation and top soil, which begins a process of path erosion*

The Dinorwig power station is a pumped-storage hydroelectric power station which was built to help meet the demands of the National Grid during sudden bursts of energy requirement. To preserve the natural beauty of Snowdonia National Park, the power station itself is located deep inside the mountain Elidir Fawr, inside tunnels and caverns. This has minimised the impact to the environment whilst meeting the human demand for energy.

The Lake District, England

The Lake District is England's largest national park, at over 2292 km². The national park contains Scafell Pike, England's highest mountain, and Wastwater, its deepest lake. Terrain within the park includes regions of moorland and fell, and includes 16 lakes dammed by glacial moraines (soil and rock debris) around the end of the last ice age. The dales and fringes of the lakes provide a rich variety of habitats, including areas of ancient woodland.

Animals present

The varied landscapes of the Lake District provide homes for a wide range of wildlife. These include water voles, natterjack toads, and a number of species of bat, through to red deer and birds of prey such as the golden eagle and osprey. A number of native species are under threat including the red squirrel and the vendace, a species of fish that is only found in this region, and which appears on the IUCN endangered list.

Plants present

In the central fells, there are habitats that exist above the tree line. These are rare in the UK and they support a diverse range arctic-alpine plant communities. These include such species as purple saxifrage and alpine cinquefoil. Specialised trees have evolved in these harsh habitats, such as the dwarf juniper and dwarf willow (*Salix herbacea*). Lower regions of the Lake District are home to the sundew (*Drosera rotundifolia*), one of the UK's few carnivorous plants. Insects are trapped within its leaves by a sticky mucilage (which glistens like morning dew, hence its common name) – enzymes are then secreted to digest the insect, releasing its nutrients for the plant.

Control of human activities

The role of the Lake District National Park Authority is like that of Snowdonia – to conserve the region while enabling access for many millions of visitors each year. Through the active management of the countryside, for example, through replanting native tree species, this fragile ecosystem is being secured for generations to come.

▲ **Figure 7** *Wastwater, the deepest of all the lakes in the Lake District, surrounded by mountains*

▲ **Figure 8** *A fly caught in the carnivorous plant, <u>Drosera</u> <u>rotundifolia</u>, common name, sundew. The fly is attracted by the glistening droplets at the tips of the tentacles, which the insects then stick to. Proteolytic enzymes secreted by the tentacles digest the insect and the products are absorbed by the leaf*

Summary questions

1 State three reasons why people might want to visit environmentally sensitive areas. *(1 mark)*

2 State one unique feature of:
   a the Galapagos Islands
   b Antarctica. *(2 marks)*

3 Using a named region, state and explain how a human activity is controlled to limit its effect on the populations of flora or fauna. *(4 marks)*

4 Environmentally sensitive regions often receive many visitors each year. State and explain the social and environmental impacts on the region of this influx of people. *(6 marks)*

Practice questions

1 The diagram shows two ways in which trees are pruned in order for wood to be continually harvested from the same trees for many years. The health of the trees is not affected in any negative way.

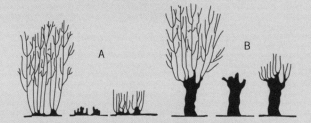

a Name the two types of tree pruning shown in the diagram. *(2 marks)*

b Explain why these forms of timber management are sustainable. *(2 marks)*

c Suggest how these forms of tree pruning help maintain biodiversity. *(3 marks)*

2 Tourism, particularly ecotourism, is a growing industry. The Antarctic Peninsula is even becoming a popular tourist destination. There are many reasons for the increase in tourist numbers such as television making people more aware of extreme environments and more flexible working hours.

a Suggest three more reasons for the increase in tourism to places like Antarctica. *(3 marks)*

The table shows the changes in the number of tourists travelling to Antarctica over the last 20 years.

Year	<1987	1990/ 1991	1992/ 1993	2002/ 2003	2006	2007
Number of visitors	<1000	4698	6500	24 281	>30 000	>40 000

b (i) Describe how the numbers of tourists visiting Antarctica have changed over the last 20 years. *(3 marks)*

(ii) Outline the advantages and disadvantages of this change in tourism to Antarctica. *(5 marks)*

(iii) Enhanced global warming has led to melting of the polar ice caps.

Suggest how this has been partly responsible for the increase in tourism. *(1 mark)*

A student carried out an investigation into the perception that people have about the impact of tourism on endangered species such as the blue whale.

The results are shown in the graphs.

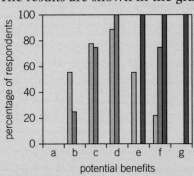

a none
b people learn about whales
c increased appreciation for whales
d people advocate for conservation
e people donate to conservation
f tourism boats carry scientists
g people get involved in research

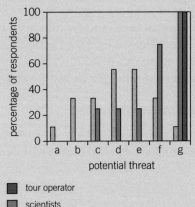

■ tour operator
■ scientists
□ tourists

a none
b travelers
c pollution (other)
d vessel noise
e oil spills
f stress
g collisions

c Discuss any conclusions that could be reached on analysis of these results. (*4 marks*)

The graph shows the changes in numbers of blue whales in the Antarctic over last hundred or so years.

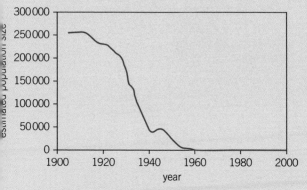

d State, giving your reason(s), whether the outlook for the blue whale population is good or not. (*3 marks*)

Peat bogs are large areas of waterlogged land that support a specialised community of plants. Peat bogs take thousands of years to form.

The flow diagram shows the main stages in the formation of a peat bog.

Bull rushes and reeds grow in the shallow water round the margins of a mineral-rich lake.

Dead plant remains accumulate at the margins and trap sediment, which begins to fill the lake.

Different plants now grow, including brown mosses, which form a floating carpet if the water level rises.

New specialised plants grown on the floating brown moss carpet, because it is mineral deficient and acidic.

Sphagnum mosses colonise, increasing the acidity further raising the bog higher, away from sources of minerals.

Plants such as heather, bog cotton, bog asphodel, and carnivorous plants colonise the Sphagnum moss and form a mature peat bog community.

a (i) Name the process summarised in the diagram that changes a lake community into a peat bog community. (*1 mark*)

(ii) Using the flow diagram, list **two abiotic** factors that play a role in determining what species of plant can grow in an area. (*2 marks*)

b Most of the minerals in a peant bog are held withing the living plants at all times, **not** in the soil.

● Plants like bog cotton and bog asphodel recycle the minerals they contain,

● The leaves of these plants turn orange as the chlorophyll within them is broken down.

● Minerals such as magnesium ions are transported from the leaves to the plants' roots for storage.

Describe **one** similarity and **two** differences in mineral recycling in a peat bog and in a **deciduous forest**. (*3 marks*)

c In Ireland in 2002, two well-preserved Iron-age bodies were found in peat bogs. Despite having been dead for over two thousand years, the bodies had not decomposed. They had skin, hair, and muscle.

Suggest why these bodies had not decomposed (*2 marks*)

d Suggest two reasons why the large scale removal of peat from bogs for use in gardens is discouraged by conservation groups. (*2 marks*)

OCR 2805/03 2010

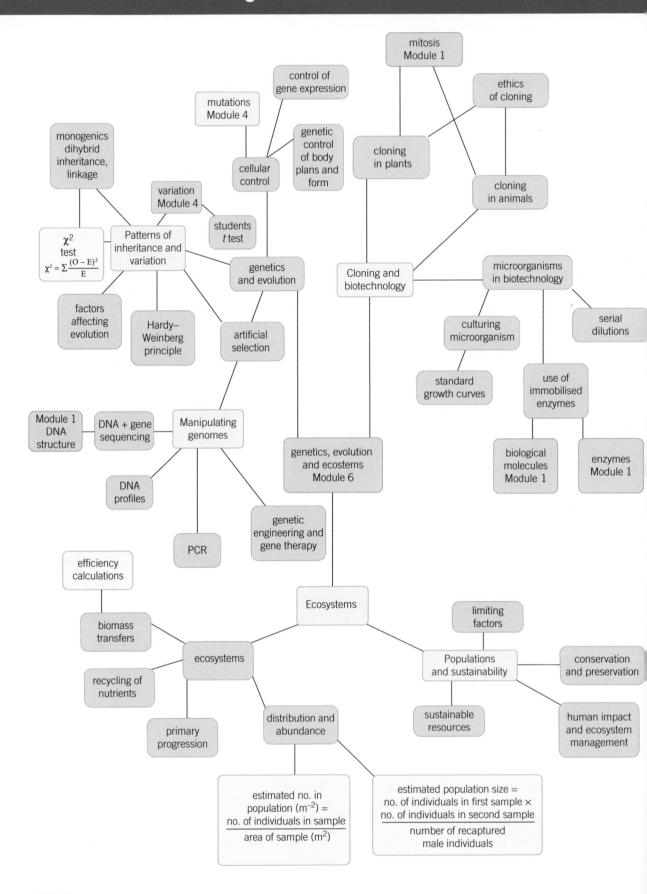

Application

DNA profiling, popularly referred to as DNA fingerprinting, has been used as a form of evidence since 1988 to help prove guilt or innocence in police investigations. The number of areas of DNA used to make the profile has been increased since the early days. Until recently DNA profiling for use in criminal trials in England and Wales involved 11 DNA regions. This has been increased to 17 areas in a new process known as DNA-17.

DNA-17 is a more sensitive test and it can be used (in conjunction with PCR) to get a DNA profile from small quantities of DNA, poor quality samples, and from older cells. This should make it more useful than ever – it will be particularly important in resolving old cases which have been impossible to solve using previous techniques. Advantages include improved discrimination between the profiles of different people (reducing the probability of a chance match between unrelated people), improved sensitivity (so less DNA needed), and improved comparability with the profiles used in the rest of the EU and beyond, where DNA-17 has been the standard for longer.

Possible disadvantages include an increased chance of partial matches from poor samples when more regions are used. Also the test is so sensitive that it is more likely to pick up traces of DNA from innocent parties

1 a Why are DNA profiles used in forensic evidence, rather than, for example, specific proteins?
  b Why is DNA profiling used rather than DNA sequencing in forensic cases and paternity testing?

2 a Summarise the importance of PCR in DNA profiling
  b A limitation in the initial technology meant that a limited number of primers were supplied. As a result, in some individuals with a particular mutation the primers did not bind so poor profiles were produced.
    i Explain what a primer is and what it does
    ii Discuss how a mutation could prevent the binding of the primers
    iii Suggest how this problem could be overcome.

3 a Consider the three advantages of the new test given here and give a scientific explanation for each of them.
    • improved discrimination between the profiles of different people: The more regions of DNA that are analysed to produce a DNA profile, the smaller the chance that two unrelated individuals will have the same pattern of introns. The more regions used in the test, therefore, the less likely it is that errors of identification will be made
    • improved sensitivity
    • improved comparability with the profiles used in the rest of the EU and beyond.
  b Suggest explanations for the disadvantages suggested:
    increased chance of partial matches from poor samples when more regions are used:
    the test is so sensitive that it is more likely to pick up traces of DNA from innocent parties.

SGM (second generation multiplex) target areas

| vWA | TH01 | D8S1179 | FGA | D21S11 | D18S51 |

SGMPlus target areas

| D2S1338 | D16S539 | D19S433 | D3S1358 |

DNA-17 target areas

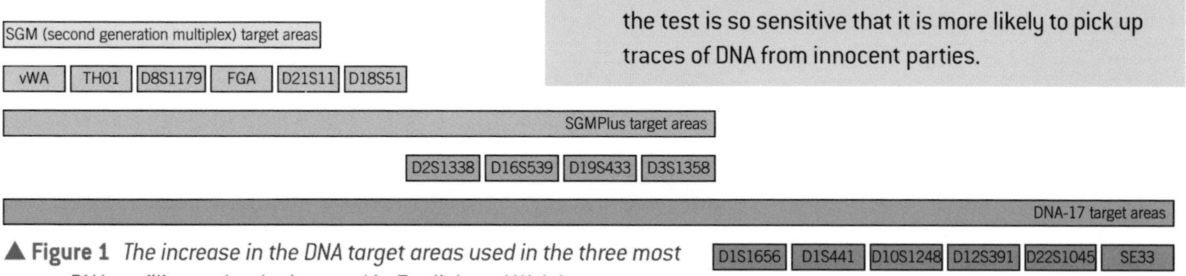

▲ **Figure 1** *The increase in the DNA target areas used in the three most recent DNA profiling technologies used in English and Welsh courts*

| D1S1656 | D1S441 | D10S1248 | D12S391 | D22S1045 | SE33 |

Extension

Investigate the story of the development of DNA profiling and the way it was first used to both prove a man innocent and to prove another man, Colin Pitchfork, guilty. Write an account of the story EITHER: for a newspaper with a view to increasing the public understanding of the science of DNA profiling and updating it to mention the introduction of DNA-17

OR: for a popular science magazine, again bringing the story up to date with the way the technology has developed since.

Module 6 Practice questions

1 Fishing regulations usually require that only fish above a certain size are caught and kept. Smaller fish are thrown back into the sea.

The North East arctic cod have been studied for many years and the average fish size has changed during this period.

The changes are summarised in the table.

Time period	Weight (kg)	Length (cm)
1930s	5.1	85
1970s	4.6	82
2000s	3.2	73

a Describe how the fish have changed over the period of study shown in the diagram. *(4 marks)*

An investigation was conducted using tank-reared Atlantic silversides. These are small, fast-growing fish that are not usually eaten.

Researchers removed 90% of the fish from the tank every year and allowed the survivors to reproduce.

The fish removed from each tank were chosen in three different ways:

1 The largest fish were removed from one tank.

2 The smallest fish were removed from another tank.

3 The fish were removed randomly from a third tank.

The results are shown in the graph.

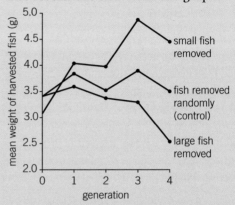

b (i) Compare the trends shown in the graph. *(4 marks)*

(ii) Explain the changes in fish size over the four generations shown by the graph. *(5 marks)*

c Suggest whether this is a form of artificial or natural selection. *(5 marks)*

2 a Define the following terms: Genome, Operon, Expression. *(6 marks)*

b Compare the roles of regulatory and structural genes. *(4 marks)*

The diagram summarises the operon responsible for the synthesis of the amino acid tryptophan.

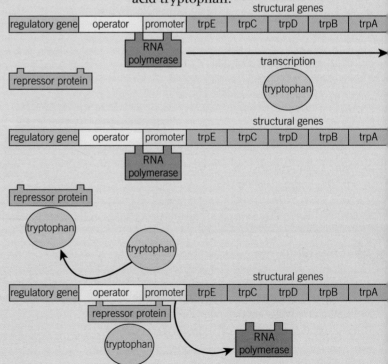

c **(i)** Compare the *structure* of the trp operon to a lac operon. (*5 marks*)

The diagram summarises the action of the repressor protein in the trp operon.

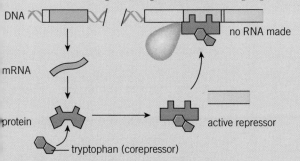

(ii) Suggest how the trp operon is switched on and off. (*4 marks*)

3 Consider the following statements:

Gene expression in prokaryotes is controlled mainly during the process of transcription.

Gene expression in eukaryotes is controlled at the transcriptional, post-transcriptional, translational, and post-translational levels.

Explain, using your knowledge of the structures of eukaryotic and prokaryotic cells, the differences in the control of gene expression. (*7 marks*)

4 **a** Describe the meaning of the following terms: Species evenness, Species richness. (*2 marks*)

The table shows the number of three species found in samples from ponds in the Indiana Dunes.

Lava species	Number of individuals in sample 1	Number of individuals in sample 2
Caddisfly larva	200	20
Dragonfly lava	425	55
Mosquito larva	375	925
Total	1000	1000

b Describe the differences in species evenness and richness between the two sites. (*3 marks*)

The richness and evenness of a community is used to calculate the Simpson's Index of Diversity.

$$D = 1 - \Sigma\left(\frac{n}{N}\right)^2$$

where:

Σ = sum of (total)

N = the total number of organisms of all species and

n = the total number of organisms of a particular species.

Two communities at different sites within the sand dunes were analysed.

The results are shown in table.

Plant species	Number of individuals (n)
Plant species on the Foredune	
marram grass	50
milkweed	10
poison ivy	10
sand cress	4
rose	1
sand cherry	3
Total	N = 78
Plant species on the mature dune	
oak tree	3
hickory tree	1
maple tree	1
beech tree	1
fern	5
moss	3
columbine	3
trillium	3
virginia creeper	4
solomon seal	3
Total	N = 27

c **(i)** Calculate the Simpson diversity index for two local communities. (*2 marks*)

(ii) Suggest reasons for the different diversity indexes. (*3 marks*)

5 It has been estimated that, at the current rate of destruction, tropical rainforests may disappear in less than 150 years.

Outline the ecological, economical, ethical, and aesthetic reasons for the conservation of rainforests. (*7 marks*)

6 There are over 40 000 plants at The Royal Botanic Gardens at Kew. These plants are susceptible to attack by insect pests, such as whitefly, and Trialeurodes spp.

Gardeners at Kew use biological control agents to help prevent pest attack.

The table provides some information regarding the biological control agents of whitefly used at Kew.

Name of control agent	Type of organism	Mode of action	Conditions required
Verticillium lecanii	fungus	infects and kills adults and larvae	high humidity
Encarsia formosa	parasitic wasp	eggs laid on larvae and adult emergence kills pest	low pest density
Macrolophus calliginosus	predatory walking bug	feeds on larvae	high pest density

a   Explain, using information from the table, why gardeners at Kew release only one control agent into one greenhouse at any particular time. *(4 marks)*

b   Suggest why *E. formosa* would not be effective at high pest densities. *(2 marks)*

c   Explain why the gardeners at Kew prefer not to use pesticides. *(4 marks)*

Some researchers investigated the effects of natural predators on another insect pest, Russian wheat aphid *(Diuraphis noxia)*. The results from this study are shown in the graph.

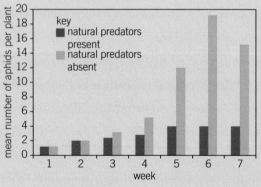

d   Describe the results of this study. *(3 marks)*

e   Despite the environmental advantages of using biological control agents, most farmers still use pesticides to control pests. Suggest why this is the case. *(3 marks)*

OCR 2805/03 2010

7   a   Explain why succession is a dynamic process. *(2 marks)*

b   Describe the changes in the following elements of an ecosystem during succession: *(6 marks)*

Species composition, Species diversity, Density and biomass of organisms, Heterotroph population

c   In an estuary, a salt marsh often develops on muddy shores between the lowest and the highest tide levels. A salt marsh is characterized by the vegetation that grows in these conditions. Succession can be seen from low tide level through the network of creeks and gulleys up to high tide level and beyond onto higher ground.

Students carried out research on two separate salt marshes, A and B. They measured the elevation range above the mean sea level where five key plant species were found. Their results are shown in the graph.

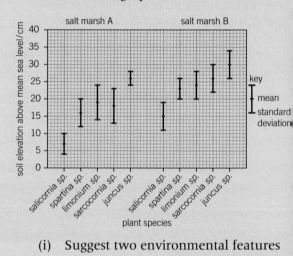

(i)   Suggest two environmental features that make a salt marsh an extreme habitat for plants to live in. *(2 marks)*

(ii)   State one similarity and one difference in the overall distribution of the five species on the two salt marshes. *(2 marks)*

(iii)   Suggest which one of the plant species shown in the graph is a pioneer species. *(1 mark)*

(iv)   Suggest how pioneer plants can alter the salt marsh environment. *(2 marks)*

d Where two or more individuals share any resource that is insufficient to satisfy all their requirements fully, then competition results.

(i) State two species, found in salt marsh A, that are likely to be involved in interspecific competition. (*1 mark*)

(ii) Name one resource for which these plants will compete. (*1 mark*)

e The students were also interested in measuring how the relative abundance of the five species changed with elevation above the mean sea level in the salt marsh.

Describe how this could be done. (*4 marks*)

OCR 2804 2009 (apart from a and b)

8 a The use of biosensors for the detection of chemicals, such as glucose and organophosphates, has proved to be a reliable alternative to other methods.

Glucose biosensors, first developed in the 1980s, enable diabetics conveniently and easily to monitor their blood glucose. The diagram shows the key components of one type of glucose biosensor.

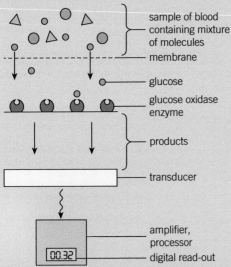

The development of organophosphate biosensors has benefits for health care and environmental monitoring.

With the increasing demand for food production, the use of organophosphates as insecticides to protect crops has increased. However, organophosphates can remain in the environment and are potentially toxic to humans and other animals.

(i) Explain how water and food supplies may contain organophosphates. (*2 marks*)

(ii) Many organophosphates are irreversible inhibitors of acetylcholinesterase.

Explain why this makes them harmful to human health. (*2 marks*)

(iii) The equation summarises the reaction catalysed by acetylcholinesterase.

$$\text{acetylcholine} + H_2O \xrightarrow{\text{acetylcholinesterase}} \text{choline} + \text{acetate} \text{ (acetic acid)}$$

Design a simple biosensor using acetylcholinesterase to detect the presence of a harmful organophosphate in a sample of river water. Include explanations for each element of your design.

You may wish to draw a labelled diagram help you answer. (*5 marks*)

(iv) Suggest the advantages of organophosphate biosensors compared with other detection methods. (*3 marks*)

b Using genetic modification, crop plants resistant to the herbicide glyphosate can be produced.

Glyphosate does not act on acetylcholinesterase but inhibits other enzyme systems.

(i) Explain why a different biosensor to that used in (a)(iii) and (iv) would need to be developed to detect glyphosate in a sample of river water. (*2 marks*)

(ii) Suggest one advantage and one disadvantage of producing glyphosate-resistant crop plants. (*2 marks*)

c Crop plants can be genetically modified for glyphosate resistance by using a restriction endonuclease and DNA ligase.

Describe the roles of these two enzymes in genetic manipulation. (*4 marks*)

OCR 2805/04 2008

SYNOPTIC CONCEPTS

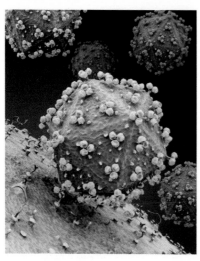

▲ **Figure 1** *The HIV virus is able to infect white blood cells by binding to specific receptor proteins on the cell's membrane*

A synoptic concept in biology is an idea or principle that is applied to more than one part of the subject. For example, cell signalling, the basis of communication within an organism is relevant to the action of local signaling, molecules like histamine, hormones in the endocrine system, transmitters in the nervous system, and the identification of cells as foreign by the immune system.

To be a good biologist, it is important that you can identify connections between different topics and apply your knowledge and understanding of each topic to answer novel questions.

In the synoptic biology examination paper, question styles will include short answer questions, practical questions, problem-solving questions, and extended response questions. These questions cover different combinations of topics from different modules, though it is impossible to cover every single one in any given set of questions.

Table 1 shows how just one learning outcome links to many other topics and modules in A Level Biology.

▼ **Table 1**

Learning outcome	Links with other modules		What you might be asked about...
2.1.5(a) the roles of membranes within cells and at the surface of cells to include the roles of membranes as sites of cell communication (cell signalling)	2.1.2(j)	how the properties of cholesterol molecules relate to their functions in living organisms	• steroid hormones are hydrophobic and can enter cells through phospholipid bilayer • fluidity of cell membranes
	2.1.3(g)	transcription and translation of genes	• the role of the nuclear membrane and the endoplasmic reticulum and Golgi apparatus in the way genes are transcribed and translated • hormones that are water soluble cannot enter cells but bind to membrane receptors and activate cAMP which acts as a secondary messenger and may result in genes being switched on and off, affecting the proteins synthesised.
	2.1.4(a)	the role of enzymes in catalysing reactions that affect metabolism	• hormones that are water soluble cannot enter cells but bind to membrane receptors and activate cAMP which acts as a secondary messenger and activates enzymes

2.1.5(a)	the roles of membranes at the surface of cells	• barrier to the movement of molecules into cells, facilitated diffusion, osmosis with links to turgor in plants and homeostasis in animals, active transport systems, site of receptors for hormones, neurotransmitters, cell recognition
2.1.5(b)	the fluid mosaic model of membrane structure	• hydrophobic core of phospholipid bilayer prevents the entry of water soluble hormones • protein pores, gated channels • osmosis with links to turgor in plants and homeostasis in animals, facilitated diffusion, active transport • surface glycoproteins as part of cell recognition system • infective mechanisms of viruses • immune system • Endo and exocytosis
2.1.5(d)(i)	the movement of molecules across membranes	• steroid hormones can diffuse through phospholipid bilayer • secondary messenger systems • osmosis with links to turgor in plant cells and homeostasis in animals • active transport systems • chemiosmosis
4.1.1(f)	the structure, different roles and modes of action of B and T lymphocytes in the specific immune response	• proteins in cell surface membrane acting as self and non-self-markers
5.1.3(d)	the structure and roles of synapses in neurotransmission	• example of cell signalling • endo and exocytosis
5.1.4(a)	endocrine communication by hormones	• example of cell signalling • secondary messengers
5.1.5(j)	the coordination of responses by the nervous and endocrine systems	• neurotransmitters and hormones involvement in cell signalling
5.1.5(k)	the effects of hormones and nervous mechanisms on heart rate	• neurotransmitters and hormones involvement in cell signalling

1 The contraction of the heart is myogenic, it is initiated by a group of muscle cells without stimulation from the nervous system. The natural rhythm of cardiac muscle is about 100–115 contractions per minute.

    a State the name given to the group of cells that initiate cardiac muscle contraction. *(1 mark)*

> Remind yourself of how the heart beat is controlled by the sino-atrial node by looking at Topic 8.5, The heart.

Answer to part **a**

SA node ✓

> For single mark answers, be clear and unambiguous in your answer – and make sure your spelling is correct.

The cells named in **a** are autorhythmic cells. Autorhythmic cell membranes have a resting potential like the membranes of neurons. The resting potential in autorhythmic cells, unlike neurons, is unstable. This leads to the membranes continually depolarizing, and subsequently repolarizing. This results in the generation of numerous action potentials per minute.

> The basis of the resting potential and action potential in neurones was explained in Topic 13.4, Nervous transmission. The same principles can be applied in a number of different places in the living world.

    b (i) Explain how resting membrane potential is maintained. *(3 marks)*

       (ii) State the number of action potentials that would be generated per minute in cardiac muscle isolated from the nervous system. *(1 mark)*

Answer to part **b**

Part **b(i)**

sodium/potassium pumps ✓

sodium ions out of muscle cells and potassium ions into muscle cells ✓

differential permeability of membrane ✓

part **b (ii)** 100–115 ✓

> Equal to natural rhythm of cardiac muscle stated in the stem of the question

The heart is connected to the nervous system by sympathetic and parasympathetic neurons. The average human resting heart rate is 60–80 beats/min.

> Check the structure and functions of the sympathetic and parasympathetic nervous systems in Topic 14.1, Hormonal communication and think about how this applies to the question.

    c Outline the effect of the sympathetic and parasympathetic branches of the nervous system on resting heart rate. *(3 marks)*

Answer to part **c**

sympathetic increases heart rate ✓

parasympathetic slows heart rate ✓

idea that parasympathetic is dominant at rest ✓

Atropine is a drug that blocks receptors in neuromuscular junctions of the parasympathetic system.

d Suggest and explain how the heart rate would change when atropine is administered. *(3 marks)*

> You met neuromuscular junctions in Topic 13.5, Synapses. Remind yourself of how they work and which neurotransmitters are involved.

> Give what is asked for in an answer and no more – you don't need a detailed description of a neuromuscular junction for 3 marks, just enough information to make the answer clear.

Answer to part **d**

heart rate would increase ✓

atropine blocks acetylcholine (in parasympathetic neuromuscular junction) from binding ✓

heart rate no longer slowed ✓

Adrenaline, a hormone released from the adrenal glands, also affects the heart rate.

> Look up hormones and the endocrine system in Topic 14.1, Hormonal communication. Make sure you know about the importance of the adrenal glands, the secretion of adrenaline, and its effects on the body.

e (i) State the precise location of the secretion of adrenaline. *(1 mark)*

> Remember that cell signalling usually involves receptors on the cell membranes. Remind yourself of the details in Topic 13.1, Coordination.

(ii) Explain why the effect of adrenaline is a form of cell signalling. *(1 mark)*

(iii) State the effect of adrenaline on the heart rate. *(1 mark)*

Adrenaline is a hydrophilic molecule and therefore is unable to enter cells.

(iv) Explain why adrenaline cannot enter cells. *(2 marks)*

> You need to understand the structure of the cell membrane and understand hydrophilic and hydrophobioc molecules to answer this part of the question, look back at Topic 3.5, Lipids.

Answer to part **e**

Part **e (i)** (adrenal) medulla ✓

Part **e (ii)** molecule released by a cell in the adrenal gland ✓

produces a response / increases rate of contraction ✓

in a cell in the heart ✓

part **e (iii)** increases heart rate ✓

Part **e (iv)** hydrophobic core of phospholipid bilayer ✓

prevents entry of hydrophilic adrenaline molecule ✓

This answer also depends on your understanding of the different types of cell signalling which you learnt about in Topic 14.5, Coordinated responses.

f Outline the way in which adrenaline brings about a response within a cell. *(6 marks)*

Answer to part **f**

adrenaline binds to receptor on cell surface membrane ✓

change in shape of receptor. G protein ✓

activates G protein ✓

G protein activates adenyl cyclase ✓

adenyl cyclase catalyzes formation of cAMP ✓

cAMP activates kinase enzymes ✓

kinase enzymes activate enzymes responsible for response ✓ (max 6)

When you answer a question that carries a lot of marks, make sure you give at least the same number of clear points as there are marks in the question, whether you present your answer as a list or as continuous prose.

You learnt about the role of the adrenal glands and their hormones in Topic 14.1, Hormonal communication.

This links to mechanisms of inflammation with the production of proteins such as histamine. Look back to Topic 12.5, Non-specific animal defences against pathogens.

g Hydrocortisone is a synthetic glucocorticoid derived from cholesterol. It has anti-inflammatory actions due to the suppression of the synthesis of proteins that have roles in the inflammatory process.

Suggest how hydrocortisone brings about its anti-inflammatory response. *(3 marks)*

Answer to part **g**

hydrophobic/lipid soluble due to cholesterol ✓

enter cells through phospholipid bilayer by diffusion ✓

switch genes off/stops transcription ✓

You learnt about cell membrane structure in Topic 5.3, Diffusion.

Check the factors affecting gene expression and how genes are switched on and off in the cell in Topic 3.10, Protein synthesis.

Glutamic acid ($C_5H_9NO_4$) is an amino acid. A salt of glutamic acid is known as glutamate. The sodium salt of glutamic acid is called monosodium glutamate (MSG) ($C_5H_8NO_4Na$). MSG occurs naturally in most foods, particularly high protein foods like milk, cheese, meat, and fish. Tomatoes and mushrooms also have high levels of glutamate.

The human body can synthesise glutamic acid and it is an essential molecule in a number of metabolic processes. Children metabolise glutamate in the same way as adults and human breast milk actually contains ten times more glutamate than cow's milk.

Glutamic acid also acts a neurotransmitter.

MSG is a commonly used as a flavour enhancer. It is added to Chinese food, canned vegetables, soups, and processed meats and many other processed foods.

MSG used to be extracted from protein-rich plants such as seaweed.

This was time-consuming process and MSG is now produced by an industrial fermentation process using bacteria.

MSG used to be blamed for causing "Chinese Restaurant Syndrome" after a doctor reported that he suffered from burning sensations along the back of his neck, chest tightness, nausea, and sweating whenever he had eaten at a Chinese restaurant.

Although a small percentage of people are sensitive to monosodium glutamate, double-blind studies have failed to show that monosodium glutamate actually causes any of the symptoms attributed to "Chinese restaurant syndrome".

Chinese food often contains a wide variety of ingredients such as seafood, peanuts, spices and herbs.

The diagram shows a molecule of MSG.

HO — molecule of MSG with NH₂ group and O⁻ Na⁺

a (i) State the name of the group of molecules that contains glutamic acid. *(1 mark)*

Glutamic acid can be used to build another type of molecule.

(ii) State the name of the group of molecules that contains this larger molecule. *(1 mark)*

(iii) State the name of the bonds formed during the synthesis of these larger molecules. *(1 mark)*

b Outline the role of a neurotransmitter. *(3 marks)*

c Explain why the production of MSG is now an example of biotechnology. *(3 marks)*

The graph shows the results of an investigation into the growth of bacteria and the production of glutamic acid by batch and fed-batch fermentations.

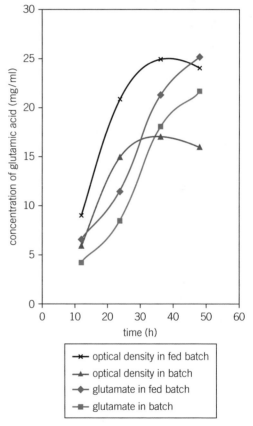

Bacterial growth was estimated by measuring the optical density (OD) of the culture using a spectrophotometer.

d (i) Describe the difference in growth rate and glutamic acid production in the graph. *(4 marks)*

(ii) Compare the processes involved during batch and fed-batch fermentation. *(3 marks)*

(iii) State the overall name given to the processes involved in the production of pure glutamic acid from the slurry removed from the fermentation vessel. *(1 mark)*

e (i) Outline how a double-blind trial would be conducted. *(4 marks)*

(ii) Discuss the arguments for and against the use of monosodium glutamate in food preparation. *(5 marks)*

(iii) Suggest why tomatoes and mushrooms are often used in cooking to flavour food. *(2 marks)*

2 The diagram shows some of the metabolic pathways that take place in a plant cell.

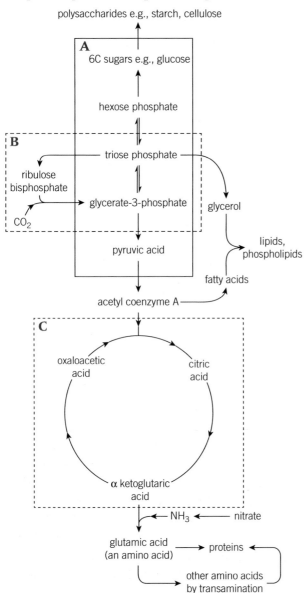

a State the names of the processes labelled A, B and C. *(3 marks)*

b Select an example from the diagram to explain what is meant by an anabolic reaction. *(2 marks)*

c Describe, using information from the diagram, how plants make the chemical components necessary to build cell walls, cell membranes and cytoplasm. *(9 marks)*

d Plants also manufacture chemicals as a protection against herbivory.

Explain why this is a compromise for the plants. *(2 marks)*

e Outline others ways that plants are adapted for adverse conditions. *(6 marks)*

OCR 2806/01 2009 (apart from (d) and (e))

3 Invertebrates, such as cockroaches and squid do not have myelin sheaths around their axons.

Giant axons that are present in cockroaches, earthworms and squid can be up to 1.0 mm in diameter.

Giant axons are essential in parts of the nervous system that control escape reflexes.

a Describe the type of behaviour that includes the escape reflex. *(2 marks)*

The diagram compares the transmission speeds of a mammal with three invertebrate organisms.

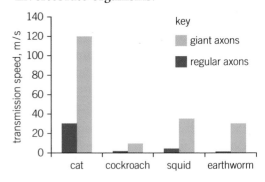

b **(i)** Describe the variation in the transmission speed of impulses shown in the diagram. *(4 marks)*

(ii) Explain effects of axon diameter and myelination on the speed of impulse transmission. *(4 marks)*

The diagram shows the changes in membrane potential during an action potential.

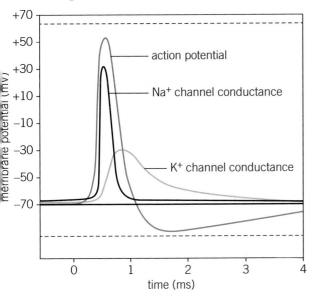

c **(i)** Explain, using information from the diagram, how an action potential is propagated. *(5 marks)*

(ii) Explain how the differential permeability of neuronal membranes is essential in maintaining resting potential. *(3 marks)*

Tetrodotoxin is a chemical produced by the puffer fish and a number of other animals. It blocks nerve impulses by blocking voltage gated sodium ion channels and is a very powerful toxin.

Some species of puffer fish are considered a delicacy in Japan and Korea and chefs in these countries have to be specially trained to prepare the fish so that the poison is removed.

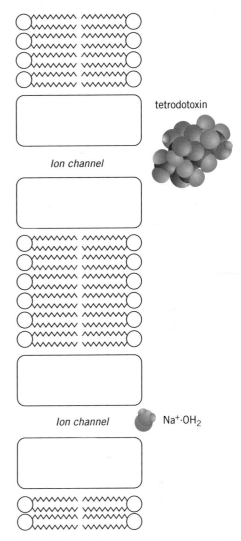

Tetrodotoxin is very specific in blocking the sodium ion voltage gated and the movement of potassium ions is not affected.

Victims of tetrodotoxin poisoning do not die as the result of a heart attack but respiratory paralysis when the diaphragm no longer contracts. The function of cardiac muscle is not affected.

d **(i)** Explain why tetrodotoxin can block sodium ion voltage gated channels but not potassium ion voltage gated channels. *(2 marks)*

(ii) Explain how paralysis of the diaphragm would lead to death. *(3 marks)*

(iii) Suggest why cardiac muscle is not affected by tetrodotoxin. *(1 mark)*

Pufferfish have undergone a mutation in the gene coding for the protein that forms the sodium ion voltage gated channel.

e (i) Explain the term mutation. (*2 marks*)

(ii) Explain how this mutation means that pufferfish are not susceptible to poisoning by tetrodotoxin. (*4 marks*)

(iii) Describe the process by which this mutation became common in pufferfish. (*4 marks*)

Tetrodotoxin is found in many organisms, including the blue-ringed octopus and some species of poisonous frogs. Toxins are usually very specific to a particular species but in this case tetrodotoxin is produced by bacteria which these organisms accumulate in their diets.

f State the name given to the process shown by these distantly related organisms all evolving to be unaffected by tetrodotoxin. (*1 mark*)

4 A short time after the death of a human being, or animal, the joints within the organism become locked in place. This process is called rigor mortis and begins with the muscles partially contracting. The graphs show how the concentration of ATP change in the hours after death and the time it takes for fully-fixed rigor mortis to develop.

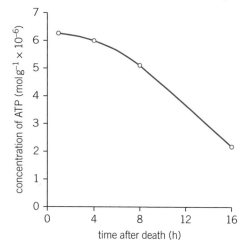

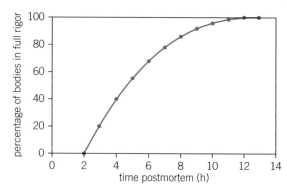

a (i) Explain why the levels of ATP level fall after death. (*2 marks*)

(ii) Describe the changes in ATP level and rigor mortis shown in the graph. (*4 marks*)

(iii) Suggest the reason for the onset of rigor mortis. (*3 marks*)

(iv) Suggest why a small amount of muscle contraction can occur after death. (*3 marks*)

The graph shows how the pH changes in muscle tissue after death.

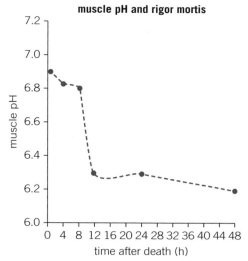

b Suggest a reason for the change in pH in muscle tissue after death. (*2 marks*)

Depending on various factors such as the temperature of the environment, rigor mortis can last up to about three days.

c Suggest an explanation for why rigor mortis is only temporary. (*4 marks*)

The diagram shows a part of the sliding filament theory.

The diagram shows a part of the sliding filament theory.

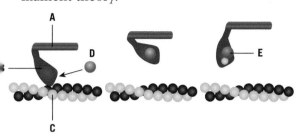

d (i) State the names of the structures A, B and C. *(3 marks)*

(ii) State the name of the molecule labelled D. *(1 mark)*

(iii) Indicate, using an arrow, in which direction structure A moves relative to structure C. *(1 mark)*

(iv) Suggest, using your knowledge of the structure of proteins, the role of the ion labelled E in muscle contraction. *(3 marks)*

(v) Outline how myosin acts as a molecular ratchet in the sliding filament theory. *(3 marks)*

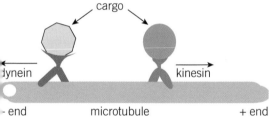

Kinesin is a motor protein that has a structure very similar to myosin. Dynein is another motor protein with a very different structure to myosin.

e Describe what determines the 3D structure of a protein and how this differs in kinesin and dynein. *(3 marks)*

Both dynein and kinesin have regions that binds to microtubules, a region that binds and hydrolyses ATP and a region that binds to an organelle.

The energy released during the hydrolysis of ATP changes their three-dimensional structures resulting in the motor proteins moving along microtubules and transporting organelles within cells. Dynein and kinesin move in opposite directions.

f (i) Compare and contrast the ways in which dyneins, kinesins and myosin lead to movement. *(4 marks)*

(ii) State two other processes, other than the movement of organelles, in which microtubules also have an important role. *(2 marks)*

Many of these questions can also be found in the AS Paper 1 and Paper 2 practice question sections of the equivalent Year 1/AS book. Here they have been organised by module for your convenience, with some new questions added.

1 **a** Explain the meaning of the terms *cation, anion* and *electrolyte*. *(3 marks)*

b Copy and complete the table *(5 marks)*

Ion	Function
calcium ions (Ca^{2+})	
phosphate ions (PO_4^{3-})	
sodium ions (Na^+)	
potassium ions (K^+)	
ammonium ions, (NH_4^+)	

2 The diagram summarises the process of DNA replication.

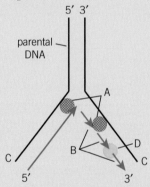

a State the names of A, B, C, and D. *(4 marks)*

The diagram outlines three different models of DNA replication.

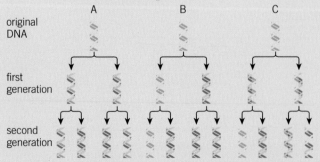

b **(i)** Outline what is meant by the term 'model'. *(2 marks)*

(ii) Identify which of the models in the diagram shows semi-conservative replication. *(1 mark)*

(iii) Discuss why the semi-conservative model is the most efficient way for DNA to be replicated. *(4 marks)*

3 The activity of an enzyme, laccase, was investigated using different concentrations of copper solution. The results are shown in the graph.

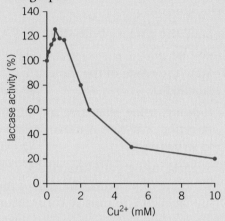

a **(i)** Describe the change in enzyme activity shown in the graph *(3 marks)*

(ii) Suggest reasons for the changes that you have described. *(4 marks)*

The activity of the same enzyme was investigated at different temperatures. The results are shown in the graph.

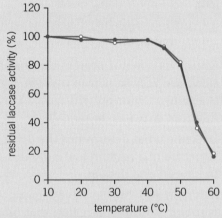

b Explain the the reason for the trend shown in the graph. *(4 marks)*

4 Consider the scientific drawings shown here.

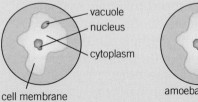

a State, giving your reasons, which is the better scientific drawing. *(4 marks)*

The photomicrograph shown was taken during the process of translation.

The magnification of the image is ×362 500.

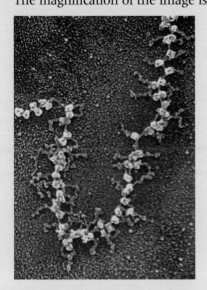

b (i) Calculate, using the photo, the mean size of a ribosome. *(3 marks)*

(ii) Suggest the name of the microscope used to capture this image. *(2 marks)*

(iii) Outline the main steps involved in translation. *(4 marks)*

5 The diagram shows four cells that have been placed in different solutions.

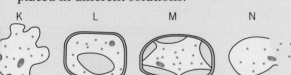

a In the table write the letter **K**, **L**, **M** or **N** next to the description that best matches the diagram. One has been done for you.

description	letter
an animal cell that has been placed in distilled water	
an animal cell that has been placed in a concentrated sugar solution	
a plant cell that has been placed in distilled water	
a plant cell that has been placed in concentrated sugar solution	**M**

(3 marks)

b Explain, using the term **water potential**, what has happened to cell **M**. *(3 marks)*

c Small non-polar substances enter cells in different ways to large or polar substance. Outline the ways in which the substances below, can enter a cell through the plasma (cell surface) membrane.

- Small, non-polar substances
- Large substances
- Polar substances *(5 marks)*

OCR F211 2009
From the AS paper 1 practice questions

6 A student carried out an investigation into the effect of the enzyme inhibitor, lead nitrate, on the activity of the enzyme amylase.

The results that were obtained are shown in the table.

percentage concentration of lead nitrate solution	transmission of light / arbitrary units			mean transmission of light / arbitrary units
	first run	second run	third run	
0	84	87	82	84.3
0.2	55	53	52	53.3
0.4	36	37	36	36.3
0.6	27	24	27	26.0
0.8	22	20	21	21.0
1.0	20	21	18	

a (i) Outline a procedure that the student could have followed to obtain the results. *(4 marks)*

(ii) Copy and complete the table *(1 mark)*

b (i) The third run for 1.0% solution of lead nitrate initially produced a reading of 70 arbitrary units.

The student discarded this result and repeated this run. Explain why. *(2 marks)*

(ii) Plot a graph using the results in the table. *(3 marks)*

c **(i)** Describe and explain the shape of the graph. *(4 marks)*

(ii) State an appropriate conclusion for these results. *(1 mark)*

The student stated in their conclusion:

'the enzyme amylase was inhibited at all concentrations of lead nitrate'

d Discuss, whether or not, the student was correct in making this statement. *(4 marks)*

e Evaluate the validity of this investigation. *(4 marks)*

f Suggest how this investigation could be improved, *(1 mark)*

From the AS paper 2 practice questions

7 **a** Copy and complete the following paragraph about cells by using the most appropriate term(s).

Cell that are not specialised but still have the ability to divide are called …………… cells. Such cells can be found in the …………….. of the long bones of mammals. These cells can …………… into other types of cell, such as erythrocytes that carry oxygen in the blood. In plants, …………………… tissues also contains cells that are not specialised. *(4 marks)*

b Sponges are simple eukaryotic multicellular organisms that live underwater on the surface of rocks. Sponges have a cellular level of organisation. This means they have no tissues.

Each cell type is specialised to perform a particular function. One type of cell found in a sponge is a collar cell. Collar cells are held in positions on the inner surface of the body of the sponge. Figure 3 is a diagram showing a vertical section through the body of a sponge and an enlarged drawing of a collar cell.

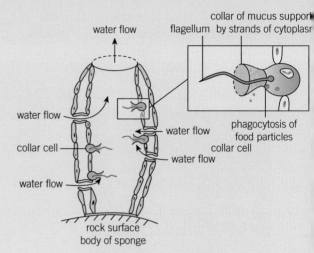

(i) Suggest **one** function of the flagellum in the collar cell. *(1 mark)*

(ii) Suggest **one** possible role for the collar of mucus in the cell *(1 mark)*

c In more advanced organisms, cells are organised into tissues consisting of one or more types of specialised cells.

Describe how cells are organised into tissues, using **xylem** and **phloem** as examples *(4 marks)*

OCR F211 2012

From the AS paper 1 practice questions

8 A student investigating how different concentrations of sucrose solution affect the size of animal cells obtained the results shown in the table.

	Concentration of sucrose solution / mol dm				
	0.05	0.10	0.20	0.40	0.8
diameter of cell 1/μm	8.4	7.2	6.6	5.7	2.3
diameter of cell 2/μm	7.8	7.3	6.8	5.7	2.5
diameter of cell 3/μm	8.1	7.4	7.0	5.5	2.4
mean diameter /μm	8.1	7.3	6.8		2.4
mean change in diameter /μm	+1.1	+0.3	−0.2		−4.

a The original mean diameter of the cells was 7.0 μm. Copy and complete the table. *(2 marks)*

b Plot a graph using the results obtained in the table. *(4 marks)*

c Describe and explain the trend shown by the graph. *(4 marks)*

d Suggest the type of cells the student used in the investigation. *(1 mark)*

From the AS paper 1 practice questions

9 A student investigating rate diffusion carried out the following procedure.

1 Prepared a petri-dish containing a layer of agar.

2 Cut a 1 cm well in the centre of the agar.

3 Placed 10 drops of a coloured solution in the well.

4 Measured the distance travelled by the colored solution from the edge of the well every 15 minutes.

The results obtained by the student are shown in the table.

time / min	distance diffused from well by coloured solution / mm
0	0
15	14
30	22
45	26
60	28
75	29

a Plot a graph using the results obtained in the table. *(4 marks)*

b Calculate, using the graph, the rate of diffusion of the solution between 10 minutes and 20 minutes. *(4 marks)*

c Describe and explain the trend shown by the graph. *(4 marks)*

The smallest units on the ruler used to measure the distances diffused by the coloured solution were 1 mm.

d (i) State the uncertainty of the measurements obtained by the student. *(1 mark)*

(ii) Calculate the percentage error for the measurement taken at 45 minutes. *(4 marks)*

(iii) Suggest how the student could have improved the precision of the measurements. *(1 mark)*

(iv) Suggest how the student could have improved the reliability of the investigation. *(2 marks)*

e Evaluate the validity of the investigation. *(4 marks)*

From the AS paper 1 practice questions

10 A student carried out an investigation into the effect of pH on the activity of the enzyme amylase.

a Suggest a hypothesis that the student made before starting the investigation. *(2 marks)*

b (i) State the independent and dependent variables in this investigation. *(2 marks)*

(ii) State two variables that should be controlled in this investigation. *(2 marks)*

The student obtained the results shown in the table.

pH		5	6	7	8	9
Time taken for starch to be broken down/min	First run	11	7	3	4	10
	Second run	10	6	4	5	9
	Third run	8	7	3	6	10

c Draw a table to present the raw data correctly. Calculate and include the mean values in the table that you produce. *(4 marks)*

d Describe how the student increased the reliability of the investigation. *(2 marks)*

e Discuss whether the results obtained by the student support your hypothesis. *(4 marks)*

From the AS paper 2 practice questions

Module 3 practice questions

1 Groundkeepers have an essential role during the winter ensuring football pitches don't get too muddy and are kept in a condition suitable to play on. It is important that they use the most suitable species of grass.

Some species of grass grow more in the winter and therefore transpire at a greater rate.

a (i) Describe what is meant by the term transpiration. *(2 marks)*

(ii) Explain why transpiration rates are affected by the rate of growth of grass. *(4 marks)*

An investigation was carried to determine the difference in transpiration rates of a winter active grass and a grass that is relatively dormant in the winter.

The results are shown in the graph.

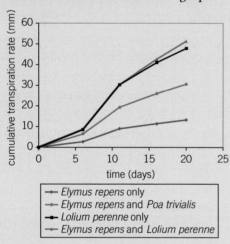

b (i) Describe the results of the investigation shown in the graph. *(4 marks)*

(ii) Suggest, giving your reasons, which grass, or combination of grasses, the groundkeeper would pick based on the results in the graph. *(3 marks)*

2 The diagram shows the diagram of a transverse section of the heart viewed from the top.

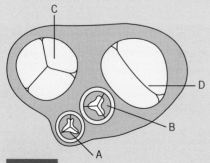

a (i) State the name of the valves labelled A, B, C and D in the diagram *(4 marks)*

(ii) Outline the functions of the four valves. *(5 marks)*

The diagram shows a normal and diseased aortic heart valve in the open and closed positions.

healthy aortic valve – closed

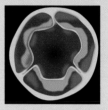
healthy aortic valve – open

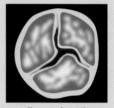
diseased aortic valve – closed

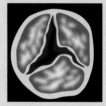
diseased aortic valve – open

b (i) Describe the differences between the normal and diseased valves. *(4 marks)*

(ii) Explain how the presence of the diseased valves would affect the flow of blood through the heart. *(5 marks)*

3 Stomatal conductance is a measure of the rate of diffusion of carbon dioxide through the stomata of a leaf.

Five-month-old seedlings grown in planting bags were not watered for 21 days.

The stomatal conductance (top graph) and leaf water potential (bottom graph) were measured during this period.

The results are shown in the graphs.

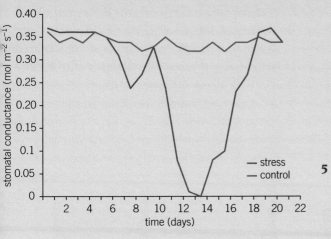

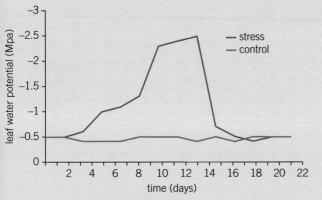

a Describe the changes in rate of diffusion of carbon dioxide and water potential over the 22 day period. (*4 marks*)

b Explain the link between the change in rate of diffusion and the change in water potential. (*4 marks*)

c Outline why controls were used in the investigations. (*2 marks*)

4 The diagram shows a section through the root of a typical plant.

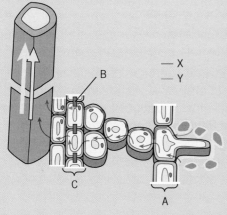

a (i) State the names of the pathways, taken by water, labelled X and Y. (*2 marks*)

   (ii) State the names of the structures labelled A, B and C. (*3 marks*)

   (iii) Describe what happens at B. (*3 marks*)

b Outline how the structure of a root hair cell is adapted to its function. (*4 marks*)

5 Various measurements of lung function are used to help diagnose lung disease and to monitor its treatment.

b State what is meant by the following terms:

   (i) vital capacity (*1 mark*)

   (ii) forced expiratory volume 1 (FEV1). (*1 mark*)

One measure of lung function is:

Percentage lung function $= \dfrac{\text{FEV 1}}{100} \times 100$

This is particularly useful in identifying possible obstructive disorders of the airways and lungs, such as asthma or chronic obstructive pulmonary disease (COPD).

- Asthma is a condition that responds to, and can be controlled by, the use of bronchodilators.
  These are drugs that dilate the airways and improve airflow.

- COPD lasts for a long period of time and is caused by progressive and permanent damage to the lung tissue.

When the value calculated for the percentage lung function is less than or equal to 70%, this indicates an obstructive disorder. A 'normal' value is approximately 80%.

The table shows data relating to three patients, **C**, **D** and **E**, before and after treatment with a bronchodilator drug.

patient	age (years)	before treatment			after treatment		
		vital capacity (dm^3)	FEV1 (dm^3)	percentage lung function	vital capacity (dm^3)	FEV1 (dm^3)	percentage lung function
C	18	5.5	3.8	69	5.6	4.5	
D	45	5.3	3.6	68	5.5	4.0	73
E	78	3.8	2.2	58	3.8	2.2	58

c (i) Calculate the percentage lung function for patient **C** after treatment with the bronchodilator drug.

Show your working and give your answer in percentage **to the nearest whole number**. *(2 marks)*

(ii) Using the information in the table and your answer to **(c)(i)**, copy and complete the table, indicating with a tick (✓) the diagnosis for each patient.

Patient	Diagnosis	
	Asthma	COPD
C		
D		
E		

(3 marks)

OCR F221 2009

From the AS paper 1 practice questions

6 a Define the following terms:

cardiac output *(2 marks)*

stroke volume *(2 marks)*

heart rate *(1 mark)*

b Copy these terms and arrange them in the correct order using the formula:

.... × .... = ... *(1 mark)*

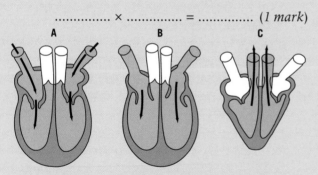

c Describe, using the diagrams in Figure 1, what is happening at each stage in the cardiac cycle shown. *(6 marks)*

From the AS paper 1 practice questions

7 Xylem vessels have two walls, a primary wall made of cellulose and a secondary wall composed of lignin. Lignin, a strong inflexible polymer, is not usually deposited uniformly but laid down in rings or spirals.

a Describe the role of lignin in xylem vessels. *(4 marks)*

b Suggest the benefit to a plant of way that lignin is deposited in the walls of xylem vessels. *(2 marks)*

From the AS paper 1 practice questions

8 The graph shows oxygen dissociation curves for both myoglobin and haemoglobin.

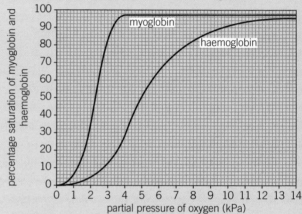

Calculate the decrease in percentage saturation of both myoglobin and haemoglobin between 4 kPa and 2 kPa partial pressure of oxygen. *(1 mark)*

OCR F224 2010

From the AS paper 2 practice questions

9 a Explain, using the term **surface area to volume ratio**, why large, active organisms need a specialised surface for gaseous exchange. *(2 marks)*

b The table describes some of the key features of the mammalian gas exchange system.

Copy and complete the table by explaining how each feature improves the efficiency of gaseous exchange. The first two have been completed for you.

Feature of gas exchange system	How feature improves efficiency of gaseous exchange system
Many alveoli	This increases the surface across which oxygen and carbon dioxide can diffuse
The epithelium of the alveoli is very thin	
There are capillaries running over the surface of the alveoli	
The lungs are surrounded by the diaphragm and intercostal muscles	

(*3 marks*)

c Outline how the diaphragm **and** intercostal muscles cause **inspiration**. (*4 marks*)

d The graph shows the trace from a spirometer recorded from a 16-year old student.

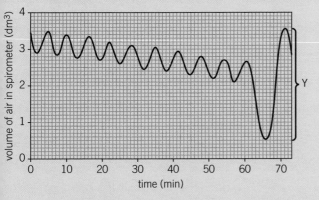

(i) Roughly copy the graph and label on the trace using the letter X, a point that indicates when the student was inhaling. (*1 mark*)

(ii) At the end of the trace the student measured their vital capacity. This is indicated by the letter Y. State the vital capacity of the student. (*1 mark*)

OCR F211 2009
From the AS paper 2 practice questions

10 The diagram shows a normal ECG trace.

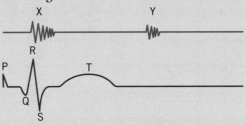

a (i) State which valves are closing at points X and Y. (*2 marks*)

(ii) Describe the function of the atrio-ventricular valves in the heart. (*2 marks*)

(iii) Explain why the right ventricular wall of the heart is less muscular than the left ventricular wall. (*5 marks*)

The graphs show an ECG trace during a heart attack (a) and during fibrillation (b).

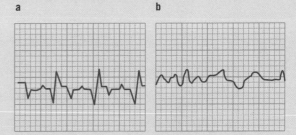

b Describe the differences between the ECG traces shown during a heart attack and fibrillation. (*3 marks*)

c 'A cardiac arrest occurs when there is a problem with the electrical activity in the heart, whereas a heart attack happens when there is a problem with the plumbing'. Discuss what you understand by this statement. (*6 marks*)

From the AS paper 2 practice questions

Module 4 practice questions

1   In 1940 it was observed that about 70% of Europeans had the ability to curl their tongues up at the sides to form a tube shape.

It was suggested that 'tongue rolling' is controlled by one gene.

a   (i)   State the name of the type of variation shown if tongue rolling is only controlled by one gene. (*1 mark*)

It was later observed that it was possible for someone to learn to roll their tongue and that only 70% of identical twins have been found to share the tongue rolling trait.

(ii) Discuss, giving your reasons, whether tongue rolling does actually follow the type of variation named in a (i).

(*4 marks*)

Put your hands together, without giving it any thought, and interlock your fingers.

It has been found that 55% of people place their left thumb on top and that 45% place their right thumb on top and the rest show no preference.

Although the majority of identical twins share this trait, it does not follow a predictable pattern of inheritance.

b   Outline the factors that influence which thumb is placed on top.   (*3 marks*)

2   Mutation plays an important role in the processes of speciation.

a   (i)   State the definition of a species.

(*2 marks*)

(ii) Explain the role of mutation in the formation of a new species. (*4 marks*)

b   (i)   Describe what is meant by the term 'selection pressure'.   (*2 marks*)

(ii) Discuss whether selection pressures are viewed as positive or negative influences on the rate of speciation.

(*3 marks*)

3   Human populations have an enormous impact on the biodiversity of ecosystems.

a   Define the term biodiversity.   (*2 marks*)

The impact of biodiversity is usually particularly apparent on farmland due to the agricultural practices in use.

b   Explain the effects of each of the following practices on biodiversity.
Deforestation
Monoculture
Fertilisers
Culling of predators   (*8 marks*)

4   Taiwan is an island with varied geography and landscape leading to the presence of a variety of habitats.

As global temperatures increase it has been predicted that many plants in Taiwan will migrate from lower to higher elevations above sea level.

A scientific study was carried out comparing the average number and Simpson's diversity index of six different species of plant against different elevations.

The results are shown in the graph.

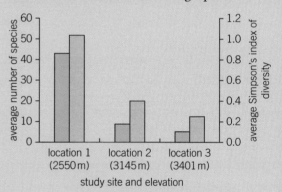

a   (i)   Describe how species diversity changes with elevation.   (*4 marks*)

(ii) Explain what the Simpson's index of diversity shows.   (*3 marks*)

It has been predicted that three species will migrate upwards to the higher zone. One species will migrate downwards to the lower zone and two species will stay where they are.

b   Suggest how this will affect the species diversity at the different elevations.

(*4 marks*)

5 Copy and complete the paragraph, filling in the missing words.

T memory cells live for a long time and are part of the.................. .... If they meet an .... for a second time, they undergo .... to form a large .... of T .... cells that destroy the pathogen. *(5 marks)*

From the AS paper 1 practice questions

6 Consider the statement:

'evolution can be summarised as change over time'

a Discuss why this statement is an over simplification as a description of evolution. *(4 marks)*

b Explain, using the finch populations observed by Darwin in the Galapagos islands the process of disruptive selection. *(6 marks)*

From the AS paper 1 practice questions

7 a Describe the difference between conservation and preservation. *(3 marks)*

b Describe the difference between 'in situ' and 'ex situ' conservation. *(4 marks)*

From the AS paper 1 practice questions

8 Which of the following statements is/are correct with respect to phylogenetic trees?

Statement 1: Phylogenetic trees depict the evolutionary relationships among groups of organisms

Statement 2: Species and their most recent common ancestor form a clade within a phylogenetic tree

Statement 3: Phylogenetic trees produced more recently show the relationship between clades and taxonomic groups

A 1, 2 and 3

B Only 1 and 2

C Only 2 and 3

D Only 1 *(1 mark)*

From the AS paper 1 practice questions

9 Variation is a fundamental characteristic of living organisms. Variation can be influenced by both genetics and the environment. The graphs below represent the two main types of variation.

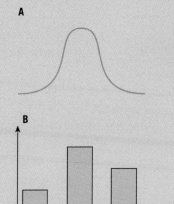

a Name the two types of variation shown in the graphs. *(2 marks)*

b Explain why evolution would not be possible without variation. *(4 marks)*

c Describe how variation arises. *(4 marks)*

d As humans, we have an anthropocentric view of evolution. This means the success of a species is measured by our own standards. Discuss whether our view of evolution is right. *(6 marks)*

From the AS paper 2 practice questions

10 a Describe the process of phagocytosis. *(5 marks)*

b Explain what is meant by the term pathogen. *(2 marks)*

c Explain why antibiotics are not prescribed to treat influenza. *(3 marks)*

d Suggest why antibiotics might still be prescribed to someone with influenza. *(3 marks)*

e Outline the different ways in which pathogens are made safe to use in vaccines. *(4 marks)*

f The diagram shows an antibody.

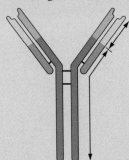

Describe how an antibody, such as the one in the diagram, has the ideal structure to carry out its role in an immune response. *(6 marks)*

OCR F224 2011

From the AS paper 2 practice questions

11 Advances in technology particularly in biochemistry have provided evidence that casts doubt on the way in which some organisms have been grouped in the five kingdom classification system.

DNA sequencing suggests that some organisms are more closely related to organisms belonging to another kingdom than other members of their own kingdom.

a Explain the meaning of the following terms:

(i) Phylogeny *(2 marks)*

(ii) Hierarchy *(2 marks)*

(iii) Taxonomy *(2 marks)*

(iv) Cladistics *(2 marks)*

b Outline the differences between kingdom and domain classification systems. *(5 marks)*

c Explain the meaning of the term 'descent with modification'. *(6 marks)*

From the AS paper 2 practice questions

12 The diagram shows the origin and development of a B lymphocyte in a tissue, X. It also shows the immune response in a lymph node following a vaccination with the measles virus.

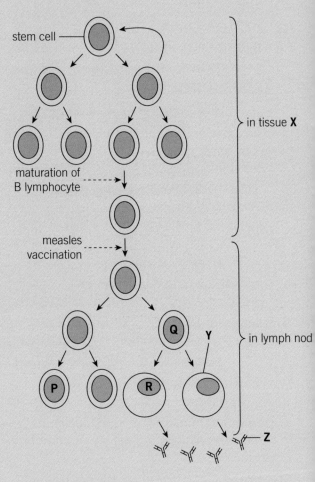

a Name tissue X, cell Y, and molecule Z. *(3 marks)*

b Molecules found on the surface of the measles virus can stimulate an immune response.

State the term given to molecules that stimulate an immune response. *(1 mark)*

c (i) Identify which of the cells P, Q, or R could become a memory cell. *(1 mark)*

(ii) State one function of memory cells. *(1 mark)*

d The graph shows the concentration of molecule Z in the blood following a vaccination for measles at day 0.

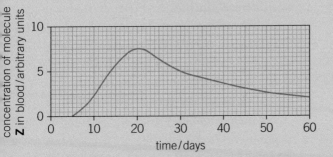

(i) Using the graph, describe how the concentration of molecule Z in the blood changes following vaccination.

(3 marks)

(ii) Copy and complete the table putting a tick (✓) in the appropriate box to indicate the type of immunity provided by this vaccination. *(1 mark)*

natural passive immunity	
natural active immunity	
artificial passive immunity	
artificial active immunity	

OCR 2802 2008

Glossary

aminoacylase enzyme used to produce pure samples of L-amino acids.

abscission the fall of leaves.

acetyl coenzyme A molecule that enters the Krebs cycle from glycolysis through a linking reaction when coenzyme A combines with an acetyl group.

acetylation addition of acetyl group.

action potential the change in the potential difference across the neurone membrane of the axon when stimulated (approximately +40 mV).

active transport movement of particles across a plasma membrane against a concentration gradient. Energy is required.

adenosine diphosphate (ADP) a nucleotide composed of a nitrogenous base (adenine), a pentose sugar and two phosphate groups. Formed by the hydrolysis of ATP, releasing a phosphate ion and energy.

adenosine triphosphate (ATP) a nucleotide composed of a nitrogenous base (adenine), a pentose sugar and three phosphate groups. The universal energy currency for cells.

alcoholic fermentation fermentation that results in the production of ethanol.

alkaloids bitter-tasting compounds found in plant leaves that may affect the metabolism of animals or insects eating them or poison them.

alleles different versions of the same gene.

allele frequency the relative frequency of a particular allele in a population at a given time.

allopatric speciation speciation that occurs as a result of a physical barrier between populations.

amino acids monomer used to build polypeptides and thus proteins.

ammonification conversion of nitrogen compounds in dead organic matter or waste into ammonium compounds by decomposers.

anabolic steroids steroid drugs used illegally by some athletes and bodybuilders to increase muscle mass.

anaerobic respiration respiration in the absence of oxygen.

antibodies y-shaped glycoproteins made by B cells of the immune system in response to the presence of an antigen.

antidiuretic hormone (ADH) hormone that increases the permeability of the distal convoluted tubule and the collecting duct to water.

antigen identifying chemical on the surface of a cell that triggers an immune response.

apical dominance the growth and dominance of the main shoot as a result of the suppression of lateral shoots by auxin.

apoptosis programmed and controlled cell death important in controlling the body form and in the removal of damaged or diseased cells.

artificial selection see selective breeding.

artificial twinning the process of producing monozygotic twins artificially.

aseptic techniques techniques used to culture microorganisms in sterile conditions so they are not contaminated with unwanted microorganisms.

autoimmune response response when the immune system acts against its own cells and destroys healthy tissue in the body.

autonomic nervous system part of the nervous system that is under subconscious control.

autosomal linkage genes present on the same, non-sex chromosome.

auxins plant hormones that control cell elongation, prevent leaf fall, maintain apical dominance, produce tropic responses, and stimulate the use of ethene in fruit ripening.

B effector cells B lymphocytes that divide to form plasma cell clones.

B lymphocytes (B cells) lymphocytes which mature in the bone marrow and that are involved in the production of antibodies.

B memory cells B lymphocytes that live a long time and provide immunological memory of the antibody needed against a specific antigen.

baroreceptors receptors which detect changes in pressure.

batch fermentation an industrial fermentation that runs for a set time.

biodiversity the variety of living organisms present in an area.

bioinformatics the development of the software and computing tools needed to analyse and organise raw biological data.

biomass mass of living material.

bioremediation the use of microorganisms to break down pollutants and contaminants in the soil or water.

biotic factors the living components of an ecosystem.

Bowman's capsule cup-shaped structure that contains the glomerulusand is the site of ultrafiltration in the kidney.

Calvin cycle the cyclical light independent reactions of photosynthesis.

carbohydrates organic polymers composed of the elements carbon, hydrogen and oxygen, usually in the ratio $C_x(H_2O)_y$. Also known as saccharides or sugars.

carrier a person who has one copy of a recessive allele coding for a genetically inherited condition.

carrying capacity the maximum population size that an environment can support.

cell cycle the highly ordered sequence of events that takes place in a cell, resulting in division of the nucleus and the formation of two genetically identical daughter cells.

cell wall a strong but flexible layer that surrounds some cell-types.

central nervous system (CNS) consists of the brain and spinal cord.

chemiosmosis the synthesis of ATP driven by a flow of protons across a membrane.

chemoreceptors receptors which detect chemical changes.

chi-square formula formula used to determine the significance of the difference between observed and expected count data.

chlorophyll green pigment that captures light in photosynthesis.

chloroplasts organelles that are responsible for photosynthesis in plant cells. Contain chlorophyll pigments, which are the site of the light reactions of photosynthesis.

chromatids two identical copies of DNA (a chromosome) held together at a centromere.

chromatin uncondensed DNA in a complex with histones.

chromosomes structures of condensed and coiled DNA in the form of chromatin. Chromosomes become visible under the light microscope when cells are preparing to divide.

citrate six carbon molecule formed in Krebs cycle by the combination of oxaloacetate and acetyl coenzyme A.

climax community final stage in succession, where the community is said to be in a stable state.

clones the offspring produced as a result of cloning.

cloning a way of producing offspring by asexual reproduction.

codominance when different alleles of a gene are equally dominant and both are expressed in the phenotype.

codon a three-base sequence of DNA or RNA that codes for an amino acid.

coenzyme A coenzyme with important roles in the oxidation of pyruvate in Krebs cycle and in the synthesis and oxidation of fatty acids.

collecting duct final part of the tubule that passes through the renal medulla and the place where hypertonic urine is produced if needed. The permeability of the walls is affected by ADH levels and it is the main site of water balancing.

communicable diseases diseases that can be passed from one organism to another, of the same or different species.

competitive inhibitor an inhibitor that competes with substrate to bind to active site on an enzyme.

complementary base pairing specific hydrogen bonding between nucleic acid bases. Adenine (A) binds to thymine (T) or uracil (U) and cytosine (C) binds to guanine (G).

computational biology the study of biology using computational techniques to analyse large amounts of data.

conservation the maintenance of biodiversity.

consumer organism that obtains its energy by feeding on another organism.

continuous fermentation an industrial fermentation where culture broth is removed continuously and more nutrient medium added.

continuous variation a characteristic that can take any value within a range, e.g. height.

correlation coefficient statistical test used to consider the relationship between two sets of data.

cortex the dark outer layer of the kidney containing the Bowman's capsules and glomeruli.

cristae fold of inner mitochondrial membranes, increases the surface area where reactions of the electron transfer chain can take place.

culture growing living matter in vitro, for example, microorganisms in specifically prepared nutrient medium.

cyclic AMP cyclic adenosine monophosphate, a molecule that acts as an important second messenger in many biological systems.

cyclic photophosphorylation synthesis of ATP involving only photosystem I.

cytolysis the bursting of an animal cell caused by increasing hydrostatic pressure as water enters by osmosis.

cytoplasm internal fluid of cells, composed of cytosol (water, salts and organic molecules), organelles and cytoskeleton.

cytoskeleton a network of fibres in the cytoplasm of a eukaryotic cell.

deamination the removal of the amino group from amino acids.

decarboxylation removal of carbon dioxide.

deciduous plants plants that lose all of their leaves for part of the year.

decomposer organism that breaks down dead organisms releasing nutrients back into the ecosystem.

decomposition chemical reaction in which a compound is broken down into simpler compounds or into its constituent elements.

dehydrogenation the removal of a hydrogen atom.

deletion a mutations where one or more nucleotides are deleted and lost from the DNA strand.

denatured (denaturation) change in tertiary structure of a protein or enzyme, resulting in loss of normal function.

denitrification conversion of nitrates to nitrogen gas.

deoxyribonucleic acid (DNA) the molecule responsible for the storage of genetic information.

depolarisation a change in potential difference from negative to positive across the membrane of a neurone.

detoxification removal or breakdown of toxins.

detritivore organism which speeds up decay by breaking down detritus into smaller pieces.

diabetes mellitus medical condition which affects a person's ability to control their blood glucose concentration.

digenic inheritance a characteristic controlled by two genes.

dihybrid inheritance A characteristic inherited on two genes.

diploid normal chromosome number; two chromosomes of each type – one inherited from each parent.

directional selection natural selection that favours one extreme phenotype.

discontinuous variation a characteristic that can only result in certain discrete values, for example, blood type.

disruptive selection natural selection that favours both extremes of a given phenotype.

distal convoluted tubule the second twisted section of the nephron where the permeability of the walls varies in response to ADH levels in the blood.

DNA polymerase enzyme that catalyses the formation of phosphodiester bonds between adjacent nucleotides in DNA replication.

DNA profiling producing an image of the patterns in the non-coding DNA of an individual.

DNA replication the semi-conservative process of the production of identical copies of DNA molecules.

DNA sequencing working out the sequence of bases in a strand of DNA.

dominant allele version of the gene that will always be expressed if present.

dominant species the most abundant species in an ecosystem.

ecological efficiency efficiency with which energy or biomass is transferred from one trophic level to the next.

ecosystem all the interacting living organisms and the non living conditions in an area.

ecotourism tourism directed towards natural environments, intended to support conservation efforts.

ectotherms animals that use their surroundings to warm their bodies so their core temperature is heavily dependent on the environment.

effector muscle or gland which carries out body's response to a stimulus.

electron carriers proteins that accept and release electrons.

electron microscopy microscopy using a microscope that employs a beam of electrons to illuminate the specimen. As electrons have a much smaller wavelength than light they produce images with higher resolutions than light microscopes.

electrophoresis a type of chromatography that relies on the way charged particles move through a gel under the influence of an electric current. It is used to separate nucleic acid fragments or proteins.

electroporation the use of a very tiny electric current to transfer genetically engineered plasmids into bacteria or to get DNA fragments directly into eukaryotic cells.

end-product inhibition the product of a reaction inhibits the enzyme required for the reaction.

endocrine glands group of specialised cells which secrete hormones.

endothermic reactions that absorb energy.

endotherms animals that rely on their metabolic processes to warm their bodies and maintain their core temperature.

enucleated with the nucleus removed.

enzymes biological catalysts that interact with substrate molecules to facilitate chemical reactions. Usually globular proteins.

epigenetics external control of genetic regulation.

epistasis the effect of one gene on the expression of another gene.

euchromatin loosely packed DNA.

eukaryotes multicellular eukaryotic organisms like animals, plants and fungi and single-celled protoctista.

eukaryotic cells cells with a nucleus and other membrane-bound organelles.

excretion the removal of the waste products of metabolism from the body.

exothermic reactions that release energy.

facilitated diffusion diffusion across a plasma membrane through protein channels.

facultative anaerobes organisms that can respire anaerobically or aerobically.

FAD coenzyme that acts as a hydrogen acceptor in Krebs cycle.

fermentation anaerobic respiration without the involvement of an electron transport chain.

forensics the application of science to the law, commonly in solving crimes.

founder effect when a few individuals of a species colonise a new area, their offspring initially experience a loss in genetic variation, and rare alleles can become much more common in the population.

gametes haploid sex cells produced by meiosis in organisms that reproduce sexually.

gene a section of DNA that contains the complete sequence of bases (codons) to code for a protein.

gene banks store of genetic material.

gene pool sum total of all the genes in a population at a given time.

genetic code the sequences of bases in DNA are the 'instructions' for the sequences of amino acids in the production of proteins.

genetic drift random change of allele frequency.

genome all of the genetic material of an organism.

genotype genetic makeup of an organism.

geotropism the growth response of plants to gravity.

germ line cell gene therapy inserting a healthy allele into the germ cells or into a very early embryo.

gibberellins plant hormones that cause stem elongation, trigger the mobilisation of food stores in a seed at germination and stimulate pollen tube growth in fertilisation.

Glomerular filtration rate (GFR) a test used to estimate the volume of blood filtered by the glomeruli each minute, used to indicate a loss of function in the kidneys.

glucoamylase enzyme used to convert dextrins to glucose.

gluconeogenesis production of glucose from non-carbohydrate sources.

glucose a monosaccharide with the chemical formula $C_6H_{12}O_6$. One of the main products of photosynthesis in plants.

glucose isomerase enzyme used to produce fructose from glucose.

glycerate-3-phosphate compound formed in Calvin cycle after carbon fixation.

glycogenesis production of glycogen from glucose.

glycogenolysis process in which glycogen stored in the liver and muscle cells is broken down into glucose.

haemoglobin the red, oxygen-carrying pigment of red blood cells.

Hardy-Weinberg equation formula used to calculate the frequency of alleles in a population.

hepatocytes liver cells.

herbivory the process of animals eating plants.

heterochromatin tightly packed DNA.

heterozygous two different alleles for a characteristic.

hexose bisphosphate the compound that results from the phosphorylation of glucose in glycolysis.

high throughput sequencing new methods of sequencing DNA that are automated, very rapid and much cheaper than the original techniques.

homeobox genes (Hox genes) genes responsible for the development of body plans.

homeodomain a conserved motif of 60 amino acids found in all homeobox proteins. It is the part of the protein that binds to DNA allowing the protein to act as a transcriptional regulator.

homeostasis the maintenance of a stable equilibrium in the conditions inside the body.

homozygous two identical alleles for a characteristic.

hormone chemical messengers which travel around the body in the blood stream.

hox genes homeobox genes.

humus organic component of soil formed by the decomposition of leaves and other plant material by soil microorganisms.

hydrophilic the physical property of a molecule that is attracted to water.

hydrophobic the physical property of a molecule that is repelled by water.

hypothalamus the region of the brain above the pituitary gland that contains osmoreceptors involved in osmoregulation.

immobilised enzymes enzymes that are attached to an inert support system over which the substrate passes and is converted to product.

inbreeding breeding between closely related organisms.

inbreeding depression reduced biological fitness due to inbreeding.

inner mitochondrial membrane the inner most of the two mitochondrial membranes. Separates the mitochondrial matrix from the intermembrane space. Is the site where the electron transport chain takes place.

insertion a mutation where one or more extra nucleotides are inserted into a DNA strand.

insulin a globular protein hormone involved in the regulation of blood glucose concentration.

interspecific competition competition between organisms of different species.

interspecific variation the differences between organisms of different species.

intraspecific competition competition between organisms of the same species.

intraspecific variation the differences between organisms of the same species.

introns regions of non-coding DNA or RNA.

ion an atom or molecule with an overall electric charge because the total number of electrons is not equal to the total number of protons. See anion and cation.

ionic bond a chemical bond that involves the donating of an electron from one atom to another, forming positive and negative ions held together by the attraction of the opposite charges.

islets of Langerhans specialised cells within the pancreas responsible for producing insulin and glucagon.

kingdom the second biggest and broadest taxonomic group.

lac operon operon responsible for the metabolism of lactose.

lactase enzyme used to hydrolyse lactose to glucose and galactose to produce lactose-free milk.

lactose a disaccharide made up of a galactose and glucose monosaccharide.

lactate dehydrogenase enzyme used in the conversion of pyruvate to lactate.

lactate fermentation fermentation that results in the production of lactate.

lamellae membranous channels which join grana together in a chloroplast.

light harvesting system a group of protein and chlorophyll molecules found in the thylakoid membranes of the chloroplasts in a plant cell.

light microscope an instrument that uses visible light and glass lenses to enable the user to see objects magnified many times.

limiting factor factor which limits the rate of a process.

line breeding form of inbreeding using less closely related organisms.

linked genes genes present on the same chromosome.

lipids non-polar macromolecules containing the elements carbon, hydrogen and oxygen. Commonly known as fats (solid at room temperature) and oils (liquid at room temperature).

loop of Henle a long loop of nephron that creates a steep concentration gradient across the medulla.

lymphocytes white blood cells that make up the specific immune system.

mature mRNA mRNA after the removal of introns and any other post-transcriptional changes.

medulla the lighter inner layer of the kidney made up of the loops of Henle.

meiosis form of cell division where the nucleus divides twice (meiosis I and meiosis II) resulting in a halving of the chromosome number and producing four haploid cells from one diploid cell.

messenger (m)RNA short strand of RNA produced by transcription from the DNA template strand. It has a base sequence complementary to the DNA from which it is transcribed, except it has uracil (U) in place of thymine (T).

methylation addition of methyl group.

micropropagation the process of making very large numbers of genetically identical offspring from a single parent plant using tissue culture techniques.

mitochondrial DNA DNA present within the matrix of mitochondria.

mitochondrial matrix the part of the mitochondria enclosed by the inner mitochondrial membrane which contains enzymes for the Krebs cycle and the link reaction.

mitosis nuclear division stage in the mitotic phase of the cell cycle.

monoclonal antibodies antibodies from a single clone of cells that are produced to target particular cells or chemicals in the body.

monogenic inheritance a characteristic inherited on a single gene.

monozygotic twins twins formed from a single fertilized egg.

mRNA see messenger (m)RNA.

multiple alleles a gene with more than two possible alleles.

mutagens chemical or physical agent which causes mutation.

mutation a change in the genetic material which may affect the phenotype of the organism.

myelin sheath membrane rich in lipid which surrounds the axon of some neurones, speeding up impulse transmission.

myofibril long cylindrical organelles found in muscle which are made of protein and specialised for contraction.

NAD a coenzyme found in all living cells involved in cellular respiration.

NADP coenzyme which acts as final electron acceptor in photosynthesis.

natural selection the process by which organisms best suited to their environment survive and reproduce, passing on their characteristics to their offspring through their genes.

nephrons tubules that make up the main functional structures of the kidneys.

neurone specialised cell which transmits impulses in the form of action potentials.

neurotransmitter chemical involved in communication across a synapse between adjacent neurones or a neurone and muscle cell.

nitrification conversion of ammonium compounds into nitrites and nitrates.

nitrile hydratase enzyme used to convert acrylonitrile to acrylamide for use in the plastics industry.

nitrogen fixation conversion of nitrogen gas to ammonium compounds.

non-competitive inhibitor an inhibitor that binds to an enzyme at an allosteric site.

non-cyclic photophosphorylation the synthesis of ATP and reduced NADP involving photosystems I and II.

nucleic acids large polymers formed from nucleotides. Contain the elements carbon, hydrogen, nitrogen , phosphorus, and oxygen.

nucleotides the monomers used to form nucleic acids. Made up of a pentose monosaccharide, a phosphate group and a nitrogenous base.

obligate aerobes organisms that can only respire aerobically.

obligate anaerobes organisms that cannot live in environments containing oxygen.

operon group of genes expressed together.

ornithine cycle a series of enzyme controlled reactions in the liver converting ammonia formed by deamination of amino acids into urea.

osmoreceptors sensory receptors that respond to changes in the water potential of the blood.

osmoregulation the balancing and control of the water potential of the blood.

osmosis diffusion of water through a partially permeable membrane down a water potential gradient. A passive process.

outbreeding breeding of distantly related organisms.

outer mitochondrial membrane the membrane that separates the contents of the mitochondrion from the rest of the cell, creating a cellular compartment with ideal conditions for aerobic respiration.

oxaloacetate four carbon molecule present at the beginning of Krebs cycle that combines with acetyl coenzyme A to form citrate.

oxidation removal of electrons or hydrogen.

partially permeable membrane that allows some substances to cross but not others.

passive transport transport that is a passive process (does not require energy) and does not use energy from cellular respiration.

pathogens microorganisms that cause disease.

pelvis the central chamber of the kidney where urine collects before passing out down the ureter.

penicillin the first widely used, safe antibiotic, derived from a mould, *Penicillium notatum*.

penicillin acylase enzymes used to make semi-synthetic penicillins from naturally produced penicillins.

peripheral nervous system (PNS) consists of all the neurones that connect the CNS to the rest of the body.

phagocytosis process by which white blood cells called phagocytes recognise non-self cells, engulf them digest them within a vesicle called a phagolysosome.

pharming the use of genetically modified animals to produce pharmaceuticals.

phenotype observable characteristics of an organism.

phospholipids modified triglycerides, where one fatty acid has been replaced with a phosphate group.

phosphorylation the addition of phosphate group to a molecule.

photosynthesis synthesis of complex organic molecules using light.

photosytem protein complexes involved in the absorption of light and electron transfers in photosynthesis.

phototropism the growth response of plants to unilateral light.

phylogeny the evolutionary relationships between organisms.

pigment molecules that absorb specific wavelengths of light.

pioneer species the first organisms to colonise an area.

plagioclimax stage in succession where artificial or natural factors prevent the natural climax community from forming.

pluripotent a stem cell that can differentiate into any type of cell, but not form a whole organism.

polymerase chain reaction a process by which a small sample of DNA can be amplified using specific enzymes and temperature changes.

polymorphic allele a gene with more than two possible alleles.

posterior pituitary gland the posterior part of the pituitary gland in the brain where ADH is stored ready for release into the blood.

predation the capturing of prey in order to sustain life.

preservation protection of an area by restricting or banning human use – so that the ecosystem is kept exactly as it is.

primary or pre mRNA the mRNA transcribed from the DNA before any post-transcriptional regulation to remove introns etc.

producer organism that converts light energy into chemical energy.

prokaryotes single-celled prokaryotic organisms from the kingdom Prokaryotae.

prokaryotic cells cells with no membrane-bound nucleus or organelles.

proteins one or more polypeptides arranged as a complex macromolecule.

protista biological kingdom containing unicellular eukaryotes.

proximal convoluted tubule the first twisted section of the nephron after the Bowman's capsule where many substances are reabsorbed into the blood.

pyruvate the three carbon product of glycolysis that feeds into Krebs cycle in the presence of oxygen.

receptors extrinsic glycoproteins that bind chemical signals, triggering a response by the cell.

recessive allele version of a gene that will only be expressed if two copies of this allele are present in an organism.

recombinant new combination of alleles / DNA from two sources.

recombination frequency proportion of recombinant offspring resulting from a cross.

reflex action involuntary response to a sensory stimulus.

regulatory gene a gene that codes for proteins involved with DNA regulation.

renal dialysis a process where the functions of the kidney are carried out artificially to maintain the salt and water balance of the blood.

repolarisation a change in potential difference from positive back to negative across the membrane of a neurone.

repressor protein protein that binds to operator affecting the rate of transcription.

respiration breakdown of complex organic molecules linked to the synthesis of ATP.

respiratory quotient ratio of carbon dioxide produced to oxygen used in respiration.

respiratory substrates organic molecules broken down in respiration.

resolution the shortest distance between two objects that are still seen as separate objects.

response the way a body reacts to a stimulus.

resting potential the potential difference across the membrane of the axon of a neurone at rest (normally about $-65mV$).

restriction endonucleases enzymes that chop a strand of DNA into small pieces.

ribonucleic acid (RNA) Polynucleotide molecules involved in the copying and transfer of genetic information from DNA. The monomers are nucleotides consisting of a ribose sugar and one of four bases; uracil (U), cytosine (C), adenine (A), and guanine (G).

ribulose bisphosphate five carbon molecule at beginning of Calvin cycle.

ribulose bisphosphate carboxylase (RuBisCo) key enzyme involved in the first step of carbon fixation in photosynthesis.

saprophytic/saprotrophic organisms that acquire nutrients by absorption – mainly of decaying material.

sarcomere functional unit of myofibril.

scanning electron microscopy (SEM) an electron microscope in which a beam of electrons is sent across the surface of a specimen and the reflected electrons are focused to produce a three-dimensional image of the specimen surface.

selective breeding selection of individuals for breeding with desirable characteristics.

selective reabsorption the reabsorption of selected substances ie those needed by the body in the kidney tubules.

sensory receptor specialised cell which detects a stimulus.

seral stages the steps in succession.

sex linked genes genes carried on the sex chromosomes.

sliding filament model movement of actin and myosin filaments in relation to each other to cause contraction.

Simpson's Index of Diversity (D) a measure of biodiversity between 0 and 1 that takes into account both species richness and species evenness.

somatic cell gene therapy replacing a faulty gene with a healthy allele in affected somatic cells.

somatic cell nuclear transfer a method of producing a clone from an adult animal by transferring the nucleus from an adult cell to an enucleated egg cell and stimulating development.

somatic nervous system part of the nervous system that is under conscious control.

Spearman's rank correlation coefficient a specific type of correlation test that compares the ranked orders of two datasets in order to consider their relationships.

speciation the formation of new species.

species the smallest and most specific taxonomic group.

stabilising selection natural selection that favours average phenotypes.

starch a polysaccharide formed from alpha glucose molecules either joined to form amylose or amylopectin.

stem cells undifferentiated cells with the potential to differentiate into a variety of the specialised cell types of the organism.

stimulus detectable change in external or internal environment of an organism.

structural genes genes that code for structural proteins or enzymes not involved in DNA regulation.

Student's *t* test statistical test used to compare the means of data values of two populations.

substrate a substance used, or acted on, by another process or substance. For example a reactant in an enzyme-catalysed reaction.

substitution a mutation where one or more nucleotides are substituted for another in a DNA strand.

substrate level phosphorylation synthesis of ATP by transfer of phosphate group from another molecule.

summation build up of neurotransmitter in a synapse to sufficient levels to trigger an action potential.

sustainable resource a renewable resource which is being economically exploited in such a way that it will not diminish or run out.

sympatric speciation speciation that occurs when there is no physical barrier between populations.

synapse the junction (small gap) between two neurones, or a neurone and an effector.

synthetic biology the design and construction of novel biological pathways, organisms or devices, or the redesign of existing natural biological systems.

T helper cells T lymphocytes with CD4 receptors on their cell-surface membranes, which bind to antigens on antigen-presenting cells and produce interleukins, a type of cytokine.

T killer cells T lymphocytes that destroy pathogens carrying a specific antigen with perforin.

T lymphocytes lymphocytes which mature in the thymus gland and that both stimulate the B lymphocytes and directly kill pathogens.

T memory cells T lymphocytes that live a long time and are part of the immunological memory.

T regulator cells T lymphocytes that suppress and control the immune system, stopping the response once a pathogen has been destroyed and preventing an autoimmune response.

tannins bitter tasting chemicals produced to prevent animals eating plant leaves; toxic to many insects.

target cells specific cells which hormones act on.

taxonomic group the hierarchical groups of classification – domain, kingdom, phylum, class, order, family, genus, species.

terpenoids chemicals found in plant leaves that may act as toxins to insects or fungi attacking the leaves.

test cross a cross used to determine genotype, involving a backcross with a homozygous recessive parent.

thermoregulation the maintenance of a relatively constant core temperature.

thin layer chromatography a technique for separating different pigments through the rate at which they move across an inert surface carried by a solvent.

thylakoid series of membranous compartments in a chloroplast that contain chlorophyll and molecules needed for light-dependent reaction.

tissue a collection of differentiated cells that have a specialised function or functions in an organism.

totipotent a stem cell that can differentiate into any type of cell and form a whole organism.

transcription the process of copying sections of DNA base sequence to produce smaller molecules of mRNA, which can be transported out of the nucleus via the nuclear pores to the site of protein synthesis.

transcription factors proteins that affect the rate of transcription.

transfer (t)RNA form of RNA that carries an amino acid specific to its anticodon to the correct position along mRNA during translation.

transmission electron microscopy (TEM) an electron microscope in which a beam of electrons is transmitted through a specimen and focused to produce an image.

triglyceride a lipid composed of one glycerol molecule and three fatty acids.

triose phosphate a molecule that is an intermediate in both photosynthesis and respiration and acts as a starting material for the synthesis of carbohydrates, lipids and amino acids.

trophic level stage in a food chain.

tropism a growth response by a plant in response to a unidirectional stimulus.

ultrafiltration the process by which blood plasma is filtered through the walls of the Bowman's capsule under pressure.

urea nitrogenous waste produced from the deamination of excess amino acids in the liver.

ureters tubes carrying urine from the kidneys to the bladder.

urethra tube carrying urine from the bladder to the outside of the body.

vector a living or non-living factor that transmits a pathogen from one organism to another, e.g. malaria mosquito.

vector (in genetic modification) a means of inserting DNA from one organisms into the cells of another organism.

vegetative propagation the artificial production of natural clones for use in horticulture and agriculture.

water potential (Ψ) measure of the quantity of water compared to solutes, measured as the pressure created by the water molecules in kilopascals (kPa).

wild type the allele that codes for the most common phenotype in a natural population.

zygote the initial diploid cell formed when two gametes are joined by means of sexual reproduction. Earliest stage of embryonic development.

Answers

Download all answers to end-of-topic questions
for free from our website at:

www.oxfordsecondary.co.uk/ocrbiologyanswers.

Alternately you can scan our unique QR code
which will take you straight to the answers page.

Appendix (Statistics data tables)

▼ *Table of values of* t

Degree of freedom (df)	p values			
	0.10	0.05	0.01	0.001
1	6.31	12.71	63.66	636.60
2	2.92	4.30	9.92	31.60
3	2.35	3.18	5.84	12.92
4	2.13	2.78	4.60	8.61
5	2.02	2.57	4.03	6.87
6	1.94	2.45	3.71	5.96
7	1.89	2.36	3.50	5.41
8	1.86	2.31	3.36	5.04
9	1.83	2.26	3.25	4.78
10	1.81	2.23	3.17	4.59
12	1.78	2.18	3.05	4.32
14	1.76	2.15	2.98	4.14
16	1.75	2.12	2.92	4.02
18	1.73	2.10	2.88	3.92
20	1.72	2.09	2.85	3.85
α	1.64	1.96	2.58	3.29

▼ *Critical values for Spearman's rank correlation coefficient,* r_s

	p = 0.1	p = 0.05	p = 0.02	p = 0.01			p = 0.1	p = 0.05	p = 0.02	p = 0.01
	5%	$2\frac{1}{2}$%	1%	$\frac{1}{2}$%	1-Tail Test		5%	$2\frac{1}{2}$%	1%	$\frac{1}{2}$%
	10%	5%	2%	1%	2-Tail Test		10%	5%	2%	1%
n						n				
1	–	–	–	–		31	0.3012	0.3560	0.4185	0.4593
2	–	–	–	–		32	0.2962	0.3504	0.4117	0.4523
3	–	–	–	–		33	0.2914	0.3449	0.4054	0.4455
4	1.0000	–	–	–		34	0.2871	0.3396	0.3995	0.4390
5	0.9000	1.0000	1.0000	–		35	0.2829	0.3347	0.3936	0.4328
6	0.8286	0.8857	0.9429	1.0000		36	0.2788	0.3300	0.3882	0.4268
7	0.7143	0.7857	0.8929	0.9286		37	0.2748	0.3253	0.3829	0.4211
8	0.6429	0.7381	0.8333	0.8810		38	0.2710	0.3209	0.3778	0.4155
9	0.6000	0.7000	0.7833	0.8333		39	0.2674	0.3168	0.3729	0.4103
10	0.5636	0.6485	0.7455	0.7939		40	0.2640	0.3128	0.3681	0.4051
11	0.5364	0.6182	0.7091	0.7545		41	0.2606	0.3087	0.3636	0.4002
12	0.5035	0.5874	0.6783	0.7273		42	0.2574	0.3051	0.3594	0.3955
13	0.4835	0.5604	0.6484	0.7033		43	0.2543	0.3014	0.3550	0.3908
14	0.4637	0.5385	0.6264	0.6791		44	0.2513	0.2978	0.3511	0.3865
15	0.4464	0.5214	0.6036	0.6536		45	0.2484	0.2945	0.3470	0.3822
16	0.4294	0.5029	0.5824	0.6353		46	0.2456	0.2913	0.3433	0.3781
17	0.4142	0.4877	0.5662	0.6176		47	0.2429	0.2880	0.3396	0.3741
18	0.4014	0.4716	0.5501	0.5996		48	0.2403	0.2850	0.3361	0.3702
19	0.3912	0.4596	0.5351	0.5842		49	0.2378	0.2820	0.3326	0.3664
20	0.3805	0.4466	0.5218	0.5699		50	0.2353	0.2791	0.3293	0.3628

▼ *continued*

	p = 0.1	p = 0.05	p = 0.02	p = 0.01
21	0.3701	0.4364	0.5091	0.5558
22	0.3608	0.4252	0.4975	0.5438
23	0.3528	0.4160	0.4862	0.5316
24	0.3443	0.4070	0.4757	0.5209
25	0.3369	0.3977	0.4662	0.5108
26	0.3306	0.3901	0.4571	0.5009
27	0.3242	0.3828	0.4487	0.4915
28	0.3180	0.3755	0.4401	0.4828
29	0.3118	0.3685	0.4325	0.4749
30	0.3063	0.3624	0.4251	0.4670

	p = 0.1	p = 0.05	p = 0.02	p = 0.01
51	0.2329	0.2764	0.3260	0.3592
52	0.2307	0.2736	0.3228	0.3558
53	0.2284	0.2710	0.3198	0.3524
54	0.2262	0.2685	0.3168	0.3492
55	0.2242	0.2659	0.3139	0.3460
56	0.2221	0.2636	0.3111	0.3429
57	0.2201	0.2612	0.3083	0.3400
58	0.2181	0.2589	0.3057	0.3370
59	0.2162	0.2567	0.3030	0.3342
60	0.2144	0.2545	0.3005	0.3314

▼ *Table of values of chi-squared*

df	p values								df
	0.99	0.95	0.90	0.50	0.10	0.05	0.01	0.001	
1	0.0001	0.0039	0.016	0.46	2.71	3.84	6.63	10.83	1
2	0.02	0.10	0.21	1.39	4.60	5.99	9.21	13.82	2
3	0.12	0.35	0.58	2.37	6.25	7.81	11.34	16.27	3
4	0.30	0.71	1.06	3.36	7.78	9.49	13.28	18.46	4
5	0.55	1.14	1.61	4.35	9.24	11.07	15.09	20.52	5
6	0.87	1.64	2.20	5.35	10.64	12.59	16.81	22.46	6
7	1.24	2.17	2.83	6.35	12.02	14.07	18.48	24.32	7
8	1.65	2.73	3.49	7.34	13.36	15.51	20.09	26.12	8
9	2.09	3.32	4.17	8.34	14.68	16.92	21.67	27.88	9
10	2.56	3.94	4.86	9.34	15.99	18.31	23.21	29.59	10
11	3.05	4.58	5.58	10.34	17.28	19.68	24.72	31.26	11
12	3.57	5.23	6.30	11.34	18.55	21.03	26.22	32.91	12
13	4.11	5.89	7.04	12.34	19.81	22.36	27.69	34.53	13
14	4.66	6.57	7.79	13.34	21.06	23.68	29.14	36.12	14
15	5.23	7.26	8.55	14.34	22.31	25.00	30.58	37.70	15
16	5.81	7.96	9.31	15.34	23.54	26.30	32.00	39.29	16
17	6.41	8.67	10.08	16.34	24.77	27.59	33.41	40.75	17
18	7.02	9.39	10.86	17.34	25.99	28.87	34.80	42.31	18
19	7.63	10.12	11.65	18.34	27.20	30.14	36.19	43.82	19
20	8.26	10.85	12.44	19.34	28.41	31.41	37.57	45.32	20
21	8.90	11.59	13.24	20.34	29.62	32.67	38.93	46.80	21
22	9.54	12.34	14.04	21.34	30.81	33.92	40.29	48.27	22
23	10.20	13.09	14.85	22.34	32.01	35.17	41.64	49.73	23
24	10.86	13.85	15.66	23.34	33.20	36.42	42.98	51.18	24
25	11.52	14.61	16.47	24.34	34.38	37.65	44.31	52.62	25
26	12.20	15.38	17.29	25.34	35.56	38.88	45.64	54.05	26
27	12.88	16.15	18.11	26.34	36.74	40.11	46.96	55.48	27
28	13.56	16.93	18.94	27.34	37.92	41.34	48.28	56.89	28
29	14.26	17.71	19.77	28.34	39.09	42.56	49.59	58.30	29
30	14.95	18.49	20.60	29.34	40.26	43.77	50.89	59.70	30
40	22.16	26.51	29.05	39.34	51.81	55.76	63.69	73.40	40
60	37.48	43.19	46.46	59.33	74.40	79.08	88.38	99.61	60
80	53.54	60.39	64.28	79.33	96.58	101.88	112.33	124.84	80
100	70.06	77.93	82.36	99.33	118.50	124.34	135.81	149.45	100

Acknowledgements

COVER: PHILIPPE PSAILA/SCIENCE PHOTO LIBRARY; **p2-3**: Bruce
Rolff/Shutterstock; **p4**(B): Vishnevskiy Vasily/Shutterstock; **p4**(T):
ia Torlin/Shutterstock; **p5**: Simon Fraser/Science Photo Library; **p6**:
Steve Gschmeissner/Science Photo Library; **p7**: Thomas Deerinck,
NCMIR/Science Photo Library; **p8**: Blueringmedia/Shutterstock;
p9: Anatomical Travelogue/Science Photo Library; **p21**: John
Greim/Science Photo Library; **p25**(C): Cginspiration/Shutterstock;
p25(L): Science Photo Library; **p25**(R): Oliversved/Shutterstock;
p26: Natural History Museum, London/Science Photo Library;
p32(C): Innerspace Imaging/Science Photo Library; **p32**(L): Eric
Grave/Science Photo Library; **p32**(R): Science Photo Library; **p33**:
Biology Media/Science Photo Library; **p34**: Microscape/Science
Photo Library; **p43**(L): Simon Fraser/Science Photo Library; **p43**(R):
Zephyr/Science Photo Library; **p49**(L): Jubal Harshaw/Shutterstock;
p49(R): Conge, ISM/Science Photo Library; **p50**(B): Vetpathologist/
Shutterstock; **p50**(T): Steve Gschmeissner/Science Photo Library;
p52(B): Laguna Design/Science Photo Library; **p52**(T): Molekuul.
be/Shutterstock; **p55**: Dmitry Lobanov/Shutterstock; **p56**: Dmitry
Lobanov/Shutterstock; **p58**: Sebastian Kaulitzki/Shutterstock; **p68**:
Visceralimage/Shutterstock; **p71**: Rebvt/Shutterstock; **p72**(L): Dirkr/
Shutterstock; **p72**(R): Pixelheld/Shutterstock; **p73**(B): Footage.Pro/
Shutterstock; **p73**(T): Zhukov Oleg/Shutterstock; **p74**: Vesilvio/
iStockphoto; **p75**(B): Arcalu/Shutterstock; **p75**(T): Elroyspelbos/
Shutterstock; **p80**(B): Biophoto Associates/Science Photo Library;
p80(T): Michael Abbey/Science Photo Library; **p84**(B): Ann Fullick;
p84(TL): Ann Fullick; **p84**(TR): Ann Fullick; **p85**(L): Manfred Kage/
Science Photo Library; **p85**(R): Ray Simons/Science Photo Library;
p93(B): Bestv/Shutterstock; **p93**(T): Kazakov Maksim/Shutterstock;
p100: Dr E Walker/Science Photo Library; **p103**: Kkimages/
Shutterstock; **p105**: Anthony Short; **p108**: Anthony Short; **p110**(B):
Steve Gschmeissner/Science Photo Library; **p110**(T): Anthony Short;
p111(B): Anthony Short; **p111**(T): Anthony Short; **p112**: Alexander
G Volkov; **p116**: Nigel Cattlin/Science Photo Library; **p119**: Geoff
Kidd/Science Photo Library; **p131**: Sinclair Stammers/Science
Photo Library; **p153**: Henner Damke/Shutterstock; **p159**: CNRI/
Science Photo Library; **p166-167**: Vinogradov Illya/Shutterstock;
p176(BR): Danny Gys/Reporters/Science Photo Library; **p176**(CL):
David Scharf/Science Photo Library; **p176**(TR): Eye of Science/
Science Photo Library; **p179**(B): Paul D Stewart/Science Photo
Library; **p179**(T): John Moss/Science Photo Library; **p182**: Sheila
Terry/Science Photo Library; **p183**(B): Image Point Fr/Shutterstock;
p183(C): Oak Ridge National Laboratory/US Department of Energy/
Science Photo Library; **p183**(T): Veda J Gonzalez/Shutterstock; **p184**:
Blacqbook/iStockphoto; **p188**(B): Varts/Shutterstock; **p188**(C): Le
Do/Shutterstock; **p188**(T): Lianem/Shutterstock; **p189**: Biophoto
Associates/Science Photo Library; **p190**: Eveleen/Shutterstock;
p191: Wally Eberhart, Visuals Unlimited/Science Photo Library;
p197: Robert J Erwin/Science Photo Library; **p199**(B): Capture
Light/Shutterstock; **p199**(CB): Milsiart/Shutterstock; **p199**(CT):
Mikkel Bigandt/Shutterstock; **p199**(T): Eric Isselee/Shutterstock;
p201: Theodore Mattas/Shutterstock; **p202**: Science Photo Library;
p203(B): Norman Bateman/Shutterstock; **p203**(CB): Iteachphoto/
iStockphoto; **p203**(CT): Jim Zipp/Science Photo Library; **p203**(T):
Michael W Tweedie/Science Photo Library; **p205**: Gary Hincks/
Science Photo Library; **p206**: Luiz A Rocha/Shutterstock; **p212**:
Brooke Fasani Auchincloss/Corbis; **p214**(B): James King-Holmes/
Science Photo Library; **p214**(T): Aiaikawa/Shutterstock; **p217**: Steve
Gschmeissner/Science Photo Library; **p218**: MRC Laboratory of
Molecular Biology; **p219**: Genome Research Limited; **p221**: Mauro
Fermariello/Science Photo Library; **p224**: Thomas Deerinck, NCMIR/
Science Photo Library; **p225**: Metin Kiyak/iStockphoto; **p230**:
Sinclair Stammers/Science Photo Library; **p231**: Thomas Deerinck,
NCMIR/Science Photo Library; **p232**: Valentyn Volkov/Shutterstock;
p234: Philippe Plailly/Science Photo Library; **p238**(B): Science Photo
Library; **p238**(T): Ilya Sirota/Shutterstock; **p239**: Photology1971/
Shutterstock; **p241**(B): Stephen Coburn/Shutterstock; **p241**(CB):
My Good Images/Shutterstock; **p241**(CT): Drozdowski/Shutterstock;
p241(T): Anyaivanova/Shutterstock; p243: Lebendkulturen.De/
Shutterstock; **p248**: Thomas Deerinck, NCMIR/Science Photo
Library; **p249**: Dream79/Shutterstock; **p252**: Maximilian Stock Ltd/
Science Photo Library; **p253**: Danny E Hooks/Shutterstock; **p254**(B):
Maximilian Stock Ltd/Science Photo Library; **p254**(T): Michal
Kowalski/Shutterstock; **p255**: Geoff Tompkinson/Science Photo
Library; **p259**: Maximilian Stock Ltd/Science Photo Library; **p261**:
Francis Leroy, Biocosmos/Science Photo Library; **p268**: Gary Yim/
Shutterstock; **p269**(B): Sally Wallis/Shutterstock; **p269**(T): Gerry
Bishop/Visuals Unlimited, Inc/Science Photo Library; **p274**: Fred
Cardoso/Shutterstock; **p275**: Fred Mcconnaughey/Science Photo
Library; **p276**(B): Dario Lo Presti/Shutterstock; **p276**(T): Schankz/

Shutterstock; **p277**: Dr Jeremy Burgess/Science Photo Library;
p278: Alfred Pasieka/Science Photo Library; **p279**: Parnumas Na
Phatthalung/Shutterstock; **p282**: Teresa Azevedo/Shutterstock;
p283(B): Photographybymk/Shutterstock; **p283**(T): Martin Fowler/
Shutterstock; **p284**: Cosmin Manci/Shutterstock; **p287**(B): Martyn
F Chillmaid/Science Photo Library; **p287**(T): Martyn F Chillmaid/
Science Photo Library; **p293**: Igor Kovalenko/Shutterstock;
p295(BL): Menno Schaefer/Shutterstock; **p295**(BR): Michael G
Mill/Shutterstock; **p295**(T): Sinclair Stammers/Science Photo
Library; **p296**: Cheryl E Davis/Shutterstock; **p297**: Cheryl E Davis/
Shutterstock; **p299**(B): Jeff Lepore/Science Photo Library; **p299**(T):
Bob Gibbons/Science Photo Library; **p301**(B): Dr Jeremy Burgess/
Science Photo Library; **p301**(T): Photocritical/Shutterstock; **p302**(B):
Peter Muller/Science Photo Library; **p302**(T): Dr Jeremy Burgess/
Science Photo Library; **p302**(B): Peter Chadwick/Science Photo
Library; **p303**(C): Bikeriderlondon/Shutterstock; **p303**(T): Art Wolfe/
Science Photo Library; **p304**(B): Art Wolfe, Mint Images/Science
Photo Library; **p304**(T): Creativex/Shutterstock; **p307**(B): Pavalena/
Shutterstock; **p307**(C): Jason Maehl/Shutterstock; **p307**(T): Neelsky/
Shutterstock; **p308**: Alexandra Lande/Shutterstock; **p309**: Steve
Horrell/Science Photo Library; **p310**(B): Francoise Sauze/Science
Photo Library; **p310**(T): Joe Gough/Shutterstock; **p311**: Photoiconix/
Shutterstock; **p312**: M Lustbader/Science Photo Library; **p313**(B):
David Vaughan/Science Photo Library; **p313**(T): Volt Collection/
Shutterstock; **p314**(B): Gyvafoto/Shutterstock; **p314**(T): Stocker1970/
Shutterstock; **p315**(B): Claude Nuridsany & Marie Perennou/Science
Photo Library; **p315**(T): Matthew Dixon/Shutterstock;

Artwork by Q2A Media

Although we have made every effort to trace and contact all
copyright holders before publication this has not been possible
in all cases. If notified, the publisher will rectify any errors or
omissions at the earliest opportunity.